FARM MACHINERY
AND EQUIPMENT

McGRAW-HILL PUBLICATIONS IN THE AGRICULTURAL SCIENCES

Lawrence H. Smith, *Department of Agronomy, University of Minnesota, Consulting Editor in the Plant Sciences*

Lawrence H. Smith, *Department of Agronomy, University of Minnesota, Consulting Editor in the Plant Sciences*

Adriance and Brison • PROPAGATION OF HORTICULTURAL PLANTS
Ahlgren • FORAGE CROPS
Brown • FARM ELECTRIFICATION
Campbell and Lasley • THE SCIENCE OF ANIMALS THAT SERVE MANKIND
Campbell and Marshall • THE SCIENCE OF PROVIDING MILK FOR MAN
Christopher • INTRODUCTORY HORTICULTURE
Crafts and Robbins • WEED CONTROL
Cruess • COMMERCIAL FRUIT AND VEGETABLE PRODUCTS
Eckles, Combs, and Macy • MILK AND MILK PRODUCTS
Edmond, Senn, Andrews, and Halfacre • FUNDAMENTALS OF HORTICULTURE
Gray • FARM SERVICE BUILDINGS
Jones • FARM GAS ENGINES AND TRACTORS
Kipps • PRODUCTION OF FIELD CROPS
Kohnke • SOIL PHYSICS
Kohnke and Bertrand • SOIL CONSERVATION
Krider and Carroll • SWINE PRODUCTION
Laurie, Kiplinger, and Nelson • COMMERCIAL FLOWER FORCING
Laurie and Ries • FLORICULTURE
Maynard and Loosli • ANIMAL NUTRITION
Metcalf, Flint, and Metcalf • DESTRUCTIVE AND USEFUL INSECTS
Rice, Andrews, Warwick, and Legates • BREEDING AND IMPROVEMENT OF FARM
ANIMALS
Roadhouse and Henderson • THE MARKET-MILK INDUSTRY
Smith and Wilkes • FARM MACHINERY AND EQUIPMENT
Thompson and Kelly • VEGETABLE CROPS
Thompson and Troeh • SOILS AND SOIL FERTILITY
Thorne • PRINCIPLES OF NEMATOLOGY
Treshow • ENVIRONMENT AND PLANT RESPONSE
Walker • PLANT PATHOLOGY
Wilson • GRAIN CROPS

Professor R. A. Brink was Consulting Editor of this series from 1948 until January 1, 1961.

FARM MACHINERY AND EQUIPMENT

Sixth Edition

Harris Pearson Smith, A. E.
Professor Emeritus of Agricultural Engineering
Texas A & M University

Lambert Henry Wilkes, M. S.
Professor of Agricultural Engineering
Texas A & M University

McGRAW-HILL BOOK COMPANY

New York   St. Louis   San Francisco   Auckland   Düsseldorf   Johannesburg
Kuala Lumpur   London   Mexico   Montreal   New Delhi   Panama
Paris   São Paulo   Singapore   Sydney   Tokyo   Toronto

FARM MACHINERY AND EQUIPMENT

1 2 3 4 5 6 7 8 9 0   D O D O   7 8 3 2 1 0 9 8 7 6

This book was set in Press Roman. The editors were C. Robert Zappa
and Douglas J. Marshall; the production supervisor was Charles Hess.
New drawings were done by J & R Services, Inc.
R. R. Donnelley & Sons Company was printer and binder.

Library of Congress Cataloging in Publication Data

Smith, Harris Pearson, date
    Farm machinery and equipment.

    (McGraw-Hill publications in the agricultural sciences).
    Includes index.
    1. Agricultural machinery. I. Wilkes, Lambert Henry,
joint author. II. Title.
S675.S55 1976   631.3   75-37668
ISBN 0-07-058957-7

Contents

Preface

Machines used in the production and processing of crops grown for food and fiber are constantly changing. New developments in farm equipment and new technology in farming practices have reduced farm labor requirements. Machines become obsolete and uneconomical within a few years. These factors make it necessary to revise and bring up to date developments and improvements in the various types of farm equipment in current use on the farm.

In the sixth edition of *Farm Machinery and Equipment* the authors discuss the latest developements in no-tillage and minimum tillage, planting with new seed-metering systems, weed control with chemicals and flame, and new techniques in hay baling and stacking. A new method of packaging harvested cotton is described and illustrated. Equipment for the application of insecticides, herbicides, and fungicides is described. Harvesting equipment for cotton, corn, small grains, and several other crops is discussed in considerable detail. Crop-processing and special labor-saving equipment is described. An effort has been made to present the latest information about equipment used on the farm that has proved to be economical in use and instrumental in reducing the cost of crop production.

As there is a trend in the United States to convert to the metric system of measurements, and this book is used in many areas where the metric system is used, the metric equivalent is shown in parentheses following the English measurement.

The authors wish to express their appreciation to the many farm machinery manufacturers for their cooperation in furnishing trade literature and illustrative material. Without their aid it would not have been possible to prepare a manuscript covering the various types of farm machines.

Thanks are due Professor Paul R. Chilen of the Department of Agricultural Engineering of Texas A & M University and the Pre-Employment Laboratory in Vocation Agriculture for his helpful suggestions. Thanks are also due Professor J. W. Sorenson Jr., of the Department of Agricultural Engineering of Texas A & M University for reviewing the chapter on crop-processing equipment, and Professor V. W. Edmonson of the Department of Agricultural Economics and

Sociology of Texas A & M University for reviewing the chapter on economics and management.

Appreciation is extended to Mrs. Verna Dillingham for her help in typing the material for this edition of this book. A sincere effort has been made to give other credit wherever due, and any oversight was not intentional.

Harris Pearson Smith
Lambert Henry Wilkes

Chapter 1

Farm Machinery and its Relation to Agriculture

In the beginning, all crops for human sustenance were produced and prepared by the power of human muscles. Many centuries passed before the power of animal muscles was used to relieve that of the human being. With the discovery of iron, tools were fashioned that further reduced the labor of human muscles. The transition from hand farming to this modern power-farming age was at first slow, but with the development of the steel plow, the internal-combustion engine, the farm tractor, and other modern farm machines, the movement has accelerated beyond the wildest dreams of our ancestors. The changes which occurred in the past two decades have so tremendously affected human values that one wonders what effect farm machines of the future will have on our welfare. In fact, there has been more farming progress in the last hundred years than in all the previous history of the world.[1] It might be said that American farm production is almost totally mechanized.

Satellites in the sky are taking photographs of farm lands which enable the scientists to spot diseased areas and to measure soil moisture, soil temperature, and many other factors. A satellite for this purpose was launched by the United States on January 21, 1975.

[1] *Agr. Engin.,* **34**(2):91, 1953; *Life,* **34**(1):62, 1953; **38**(1):54, 1955.

PROGRESS OF FARM MECHANIZATION

In 1855 practically 80 percent of the population of the United States lived on farms, while in 1973, more than 90 percent lived in towns, cities, and urban areas (Fig. 1-1).

> Since the peak of farm population in 1916, the trend in the number of persons living on farms has been generally downward. The depression in the 1930's brought a temporary increase, but World War II with its demand for manpower in industry and the armed forces caused a rapid loss in the farm population. The high level of nonfarm employment prevailing since 1916, together with defense mobilization following the outbreak of hostilities in Korea and Vietnam, have been conducive to a continuation of a relatively high rate of net migration from farms.[1]

In 1854, farm tools were so crude that each farm worker could produce only enough food for five to six people. By 1920, with improved horse-drawn equipment, the farm worker could support ten people. In 1955, with modern power equipment, the farm worker could support about eighteen people. In 1963 one farm worker could support about thirty people. By 1974 it was estimated that one farmer could produce enough food and fiber to support more than 55 people (Fig. 1.2).

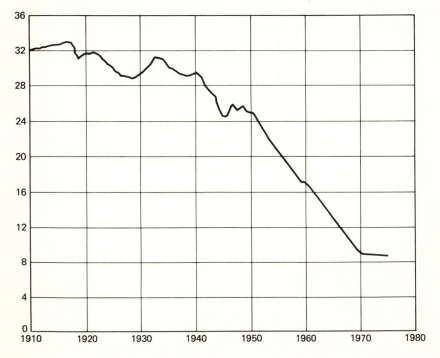

Fig. 1-1   Decline in farm population in the United States, 1910–1973.

[1] *U.S. Dept. of Agr., BAE.*

Table 1-1   Population: Total and Farm, U.S., 1959, 1960, 1965, 1970, 1972, 1973

Year	Total population	Farm population	
		Number	Percent of total
1973	210,036,000	9,472,000	4.5
1972	208,441,000	9,610,000	4.6
1970	204,335,000	9,712,000	4.8
1965	193,709,000	12,363,000	6.4
1960	180,007,000	15,635,000	8.7
1959	176,551,000	16,592,000	9.4

Source: Agricultural Statistics, 1974.

Farm equipment is being increased in size and efficiency so that farmers can produce more with less labor and cost.

A key part of the technological revolution under way in agriculture, and largely a product of it, has been the rapid increase in output per worker-hour of labor on farms. Output per worker-hour is now the greatest in history. The period from 1945 to 1973 has had rapid progress in farm mechanization and sharp increases in yields of crops and livestock because of widespread adoption of improved farming practices. These changes have made possible a great rise in total farm output, with fewer worker-hours spent at farm work.

Power Equipment Cuts Production Worker-Hours

The effect of the mechanization of agriculture is shown in the number of worker-hours required to grow and harvest an acre of wheat yielding 20 bushels. In 1830, when the grain was sown by hand and harvested by hand with a cradle, 55.7 worker-hours were required. In 1896, with the use of a horse-drawn drill and binder, 8.8 worker-hours were required, while in 1930, with the

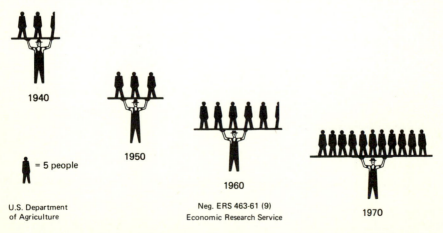

1940

= 5 people

1950

1960

U.S. Department
of Agriculture

Neg. ERS 463-61 (9)
Economic Research Service

1970

Fig. 1-2 Fewer farm workers support more people by the use of new developments and improvements in farm machinery.

tractor-drawn drill and combine, only 3.3 worker-hours were necessary.[1] Newer machines and improved practices in producing spring wheat reduced the required worker-hours in 1950 to 1.4 in South Dakota and to 1.8 in northeastern Montana and southwestern North Dakota. Where summer fallow was practiced, the worker-hour requirement per acre was 1.9 in western South Dakota and 2.6 in the central areas of the Dakotas. The per-acre tractor-hours for these areas were 0.8, 1.4, 1.5, and 1.8, respectively.[2] The difference in worker- and tractor-hours results from the use of self-propelled combines and hauling where no tractors were used. Improved machines and practices have brought about similar reductions in worker-hour requirements in the production and harvesting of most field crops, in the relation of farm output to labor input (Table 1-2).

There has been more progress in the reduction of worker-hours required to grow and harvest an acre of cotton from 1940 to 1970 than in all the previous history of the crop. This resulted from farm-mechanization practices and the use of chemicals.

Equipment Must Suit the Crops and the Types of Farming

The two major crop systems in the United States are *row crops* and *broadcast crops*. The principal row crops are corn, soybeans, cotton, potatoes, tobacco, and truck crops. Hay, rice, wheat, and the small grains are broadcast crops. Farm machinery can be profitably used with both systems, but the more uses to which a machine can be adapted, the less the initial investment in equipment. Certain types of plows and harrows for seedbed preparation have a wide application. Grain drills and combines are adapted for seeding and harvesting of wheat and the small grains, and the combine can also be used for harvesting some row crops, such as sorghum grain, corn, soybeans, and other crops, with special attachments.

Breeding Crops to Suit Machinery

Certain field crops do not readily lend themselves to machine harvesting. Originally grain sorghum had drooping heads which made it difficult to head them without cutting excessively long stems. Plant breeders have developed varieties of sorghum with straight, erect heads of uniform height that are well adapted to combining. As cotton matures, it produces long vegetative and fruiting branches with an abundance of foliage, which make it difficult to harvest the cotton bolls with machinery. Plant breeders, however, have developed types of cotton plants that are more suitable to machine harvesting. Fluffy types are suitable for the picker, and nonfluffy or stormproof types are best harvested by the stripper.

Trend toward Tractor-mounted, Pickup, and Quick-Coupler Units

When the tractor was first used for the operation of field equipment, all machines were pulled behind the tractor. Specially designed planters and

---

[1] *U.S. Dept. Agr. Misc. Rpt.* **157**, 1933.
[2] *U.S. Dept. Agr., BAE,* F. M. 92, Section 4, 1953.

Table 1-2 Labor: Indexes of Farm Production per Hour, by Enterprise Groups, 48 States, 1950-1972*
(1967 = 100)

All crops	Feed grains	Hay and forage	Food grains	Vege- tables	Fruits and nuts	Sugar crops	Cotton	Tobacco	Oil crops
39	23	55	40	57	64	38	25	66	47
38	23	58	38	59	65	39	28	67	46
42	26	60	46	63	67	43	30	67	50
43	27	63	43	64	70	49	33	67	51
45	29	58	46	67	74	51	35	71	54
48	31	63	50	70	75	53	39	75	60
52	35	65	54	76	75	59	41	80	67
56	40	71	62	80	69	67	44	77	69
65	47	77	85	82	73	67	48	81	81
66	52	81	77	88	74	75	52	81	82
71	58	84	93	89	74	79	56	87	84
73	64	89	86	93	78	82	61	88	90
77	70	89	89	92	87	82	71	93	92
82	77	92	90	97	87	97	78	97	94
85	78	93	97	96	91	90	87	99	89
92	92	96	101	99	95	88	101	95	100
95	93	99	102	99	99	92	101	96	101
100	100	100	100	100	100	100	100	100	100
106	102	102	108	101	98	113	130	98	110
112	109	145	113	106	109	115	117	98	114
110	101	148	117	106	107	121	125	104	115
122	118	152	125	106	110	130	127	101	120
126	127	158	120	109	109	139	143	105	123

* Annual data for 1910-1949 in 1964 issue of this publication (see source note).
    Source: Statistical Bulletin No. 233, Economic Research Service, U.S. Dept. of Agriculture, 1973.

cultivators were mounted on row-crop tractors about 1930. With the development of the power takeoff, other machines, such as mowers and corn pickers, were fitted to and mounted on the tractor. These first tractor-mounted units required considerable labor and time to mount and dismount. They were put together practically piece by piece and taken off in a similar manner. All lifting and adjusting was done by hand levers. Later, units were developed that could be attached as assembled units and lifted with mechanical power lifts. Two- and three-bottom plows were at first considered too heavy to be picked up or lifted as units. In 1955, however, three-bottom moldboard plows, tandem disk harrows, and many other formerly trailing machines were picked up, while making turns, by means of hydraulic power. Consequently, most farm equipment for cultural purposes is now designed for tractor attachment. The exceptions are self-propelled combines, corn pickers, and cotton pickers and strippers.

Farm Management

Farm machines designed for higher speeds, constructed of heat-treated steels, and equipped with more durable bearings will lessen operating time and will lower costs. Terracing and contouring of fields will cause changes in farming practices, both in the types of machinery used and in cropping systems. These and various other factors will materially affect the management of farm labor and equipment.

See Chap. 25 for a complete discussion of the management of farm equipment.

Rubber Tires for Farm Equipment

When the first edition of *Farm Machinery and Equipment* was published in 1929, there were no farm machines equipped with rubber tires. In the second edition in 1937, only a few machines equipped with rubber tires were shown. The third edition in 1948 showed rubber tires on most equipment. All other editions show all field equipment mounted on pneumatic rubber tires.

The various factors related to the types and use of rubber tires are given in Chap. 8.

REFERENCES

Anderson, K. W.: New Horizons in Farm Machinery Development, *Agr. Engin.,* 33(12):765, 1953.

Anonymous: Changes in Farm Production and Efficiency, A Summary Report, *Statistical Bul. 233,* Economic Research Service, U.S. Dept. of Agr., 1973.

Buchele, Wesley F.: 2003: A Farming Odyssey, *Implement & Tractor,* 88(3 and 4), 1973.

McKinney, John,: Satellites, *Progressive Farmer,* 88(2), 1973.

Wennblom, Ralph D.: The Eye Watches Your Crops, *Farm Journal,* 98(2):29, 1974.

QUESTIONS AND PROBLEMS

1   Discuss the development and progress of farm mechanization.
2   Explain how power equipment reduces man-hours in crop production.
3   Enumerate machines that can be used in producing two or more crops. List some machines that have a single use.
4   Trace the trend in the development of tractor-mounted and quick-change units.
5   Discuss the value of satellites.

Chapter 2

Materials of Construction

The strength, durability, and service of a farm implement depends largely upon the kind and quality of material used in building it. There is a tendency in the construction of implements to eliminate as many castings as possible and to use pressed and stamped steel. Where this is done, the cost of manufacturing machinery in quantities is materially reduced. The weight of the machine is reduced, but the strength and durability are retained and often improved. The success or failure of an implement frequently depends upon the material used in building it.

The materials used in the construction of farm equipment may be classified as *metallic* and *nonmetallic*. The metallic is further divided into ferrous and nonferrous materials.

NONMETALLIC MATERIALS

The nonmetallic materials are wood, rubber, leather, vegetable fibers, and plastics.

Wood

Iron, steel, and plastics have practically taken the place of wood. There are, perhaps, two reasons for this: first, steel and plastics are more durable; second, they are cheaper than good wood because of the scarcity of the latter.

Rubber

Rubber is both derived from the gum of trees and made synthetically. Special compositions of rubber are developed to obtain the properties desired for a particular application. Design engineers should have a thorough knowledge of the properties of rubber—both natural and synthetic. There are several grades of rubber materials varying in the general properties of hardness, flexibility, bonding properties, and chemical resistance. The leading use of rubber on farm equipment is in the production of implement tires and tubes. Much rubber is also used in making flat and V belts and for the insulation of ignition wires. Rubber bushings on suspended oscillating components often give an excellent service life and require no lubrication. Disks of rubber to clasp plants are used on transplanters.

Plastics

A plastic material is an organic solid, polymerized to a high molecular weight, that is capable of being molded, usually with the aid of heat or pressure or both. There are many groups and types of commercially available plastics and they are sold under many trade names.

Plastics fall into two general categories, thermoplastics and thermosettings. Thermoplastics are usually soft and pliable at normal temperatures and become hard when cold. Typical thermoplastics used on machinery include acrylics and polyethylene polyvinyl chloride (PVC). Thermosetting plastics retain a permanent form or shape when heat and pressure are applied during the forming process. Materials that fall into this category include epoxies, phenolics, polyurethane, and silicones.

A common plastic product used for seed hoppers and chemical tanks is fiber-glass-reinforced material made from acrylic or polyester and is commonly referred to as "fiber glass." It is transparent and has good resistance to weathering and most chemicals used in agriculture. Because of many properties of the plastics, they are widely used. Some uses include plow handles, bearings, tubing, conveyor belting, bristles for brushes, windows, and machine panels or hoods. Polyethylene is often used as protective covering for production of certain vegetable crops.

Leather and Vegetable Fibers

Leather is largely a belting material. Vegetable fibers are used in brushes, fabrics, and upholstery padding.

NONFERROUS METALS

The nonferrous metals are copper and its alloys (such as brass and bronze), aluminum, magnesium, lead, zinc, and tin.

Alloy

An *alloy* is a substance that has metallic properties and is composed of two or more chemical elements of which at least one is a metal. The number of possible alloys is infinite. They are made by the fusion of metals. The most common groups of alloys are bronze, brass, babbitt, alloy steels, and the aluminum alloys.

Copper

In commercial importance, copper ranks next to iron and steel because of its electrical and heat conductivity and its capacity to form useful alloys. Copper is soft enough to be rolled or hammered into thin sheets or drawn into fine wire. It is used for ignition and electric wires on engines, in generator and starting motors, and in tubing for conducting fuel from tank to carburetor.

Brass

Ordinary brass is an alloy of copper and zinc. Some commercial brasses contain small percentages of lead, tin, and iron. The percentage of copper in brass may range from 60 to 90 percent, and the percentage of zinc from 10 to 40 percent. Brass is used for making radiators, pipe, welding rods, screens for fuel lines, instrument parts, and fittings.

Bronze

Bronze is an alloy of copper and tin. However, zinc is sometimes added to reduce the cost of the alloy or to change its color and increase its malleability. The amount of tin in bronze may vary from 5 to 20 percent. Phosphor bronze, manganese bronze, and aluminum bronze are special copper alloys containing small percentages of tin, zinc, and other metals such as phosphorus, manganese, and aluminum. These are used for bearing bushings, springs, pipe fittings, valves, pump pistons, and bearings.

Babbitt

Babbitt is a tin-base alloy containing small amounts of copper and antimony. Good babbitt for automobile bearings should contain 7 percent copper, 9 percent antimony, and 84 percent tin. It is used mostly as a bearing metal.

Solder

Common solder contains about one part tin and one part lead. Hard plumbers' solder contains two parts tin and one part lead. Solder is used extensively in joining brass, copper, tin, steel, and cast iron.

Aluminum

This is a white metal with a bluish tinge which is resistant to corrosion and to many chemicals. It, however, can be dissolved by alkalies and hydrochloric acid. It is frequently alloyed with iron and copper. Aluminum is extensively used to make light castings for certain types of farm equipment and for coating chemical tanks.

Zinc

Zinc is a bluish-white, crystalline, metallic element, brittle when cold, malleable at 110 to 210°C. It is used mostly as a coating on sheet iron and die castings as a protection against corrosion.

FERROUS METALS

The ferrous metals include cast iron, wrought iron, and steel. These metals are all produced by the reduction of iron ore into pig iron and subsequent treatments of the pig iron by various manufacturing processes. The term "cast" refers to the process that is used to obtain the final form or shape of the metal. The hot molten metal is poured into a mold and allowed to cool and harden into the shape dictated by the pattern of the mold. The method is used to form many intricate and irregular-shaped parts on farm machinery. The basic differences between iron and steel include the manufacturing processes, the amount of carbon, and impurities, which in turn affect the physical properties.

Cast Iron

There are five general types of castings that are made of iron. These include gray, white, chilled, malleable, and ductile.

Gray Cast Iron

Gray cast iron is formed by allowing the molds to cool slowly in natural air. Most of the carbon occurs in the casting as graphite flakes that are responsible for the gray color noted when the part is broken. The parts formed are high in compressive strength and low in tensile strength and are relatively brittle. The wearing characteristics due to abrasion are low. To obtain rigidity required in the parts made in this manner, the castings are usually large.

White Cast Iron

Rapid cooling of the castings causes the carbon to remain in a chemically combined form and gives a characteristic white color when fractured. The parts so made are very hard and brittle; thus it is used in forming parts that are subjected to abrasion or wear. Typical parts include plates for burr and roller mills and plain bearings used on some disk harrows.

Chilled Cast Iron

Chilled cast iron is made by chilling or rapidly cooling only portions of the casting. This is accomplished by lining or making the mold with metal at the areas where chilling is desired. The hot molten metal is chilled in these areas and the casting assumes the characteristics of white cast iron, whereas the remaining portions are cooled more slowly and have the same properties as a gray cast iron. This process is used to form many machine parts that require good wear or abrasion resistance of surface or edge but where it is desired to retain toughness in the body to absorb shock loads. Chilled cast iron is used on moldboards and shares on plows and some bearings and on chain sprockets.

Malleable Cast Iron

Malleable iron is made by subjecting a white casting to the annealing or "softening" process. The casting is heated to a temperature of about $1600°F$ and held in the oven for a long period of time; then it is cooled very slowly. This heat treatment converts the combined carbon into free carbon in an amorphous condition, but not the crystalline form as in gray cast iron. The castings or parts are malleable, tough, and strong. The properties are similar to those of low-carbon steel, but the cost is considerably less. Typical farm-machinery parts made of malleable iron include mower guards, ledger plates, control pedals, and chains.

Ductile Cast Iron

This is a new metal for farm-equipment parts. Patents were granted on the process of producing ductile cast iron in 1949. It is a high-grade iron produced by the ladle addition of magnesium alloy to molten iron prepared to produce gray cast iron. The magnesium acts as a desulfurizer, and when added in controlled amounts, it produces spheroidal carbon instead of flake carbon (graphite).

Ductile cast iron has many applications in farm equipment, such as sprockets, gears, chilled plowshares, mower guards, parts for hay-baler knotter mechanism, and tail-wheel mounting brackets for plows.

Ductile cast iron can be welded similarly to gray cast iron. It requires, however, a special reverse-polarity arc rod designated *Ni-rod* 55. This rod deposits a bead with 8 percent elongation and with tensile properties of over 60,000 lb/in^2.

Wrought Iron

Wrought iron is nearly pure iron, with some slag, and is used in forge work, as it is readily welded and easy to work. Wrought iron has very little carbon in it, ranging from 0.05 to 0.10 of 1 percent. It is expensive, however, and a mild steel is used to a considerable extent in place of it. The commercial form is obtained by rolling the hot iron into bars or plates from which nails, bolts, nuts, wire, chains, and many other products are made.

Steel Alloys

A steel alloy is a mixture of two or more metals. The mixture is composed largely of steel with small amounts of one or more alloy metals. The more common alloy elements used in steel are boron, manganese, nickel, vanadium, tungsten, and chromium.

Steel

Steel is made from pig iron with manufacturing processes different from those used for making cast iron. The carbon content is lower and more carefully controlled. There are several ways that steel may be classified. Some of these include (1) the manufacturing processes (bessemer steel, open-hearth steel, and electric steel), which affect the quality, (2) the carbon content, (3) alloy steel, where other metals are added, (4) uses such as structural or tool steel, and (5) methods of forming, such as rolled, forced, and cast.

Carbon Steel   Steel is an alloy of iron and carbon having a carbon content generally below 1.5 percent. Carbon is very important and its presence controls the hardness of the finished steel product. The proportion of carbon is carefully controlled in the manufacturing processes, since the carbon also influences the stiffness and brittleness. Carbon steel is generally classed as low-carbon (carbon content not exceeding 0.25 percent), medium carbon (0.25 to 0.50 percent carbon), and high-carbon (over 0.50 percent carbon).

Low-carbon steels are used extensively in the construction of farm machinery. Practically all of the structural parts are made of low-carbon steel. This material is malleable and easy to cut and weld. Medium-carbon steels are used for parts and components requiring greater strength and hardness, such as shafting and connecting rods. High-carbon steel is very hard and is used for making tools, ball and roller bearings, and cutting tools.

Alloy Steel   The physical properties of steel can be changed and improved to meet the requirements of special applications and uses by the addition of special alloy metals. The more common alloy elements used in steel include boron, manganese, nickel, tungsten, and chromium.

Boron Steel   This contains a small amount of boron. The boron acts to increase the hardening ability of the steel, that is, its ability to harden deeply when heat-treated by quenching and tempering. It is used for axle shafts, wheel spindles, steering-knuckle arms, cap screws, and studs.

Manganese Steel   This usually contains 11 to 14 percent manganese and from 0.8 to 1.5 percent carbon and has properties of extreme hardness and ductility. It is usually cast for the desired shape and finished by grinding. It is used in feed grinders and machine parts subject to severe wear.

Nickel Steel   Steel containing from 2 to 5 percent nickel and from 0.10 to 0.50 percent carbon is strong, tough, and ductile. Nickel steels are used in making parts that are subjected to repeated shocks and stresses.

Vanadium Steel   When less than 0.20 percent vanadium is added to steel, the resulting alloy is given additional tensile strength and elasticity comparable to the low- and medium-carbon steels with a corresponding loss of ductility.

Chrome-Vanadium Steels   These contain about 0.5 to 1.5 percent chromium, 0.15 to 0.30 percent vanadium, and 0.15 to 1.10 percent carbon. These steels are used extensively in making machinery castings, forgings, springs, shafting, gears, and pins.

Tungsten Steel   Steels containing from 3 to 18 percent tungsten and from 0.2 to 1.5 percent carbon are used for dies and high-speed cutting tools.

Molybdenum Steel   This steel has properties similar to tungsten steel.

Chrome Steel   Chrome steels usually contain from 0.50 to 2.0 percent chromium and from 0.10 to 1.50 percent carbon. Chromium steels are used in making high-grade balls, rollers, and races for ball and roller bearings. Chrome steels containing from 14 to 18 percent chromium produce a variety of stainless steel.

Chrome-Nickel Steel   The average chrome-nickel steel contains about 0.30 to 2.0 percent chromium, from 1.0 to 4.0 percent nickel, and from 0.10 to 0.60 percent carbon. Heat treatment increases its tensile strength, elasticity, and endurance limits. It is tough and ductile. Chrome-nickel steel is used in making gears, forgings, crankshafts, connecting rods, and machine parts.

When chrome-nickel steel contains from 16 to 19 percent chromium, 7 to 10 percent nickel, and less than 0.15 percent carbon, it is generally called *stainless steel*. The commonly called 18-8 *stainless* falls in this group.

Tool Steel

The term *tool steel* is used in designating a high-carbon steel that is used for making tools. It has the property of becoming extremely hard by quenching from a temperature of 1400 to 1800°F. It can then be treated to obtain any degree of hardness by heating at lower temperatures.

Soft-center Steel

Soft-center steel consists of three layers of steel, as shown in Fig. 2-1. Two layers of hard steel are placed on each side and welded to an inner layer of soft steel. In this manner, a hard surface is obtained, without brittleness. Soft-center steel is used in the making of plow bottoms. Filing a slight notch in the edge of the metal will reveal the three layers.

Clad steels or *bimetal* steels are made by permanently bonding a layer of nickel, inconel, or monel metal to a heavier base layer of steel by hot rolling. The cladding layer may range in thickness from $\frac{3}{16}$ in up, with the cladding amounting to about 10 to 20 percent of the total plate thickness.

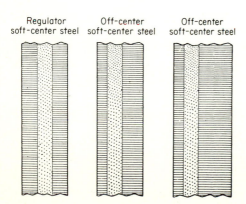

Regulator soft-center steel   Off-center soft-center steel   Off-center soft-center steel

Fig. 2-1   Different types of soft-center steel.

Shapes of Steel

Steel that is formed into angles, channels, tee bars, I beams, Z bars, U bars, and hollow squares, as shown in Fig. 2-2, is known as *structural steel*. Solid bars are furnished in many shapes, such as round, half-round, oval, half-oval, square, hexagon, and flat-rectangle strips. Various sizes of round and square tubing are available. Many special parts are formed from flat-rolled carbon steel and stainless sheets and plates. A few of these shapes are shown in Fig. 2-3.

Hardening of Finished Steels

In many cases where long-life service is desired, extremely hard steels cannot be forged and machined to the required shape and finish. Under these conditions a softer steel is shaped and finished, then given a hardening treatment. The most common hardening processes are casehardening and hardening by heat treatment.

Casehardening   This is a process of hardening a ferrous alloy so that the surface layer or case is made substantially harder than the interior or core (Fig. 2-4). Casehardening can be done by several processes, such as carburizing and quenching, carbonitriding, nitriding, cyaniding, induction hardening, and flame hardening.

Carburizing is a process in which steel is packed in charred peach pits or charcoal and heated at about 1600°F for a long enough period to give the desired depth of hardness. It is then removed, quenched, and tempered to give the desired hardness.

Nitriding is a process of casehardening by placing the finished heat-treated steel in an airtight box and heating to 1000°F as ammonia gas is injected into the chamber.

Carbonitriding is a process of hardening steel by the addition of a carbon-rich gas as well as ammonia.

Cyaniding is a process where the steel is dipped into a molten bath of potassium cyanide for a short time. Some carbon and nitrogen are absorbed by the steel, which results in the hardening of a thin surface layer.

Induction hardening is accomplished by the use of a high-frequency alternating electric current for a short period. A current is induced on the surface of the steel, which causes localized heating. After heating, the surface is flooded with water to quench and harden it.

Flame hardening is a process in which an oxyacetylene torch is used to heat the surface quickly to a temperature above the critical temperature, after which the surface is quenched with water.

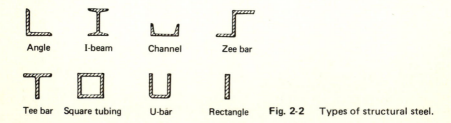

Angle   I-beam   Channel   Zee bar

Tee bar   Square tubing   U-bar   Rectangle   **Fig. 2-2**   Types of structural steel.

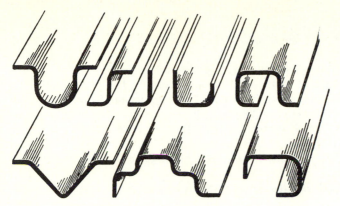

Fig. 2-3   Sheet metal can be shaped or rolled into many shapes.

Hardening by Heat Treatment   *Heat treatment* is a term used to describe the application of heating and cooling processes to steel, through a range of temperatures, to improve the structure and produce desirable characteristics. Such treatments include annealing, hardening, tempering, and casehardening.

Plow beams, plow disks, and disk-harrow blades are examples of parts of agricultural machines that are heat-treated in order to make more serviceable implements.

Hard Facing or Surfacing   The application of a hard surface, or face, by welding is not to be confused with the hardening of finished surfaces. *Hard facing*, or surfacing by welding, is the addition of a hard metal over the base metal by applying a welding-rod deposit to provide a final surface that is harder than the original surface.

Hard facings are applied to parts for wear resistance, heat resistance, corrosion resistance, or combinations of the three. Most hard facing is done to prevent wear. In hard-facing parts, it is essential that the correct hardening material be selected to suit the base metal.

There are possibly hundreds of different hard-facing alloys available, and these are manufactured in three forms: as welding rods, as insert shapes, and in powdered forms. There are many types of welding rods. The rods used with the oxyacetylene torch are not coated. They are heated and dipped into a special flux. Electric rods usually have a flux coating.

Inserts and filler bars are welded on surfaces where extra-heavy hard facing is required.

Hard-facing powders are spread over the base metal, which is heated to the melting point to embed the powders firmly.

Fig. 2-4   Casehardened steel.

REFERENCES

Brady, G. S.: *Materials Handbook,* McGraw-Hill Book Company, New York, 1944.

Clapp, H. W., and D. S. Clark: *Engineering Materials and Processes, Metals and Plastics,* International Textbook Company, Scranton, Pa., 1949.

DuMond, T. C.: *Engineering Materials Manual, Materials and Methods,* Reinhold Publishing Corporation, New York, 1951.

Geiger, H. L., and H. W. Northrup: A New Metal for Farm Tool Components, *Agr. Engin.,* **32**(3):143-147, 1951.

Hessenthaler, W. H.: Machine Laying of Polyethylene Mulch, *Agr. Engin.,* **39**(4):235, 1958.

Lyman, Taylor: *Metals Handbook,* The American Society for Metals, Cleveland, 1948.

Marks, Lionel S. (ed.): *Mechanical Engineers' Handbook,* 6th ed., McGraw-Hill Book Company, New York, 1958.

Oberg, Erik, and F. D. Jones: *Machinery's Handbook,* The Industrial Press, New York, 1949.

Ryerson Steels, Joseph T. Ryerson & Son, Inc., St. Louis, 1953-1954.

Smith, Ronald B.: *Materials of Construction,* McGraw-Hill Book Company, New York, 1973.

QUESTIONS AND PROBLEMS

1 Classify and give examples of construction materials.
2 Discuss the various nonmetallic materials and give uses of each.
3 Discuss the use of plastics in farm equipment.
4 Name the nonferrous metals and give uses of each.
5 Define an alloy.
6 What are the ferrous metals?
7 Explain the differences in the various types of castings.
8 How is malleable cast iron produced?
9 Discuss the influence of carbon content on the hardness of steel.
10 Discuss the differences in structure and metals in soft-center steel and in the clad steels.
11 Describe the various methods of hardening finished steels.
12 Discuss hard facing of metals.
13 Name the common steel alloys and their uses.
14 What is stainless steel?

Chapter 3

Mechanics

A clear conception of the fundamental principles of mechanics, as well as of their practical application to machinery, is necessary to a comprehensive study of farm machinery. *Mechanics* is the science that treats of forces and their effect.

Force

Force is the action of one body upon another which tends to produce or destroy motion in the body acted upon. Force may vary in magnitude and in method of application. In general, force is associated with muscular exertion. This, however, does not completely cover the scope and action of force, because flow of an electric current, freezing of a liquid, and ignition of explosives may exert a certain amount of force. In order for different forces to be compared, they must all be in terms of the same unit. One such unit is called the *pound weight*.

Work

Whenever a force is exerted to the extent that motion is produced, work is performed. Work is measured by the product of the force times the distance moved, and it can be expressed in several combinations of units of weight (force) and distance, as inch-pounds and foot-pounds. A foot-pound of work is done when a body is moved 1 ft against a force of 1 lb weight. A centimeter-gram of

work is accomplished when an object is moved 1 cm against 1 g of force. The
amount of work required to place a 100-lb bag of grain on a wagon that has a
box 4 ft from the ground can be determined by multiplying the weight, 100 lb,
by the height, 4 ft, which will equal 400 ft·lb of work done to place the bag of
grain upon the wagon, or

Work = force x distance

or

$$W = F \times D$$
$$W = 100 \times 4 = 400 \text{ ft·lb of work}$$

If a force moves in a circular direction to produce a twisting action, this
rotating force is termed *torque*. For example, a belt which exerts a force to turn
a pulley and thus transmits power through a shaft gives the shaft a twisting
action or a torque force. The pull on the belt in pounds multiplied by the radius
of the pulley equals the torque in foot-pounds or, rather, pound-feet
(gram-centimeters).

A force which produces the same effect upon a body as two or more forces
acting together is called their *resultant*. The separate forces which can be so
combined are called *components*. The finding of the resultant of two or more
forces is called the *composition* of forces.

The *moment* of a force with respect to a point is the product of the force
multiplied by the perpendicular distance from the given point to the direction of
the force. In Fig. 3-1, the moment of the force P with relation to the point A is
P times AB. The perpendicular distance is called the *lever arm* of the force. The
moment is a measure of the tendency of the force to produce rotation about the
given point, which is termed the *center of moments*. If the force is measured in
pounds and the distance in inches, the moment is expressed in pound-inches; if
measured in grams and meters the expression would be gram-meters. If P is a
force of 10 lb and 20 in (50.8 cm) from A, its moment about A is 200 lb in.

Power

Power is the rate of doing work. To determine the power used or transmitted by
a machine, it is necessary to measure the force, also the distance through which

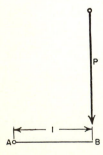

Fig. 3-1   The moment of force.

the force acts and the length of time required for the force to act through this distance. The units of power ordinarily used in America are the *foot-pound per second*, the *foot-pound per minute*, and the *horsepower*.

If a body is moved 1 ft/sec against a force of 1-lb weight, the rate of work is 1 ft·lb/s. If it moves 1 ft/min against the same force, the rate is 1 ft·lb/min. If it moves so that 33,000 ft·lb are done each minute, the rate is 1 hp. The horsepower is based on the rate at which a 1500-lb horse can do work. If such a horse pulls 150 lb, 10 percent of its weight, and moves at the rate of 220 ft (67.1 m)/min, or 2½ mi/h (4.0 km/h), it would do 33,000 ft·lb of work per minute, this being equal to 150 times 220, or 33,000 ft·lb, or 1 hp.

Energy

Energy is defined as the capacity for doing work. When a 1-lb (454-g) weight has been raised 1 ft (30.48 cm), it is said to have 1 ft·lb (0.138 m·kg) of greater potential energy than it had in its original position. The energy possessed by a body, such as a tractor, due to its motion is termed *kinetic* energy. *Inertia* is the property of a body which causes it to tend to continue in its present state of rest or motion unless acted upon by some force such as a brake.

Simple Machines

A machine is a device that gives a mechanical advantage which facilitates the doing of work. The term is usually associated with such tools as grain binders, threshing machines, mowing machines, and so forth. But really, such machines are made up of many simple machines. There are six simple machines, namely:

1	The lever	4	The inclined plane
2	The wheel and axle	5	The screw
3	The pulley	6	The wedge

Any simple machine is capable of transmitting work done upon it to some other body. The *mechanical advantage* of a machine is the ratio of the force delivered by the machine to the force applied. The force which operates the machine is called the *applied* force. The *efficiency* of the machine is the ratio of the work accomplished by the machine to the work applied to the machine. If the efficiency of a machine could be 100 percent, perpetual motion would exist. Since there is always a loss due to friction, the efficiency of the machine falls below 100 percent.

Lever

The lever is a rigid bar, straight or curved, which rotates about a fixed point called the *fulcrum*. It has an applied force and a resisting force that are well defined by their names. The lever arms for a straight bar are the parts or ends on each side of the fulcrum if the forces act perpendicular to the bar. The mechanical advantage of the lever is the ratio of the length of the lever arm of the applied force to the length of the arm of the resistance force, or

Weight x weight arm = applied force x force arm

Levers are of three classes (Fig. 3-2). In the lever of the first class, the applied force is at one end and the resisting force, or force exerted by the object to be moved, at the other. The fulcrum, or fixed point, is placed between the applied and the resisting forces. Such a lever may have a mechanical advantage of any value, depending directly upon the length of the lever arm between the fulcrum and the point of applied force as compared with the length of the lever arm between the fulcrum and the point of resisting force. The majority of levers found on farm machinery will fall in this class.

Levers of the second class have the applied force at one end, the fulcrum at the other, and the resisting force between them. This class of levers will have a mechanical advantage that will always be greater than unity. As in the case of the lever of the first class, a lever of the second class sacrifices speed and distance for a gain in pull or force.

A lever of the third class has the resisting force at one end, the fulcrum at the other, and the applied force between them. The mechanical advantage of this kind of lever is always less than unity, and unlike the two previous classes, force is sacrificed for a gain in speed and distance. An ordinary crane is a lever of this kind.

Wheel and Axle

This is a modification of the lever and acts on the same principle, but the forces operate constantly (Fig. 3-3). The center of the axle corresponds to the fulcrum,

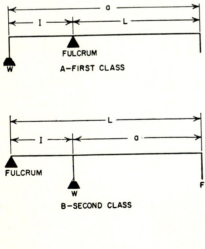

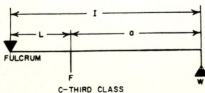

Fig. 3-2   The three classes of levers.

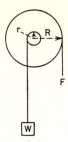

Fig. 3-3   Wheel and axle.

the radius of the axle to the short arm, and the radius of the wheel to the long arm. The mechanical advantage is expressed by the equation

$$F \times R = W \times r$$

where W = weight
F = force applied
R = radius of wheel
r = radius of axle

Pulley

A pully consists of a grooved wheel turning freely in a frame called a *block*; it is a lever of the first or second class. There are several different applications of pulleys, depending on their arrangement. A single fixed pulley affords no mechanical advantage except to change the direction of motion. When one or more fixed pulleys and one or more movable pulleys (Fig. 3-4) are used in combination, they form the *block and tackle*. The mechanical advantage varies directly as the number of ropes that support the movable pulley and the weight,

$$w \times h = F \times 3h$$

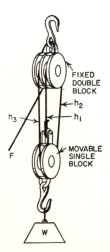

Fig. 3-4   Block and tackle.

or $\dfrac{w}{F}$ = 3, theoretical mechanical advantage

where w = weight
  h = distance weight moves
  F = force applied
  3 = number of ropes supporting w

    The *differential pulley* (Fig. 3-5) is a modification of a block and tackle but differs in that the two pulleys D and C are of different radii and rotate as one piece about a fixed axis B. The endless chain passes under and supports the movable pulley G and any weight attached to it. To raise a load, force is applied downward to chain F, which will rotate pulleys C, D, and G, causing the chain to wind up on the larger fixed pulley D and unwind on the smaller fixed pulley C, thus raising movable pulley G. In operation, consider that point D of the section of chain DH moves up through an arc whose length is equal to BD. At the same time the point C of the section of chain CA will move downward an arc, a distance equal to BC. The length of the chain loop $DHAC$ will be shortened to $BD - BC$, which will cause pulley G to be raised half this amount, P, the pulley force, is then applied to the section of chain EF, and the weight W is lifted at G. The mechanical advantage will be

$$P \times BD = W \times \tfrac{1}{2}(BD - BC)$$

Figure 3-6 shows a geared differential hoist.

Inclined Plane

The inclined plane, shown in Fig. 3-7, is an even surface sloping at any angle between the horizontal and the vertical. The law or principle which governs the

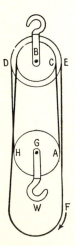

Fig. 3-5   Differential hoist.

Fig. 3-6   Geared differential hoist: (*A*) worm-geared hoist;   (*B*) planetary-geared hoist.

inclined plane in mechanics is that the force applied is increased as many times as the length of the inclined plane is greater than the elevation *H*. Briefly, it is equal to the length over the height, varying with the direction in which the force is applied. Instead of lifting the entire weight of the object vertically, part is supported on the plane and part by the force. Referring to Fig. 3-7, if the force *F* causes the weight *W* to move from *A* to *C* and parallel to the plane, the work done is *F* times *AC*, while the work done against gravity is the weight *W* times *CE* if friction is disregarded, or briefly,

$$F \times AC = W \times CE$$

If the force is parallel to the base *AE*, the advantage would be

$$F \times AE = W \times CE$$

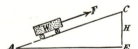

Fig. 3-7   The inclined plane.

Screw

The screw (Fig. 3-8) is the application or modification of the inclined plane combined with the lever. The threads winding around a cylinder bear the same relation to the inclined plane that a winding staircase bears to a straight one. When the screw is turned on its axis with the aid of a lever or gear, its sloping thread causes the load to move slowly in the direction of its vertical axis. The vertical distance between threads is called the *pitch* of the screw. The mechanical advantage is figured upon the condition that the applied force moves through a distance equal to the circumference of a circle whose radius is the length of the jackscrew bar or the radius of the driving gear, while the weight is being moved through a distance equal to the pitch of the screw.

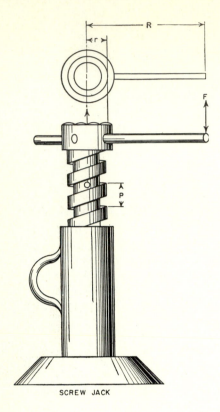

SCREW JACK   **Fig. 3-8**   Screw.

Wedge

The wedge is a modification of the inclined plane. Actually it consists of two inclined planes placed base to base (Fig. 3-9). The force pushing on the wedge into any material, such as a log, will cause forces to act perpendicular to each of the two faces of the wedge.

Fig. 3-9   Wedge.

REFERENCES

Clyde, A. W.: Mechanics of Farm Machinery, *Farm Impl. News,* January-March 1944.

Fairès, Virgil M., and Robert M. Keown: *Mechanism,* McGraw-Hill Book Company, New York, 1960.

QUESTIONS AND PROBLEMS

1   In Fig. 3-2A, a force of 80 lb is exerted at W; l = 12 in and L = 32 in. Find the value of force F required to balance the lever.

2   In Fig. 3-2B, a weight of 90 lb is supported at W; l is 10 in. The total length of L of the lever is 25 in. Find the value of F.

3   In Fig. 3-2C, F = 28 lb; L = 10 in, a = 24 in. What is the weight supported at W?

4   The radius of a roller on which is wound the lifting rope of a windlass is 4 in. What force must be exerted at the end of a crank arm 24 in in length attached to the shaft to lift one ton (2000 lb)? W = 2000; R = 24; r = 4 (Fig. 3-3).

5   In Fig. 3-4 is shown a block and tackle. The pulleys turn freely on a pin. There are three parts of rope. If 90 lb is to be lifted, what force is required at the end of the rope F?

Chapter 4

Transmission of Power

The method of transmitting power from its source to the point of use is one of the greatest problems of the farm equipment designer. The problem is relatively simple on stationary machines or when a tractor is used to operate a stationary thresher or feed grinder. The size and speed of the pulleys on the two units are approximately the same. The tractor pulley is lined up with the driven pulley on the machine and a flat belt is fitted over them and tightened by backing the tractor. Power may also be transmitted from the tractor to the machine with a shaft. But the problem is multiplied many times in the case of a self-propelled combine, where the source of power is an engine mounted on the machine. Power must be transmitted to a slow-revolving reel and to a high-speed fan. Rotating movement must be changed into back-and-forth movement for the knife on the cutter bar and to an oscillating or shaking movement for the straw rack and grain pan. All this is done by means of pulleys and belts, sprocket wheels and chain, gears, and shafts. The operation is made possible by having good bearings to support the shafts and parts. The various parts of the machine are held together by different kinds of bolts and screws. Therefore, it is well to learn something about all these units to have an appreciation of their use in the design and construction of farm equipment.

METHODS OF TRANSMITTING POWER

The methods of transmitting power in connection with farm equipment are (1) direct drive, (2) pulleys and belts, (3) sprocket wheels and chain, (4) gears, (5) shafts and universal joints, (6) hydraulic systems, and (7) flexible shafting.

Direct Drives

When a machine is driven directly from the shaft of an electric motor or internal-combustion engine, this is termed a *direct drive* or *direct connection*. Feed mills and centrifugal water pumps are often driven in this manner. There is usually a clutch or flexible coupling between the power source and the machine.

Pulleys and Belts

A belt of flexible material forming a band about two or more pulleys is a simple method of transmitting power in farm equipment. Belts can be used in many intricate patterns over several pulleys on parallel shafts as shown in Fig. 4-1. The pulleys and belts may be either flat or V-shaped.

   Flat Belts   The most common belting materials are leather, natural or synthetic rubber, and canvas (Fig. 4-2). The principal use of flat belts on field equipment is in elevator chutes to convey harvested crop material from the harvester to a trailer. These belts are made mostly from rubber and canvas belting. Leather belting is expensive, must be kept dry, and is not commonly used on farm equipment. The standard belt speed for farm tractors should be 3100 ft/min ± 100 ft/min.[1] Metal fasteners are generally used to join the two ends of flat belts (Fig. 4-3).

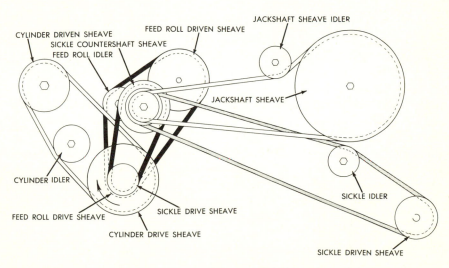

Fig. 4-1   Diagram of V-belt power transmission on a field forage harvester.

[1] ASAE-SAE Standard, ASAE S210.2 (1971).

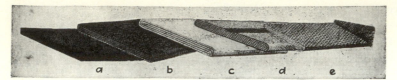

Fig. 4-2 Different kinds of flat belting: (*a*) leather; (*b*) stitched canvas; (*c*) balata; (*d*) rubber; (*e*) solid woven.

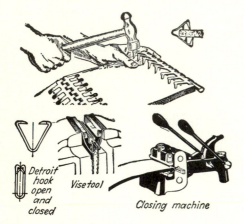

Detroit hook open and closed   Vise tool   Closing machine

Fig. 4-3 Methods of closing metal belt laces: (*above*) alligator; (*below*) clipper.

V Belts   The trapezoidal-shaped or V belts are so named because the sides of the belts are beveled to fit into the V slot of a pulley or sheave. The frictional contact between the sides of the belt and the sheave flanges results in less belt slippage with less tension than flat belts.

The sizes of V belts for agricultural machines have been standardized and referred to by letter designations of HA, HB, HC, HD, and HE, as shown in Fig. 4-4 and Table 4-1. These have been developed to cover the needs of designers and users of farm machines. Belts with these cross sections can be obtained from manufacturers of V belts. Double-sided V belts are used for transmitting power to the sheaves with both sides. Adjustable-speed belts are used on variable-speed drives that are equipped with variable-pitch sheaves (Fig. 4-5).

V belts are often used in multiple when the power being transmitted exceeds the capacity of single belts. In a multiple V belt drive it is important that all belts have the same length so that the load is evenly divided. The belts should be purchased as a matched set when replacement is required. To

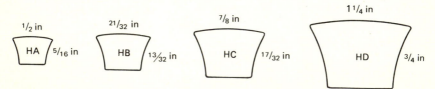

Fig. 4-4 Nominal cross-sectional sizes for standard agricultural V belts. (*Gates Rubber Company.*)

Table 4-1   Nominal Dimensions of Standard Agricultural V Belts

Cross section	Belt size designation	in		mm	
		Width	Depth	Width	Depth
V BELTS	HA	0.50	0.31	12.7	7.9
	HB	0.66	0.41	16.7	10.3
├── WIDTH ──┤	HC	0.88	0.53	22.2	13.5
	HD	1.25	0.75	31.8	19.0
DEPTH	HE	1.50	0.91	38.1	23.0
DOUBLE V BELTS	HAA	0.50	0.41	12.7	10.3
├WIDTH┤	HBB	0.66	0.53	16.7	13.5
	HCC	0.88	0.69	22.2	17.5
DEPTH	HDD	1.25	1.00	31.8	25.4
ADJUSTABLE SPEED BELTS	HI	1.00	0.50	25.4	12.7
	HJ	1.25	0.59	31.8	15.0
├── WIDTH ──┤	HK	1.50	0.69	38.1	17.5
	HL	1.75	0.78	44.4	19.8
DEPTH	HM	2.00	0.88	50.8	22.2

Source: ASAE Standard: ASAE S211.3.

overcome some of the problems encountered with single belts used in multiple, belts are available with two or more standard V belts bonded together as shown in Fig. 4-6, to give the advantages of a multiple drive in a single belt.

In operation the V belt is bent when passing around the sheave. The outer portion is under tension and the inner section is compressed. Typical construction of V belts consists of: (1) an elastic tension and a compression section; (2) load-carrying section containing low-stretch, oil-resistant fabric cover (Fig. 4-7).

Standard V belts are made endless and are made in a sufficiently wide range of lengths to meet most requirements. Most V belt drives are designed to use

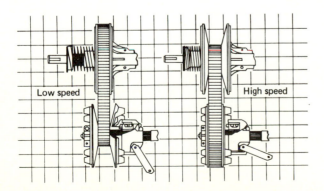

Fig. 4-5   Variable-speed drive. (*Dayco Corporation.*)

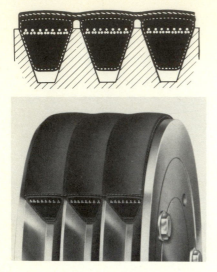

Fig. 4-6   Standard individual V belts bonded together to form a multiple drive. (*Gates Rubber Company*.)

endless belts so that they can be removed and replaced easily and adjustments made for belt stretch. In some cases, however, this is not always practical. Standard sizes of open-end V belting are used where it is necessary to transmit power around structural members or other permanent obstructions. Special detachable fasteners (Fig. 4-8) are used to connect the ends of the belt. Fig. 4-9 shows how power can be transmitted around corners.

Pulleys and Sheaves

Pulleys for flat belts are made of cast iron, wood, steel, or composition fiber. The diameter of a flat pulley is slightly larger at the center to aid in belt tracking. This is called the "crown" of the pulley. The sizes of flat pulleys are designated by the diameter and width of face.

Sheaves for V belts are made of cast iron or steel or die-pressed steel (Fig. 4-10). Information normally required for ordering a sheave should include belt size (HA, HB, etc.), number of grooves, pitch diameter, type of construction, and size and type of hub. Hubs may be obtained with a finished bore to fit the

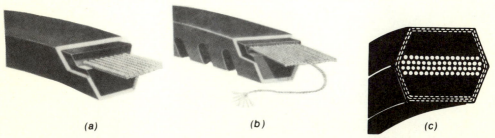

(a)   (b)   (c)

Fig. 4-7   Cross-sections of V belts: (a) textile-cord load-carrying members; (b) steel-cord load-carrying members (*T. B. Wood's Sons Company*); (c) double-angle V belt. (*Dayco Corporation*.)

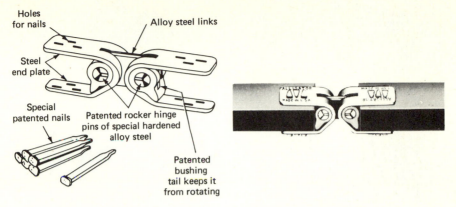

Holes
for nails

Alloy steel links

Steel
end plate

Special
patented nails

Patented rocker hinge
pins of special hardened
alloy steel

Patented
bushing
tail keeps it
from rotating

Fig. 4-8   Connector for open-end V belting. (*Flexible Steel Lacing Company.*)

shaft or with split-tapered bushings that make it possible to use the sheave on several different shaft sizes by merely changing the bushing (Fig. 4-11).

Three types of variable-pitch sheaves to obtain different speed ratios are available (Fig. 4-12). The first is referred to as a stationary control type; in it the pitch diameter is adjusted by changing the movable disc while the machine is not in operation. On the latter two types, the pitch diameter may be changed while the machine is in operation. The manually controlled sheave is adjusted with a threaded rod attached to a hand knob. Belt tension is used to control the distance between the two sheaves in the spring-load variable-pitch sheave. The latter two are often used together in a variable-speed drive, thereby reducing the necessity of changing the shaft-center distance or using an idler (Fig. 4-13).

There are two basic types of grooves available for V-belt sheaves: standard and deep (Fig. 4-14). Standard-groove sheaves are the most common and are used on drives with parallel shafts. Deep-groove sheaves are used on drives where the shafts are not parallel and the belt operates in a turn or transmits power around a corner, as shown in Fig. 4-9. The effective outside diameter of a sheave is the diameter at which the top of the belt rides in the groove. This is normally

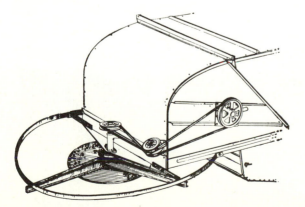

Fig. 4-9   V belts can be used to transmit power around corners.

Type A
turned steel

Web type

Pressed steel

Spoke type   Tru-run hub   Adapter ring   Hex nut

Fig. 4-10   Sheaves for V belts. (*Gates Rubber Company.*)

Fig. 4-11   Multiple-groove sheaves with two types of malleable-iron split-taper bushings. (*Gates Rubber Company.*)

(a)   (b)   (c)

Fig. 4-12   Variable-pitch sheaves: (a) stationary control pitch; (b) manually controlled;
(c) spring loaded. (*Browning Manufacturing Division, Emerson Electric Company*.)

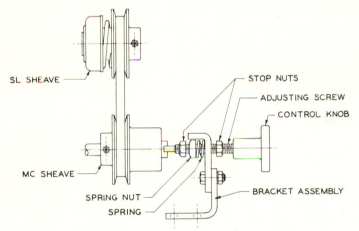

Fig. 4-13   Variable-speed drive using a manually controlled variable-pitch sheave on the
driver and a spring-loaded sheave on the driven. (*Browning Manufacturing Division, Emerson
Electric Company*.)

equal to or less than the actual outside diameter of the sheave. The pitch
diameter is the diameter formed by the neutral axis of the belt riding in the
groove.

How to Determine the Length of a Flat Belt

To find the length of a flat belt for two pulleys of unequal size, add the
diameters of the two pulleys together, divide this sum by 2, multiply by 3, and
to this product add twice the distance between the centers of the two shafts.

How to Determine the Size and Speed of Pulleys and Sheaves

To calculate the speed or size of a flat pulley: the revolutions per minute of the
driving pulley times its diameter equals the revolutions per minute of the driven

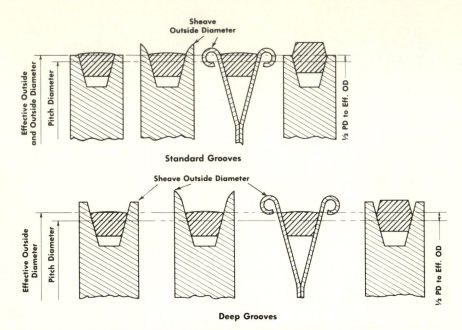

Fig. 4-14   Relationship of actual outside, effective, and pitch diameters for standard and deep-groove sheaves. (*Gates Rubber Company*.)

pulley times its diameter. If three of the quantities are known, the fourth can be easily determined.

$$S \times D \quad = \quad S \times D$$
$$\text{(Driver)} \qquad \text{(Driven)}$$

where S = r/min
$\qquad D$ = diameter

Another expression is

$$\text{Diameter of driving pulley} = \frac{\text{r/min of driven} \times \text{diam. of driven}}{\text{r/min of driver}}$$

The speed or revolutions per minute can be determined by substituting the known and unknown quantities, as

$$\text{R/min of the driving pulley} = \frac{\text{diam. of driven} \times \text{r/min of driven}}{\text{diam. of driving pulley}}$$

The same procedure is followed for determining the size and speed (r/min) of sheaves for V belts; however, the pitch diameter of the sheaves is used instead of the outside diameter. To find the speed of a belt, multiply the circumference of the pulley by the number of revolutions at any given time. This disregards slippage and creep. The speed of a belt should not exceed 5000 ft/min (1524 m/min). A good speed is 3500 to 4000 ft/min (1066 to 1220 m/min).

How to Measure the Length of a V Belt

When the drive consists of two sheaves (Fig. 4-15), the relation between the center distance between shafts and the belt length can be determined by the following formula:

$$L = 2C + 1.57(D + d) + \frac{(D - d)^2}{4C}$$

where L = effective length of belt, in (mm)
$\quad\quad C$ = distance between centers of sheaves, in (mm)
$\quad\quad D$ = effective outside diameter of large sheave, in (mm)
$\quad\quad d$ = effective outside diameter of small sheave, in (mm)

General Rules for Maintaining and Using Belts   Based on field experience and product testing, it has been established that the acceptable belt life for agricultural machines will generally fall into the range of 500 to 1000 hours or more of use when proper care is taken.

1   Maintain proper tension. Belts running loose will slip and vibrate, causing excessive wear. Too much tension places excessive load on belts and shaft bearing. If proper equipment for checking tension is not available, take up the belt drive until the belts fit snugly in the grooves. Run the drive for about 15 minutes to allow the belt to "seat." Check the grooves of the sheaves for excessive heating, which is an indication of slippage. If belts slip, retighten and run under load. Check the belt several times the first day of operation.
2   Keep pulleys and sheaves in proper alignment.
3   Belts should be stored in a cool, dark place out of direct sunlight.
4   Do not force or stretch a belt over the groove when removing or replacing.
5   Keep belt drive free of moisture, dirt, grease, and oil.
6   If belt turns or twists on straight drives, check tension and sheave alignment.
7   Do not use belt dressing on standard agricultural V belts.

Positive Belt Drive   Positive-drive belts, often referred to as *timing* belts (Fig. 4-16), were originally designed to transmit power between two shafts

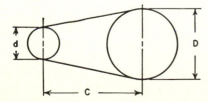

Fig. 4.15   The relation between center distance and V-belt length.

Fig. 4-16   Positive-drive belt. (*Morse Chain, Division of Borg-Warner Corporation.*)

where precision timing was required. However, this type of drive is used on many machines where high torque is transmitted at low to high speeds. The tension load is carried by steel cables or fabric cord embedded in neoprene. The teeth are molded integrally with the neoprene backing and are covered with a wear-resistant nylon duck.

Positive-drive belts are made in five stock pitches, including $\frac{1}{5}$, $\frac{3}{8}$, $\frac{1}{2}$, $\frac{7}{8}$ and $1\frac{1}{4}$ in (0.5, 1.0, 1.3, 2.2, and 3.2 cm) (Fig. 4-17). There are several stock widths within each pitch category. Specifications for ordering replacement belts should include pitch length, pitch code, and width. The relationship of the pitch diameter and the outside diameter of pulleys is illustrated in Fig. 4-18. Information required in reordering pulleys include number of grooves, pitch, width, type of hub, and diameter of shaft.

SPROCKET WHEELS AND CHAINS

Hook-link and roller are the two types of chain commonly used in transmitting power on farm equipment. Sprocket wheels are designed to fit each type of chain.

The *hook-link* chain may be made of either malleable iron or pressed steel (Fig. 4-19). Hook-link chains are used where the power requirements are low and the speed relatively slow, such as in planters. The steel hook-link chain is most extensively used. In the operation of hook-link chains, the hook of the chain link should be run with the open lip away from the sprocket wheel and leading

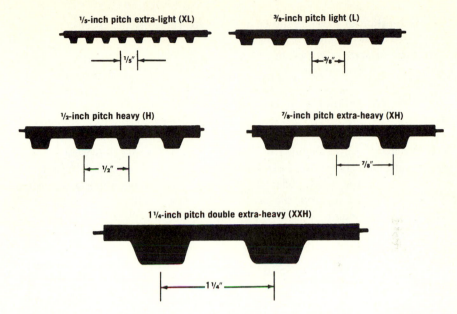

Fig. 4-17 Standard tooth dimensions for positive-drive belts. (*Dayco Corporation.*)

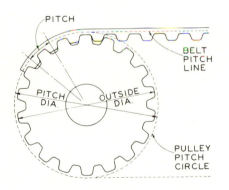

Fig. 4-18 Relationship of outside diameter to pitch diameter for positive-drive belt pulley. (*Browning Manufacturing Company.*)

Fig. 4-19 Malleable iron (*top*) and steel hook-link chain. (*FMC Corporation, Chain Division.*)

in the direction of travel, as shown in Fig. 4-20. There may be exceptions to this rule when the drive sprocket is small.

Roller chains as shown in Fig. 4-21 are made of a special high-grade steel and are used extensively on agricultural machines. The chain consists of alternate roller links and connecting pin links, as shown in Fig. 4-22. The rollers turn freely on bushings that are pressed into the roller-link plates. The pins of the pin links fit and turn freely inside the bushings of the roller links. Roller-chain assemblies are available with either cottered pin or riveted pins. A special tool may be used for disassembling riveted roller chain (Fig. 4-23). Several types of connecting links are shown in Fig. 4-22.

The size of standard roller chain is designated by the pitch, which is the distance in inches between the centers of the links, as illustrated in Fig. 4-21. Standard chain numbers include 35, 40, 50, 60, and 80, which correspond, respectively, to pitch lengths of $\frac{3}{8}$, $\frac{1}{2}$, $\frac{5}{8}$, $\frac{3}{4}$, and 1 in (1.0, 1.3, 1.6, 1.8, and 2.5 cm). The outside diameter of the roller and the inside width of each chain are approximately five-eighths of the pitch.

Standard single-strand chain is the type most commonly used on field machines; however, in special power-transmission cases, multiple-strand chains

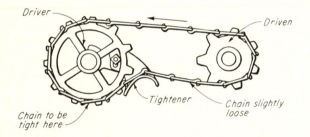

Fig. 4-20   The proper method of running a hook chain on the sprockets.

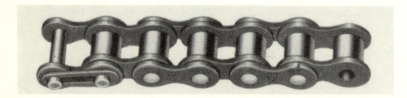

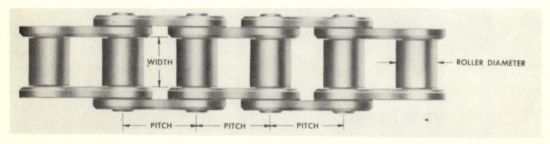

Fig. 4-21   Single-strand standard steel roller chain. (*FMC Corporation, Chain Division.*)

Cotter pin connecting link

Roller link

Offset link

Spring clip connecting link

Fig. 4-22 Connecting links for roller chain. (*Browning Manufacturing Division, Emerson Electric Company.*)

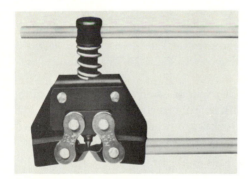

Fig. 4-23 Tool for disassembling riveted-type roller chain. (*Acme Chain Division, Rockwell International.*)

are used to transmit more power without increasing the chain pitch or speed. Double-pitch roller chains (Fig. 4-24) are used for transmitting light load at low speeds, such as unit-type planters. The pitch on these chains is twice that on a standard chain.

The sprockets used with roller chain are usually specified according to the pitch, the number of teeth, the type of hub (Fig. 4-25), and the bore. Sprockets are also available with tapered hubs similar to the ones used on sheaves (Fig. 4-11).

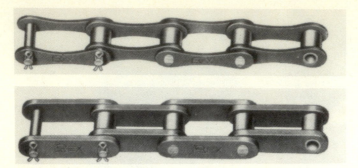

Fig. 4-24   Double-pitch roller chain: (*top*) standard-drive type; (*bottom*) conveyor series. (*Rexnord, Inc.*)

Fig. 4-25   Sprockets for standard roller chain: (*a*) type "A" plate; (*b*) type "B" steel; (*c*) type "C" cast iron. (*Morse Chain Division, Borg-Warner Corporation.*)

Gears

When the machine is rather compact and the shafts are close together, gears can be employed to transmit the power, as shown in Fig. 4-26. The various types of gears are shown in Fig. 4-27.

Often there is a combination of either spur or bevel and another type. If the power is transmitted parallel to the shaft, helical or spur gears are employed; but if the shafts are at right angles, the beveled or worm gears must be employed. The use of gears makes a more substantial construction and eliminates a great amount of lost motion; however, the cost is greater, especially in the case of repairs. It is much cheaper to replace one or two links in a chain than to replace a complete gear. When one tooth is broken and all the others remain, the gear cannot be used.

Spur gears have their shafts parallel. The teeth of the gear are parallel to the shaft. In an *internal* spur gear (Fig. 4-27C), the teeth are on the inside of the rim. An *external* spur gear (Fig. 4-27A) has teeth on the outside of the rim. For every internal spur gear, it is necessary to have an external spur gear to operate it. The combination is used to rotate the parallel shafts in the same direction at different speeds.

Beveled gears (Fig. 4-27H) have their shafts at right angles or nearly so. Where the power has to turn a corner, beveled gears are used. The teeth are on an incline which varies according to the difference in diameter of the gears

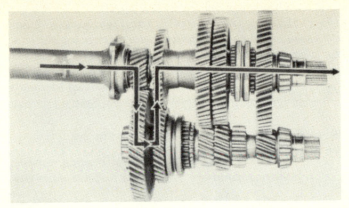

Fig. 4-26   Gears used in a tractor transmission to obtain a selection of speeds. (*Ford Motor Company*.)

meshing together. Beveled gears tend to wear so that their teeth do not fit one another closely. For this reason there should always be some method of adjustment. *Miter* gears have an equal number of teeth cut at the same angle (Fig. 4-27G).

Worm gears (Fig. 4-27F) consist of a shaft, called the *worm*, with screwlike threads which run spirally around it. This meshes with a helical spur gear called the *sector*. As the worm turns, the teeth of the sector, which fit in the grooves, are turned slowly. This type of gear is used to a limited extent in farm machinery.

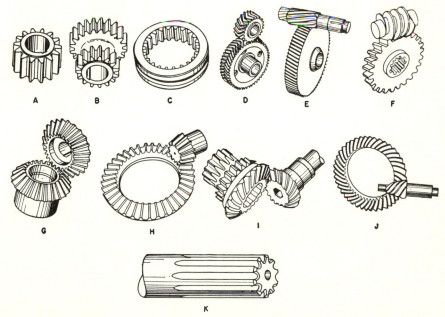

Fig. 4-27   Types of gears: (*A*) spur; (*B*) cluster; (*C*) internal spur; (*D*) herringbone; (*E*) helical; (*F*) worm and worm-wheel; (*G*) straight-bevel; (*H*) straight-bevel gear set; (*I*) spiral-bevel set used in tractors; (*J*) hypoid gear set; (*K*) spline-shaft gear.

Helical gears (Fig. 4-27E) may take the form of either spur gears or beveled gears, but they do not have straight teeth. The teeth are more or less curved so that they will remain in mesh or in contact longer than straight teeth. In the spur gear, they are called *helical spur gear*; in the beveled type, they are called *helical beveled gear*. When helical gears are used, much noise is eliminated because of the fact that the teeth remain in contact longer, giving an even, constant pressure at all times.

A *pinion* is the smaller gear of any two gears that are meshing together; it may be a spur, bevel, or helical gear.

Power-Takeoff Shafts and Universal Joints

In the operation of many farm machines, the tractor is used to move the machine forward and at the same time furnish the power for its operation. The power is transmitted from tractor to machine by means of a shaft which is usually termed a *power-takeoff shaft*. If the travel of the tractor and the machine were always in a straight line, a solid shaft could be used. Field operation requires turning of corners in harvesting broadcast crops and back-and-forth trips for row crops. Thus, the power-takeoff shaft must be equipped with at least two universal joints to permit these turns (Figs. 4-28 and 4-29). The complete shaft, including universal joints, is called a *power-takeoff drive*. Shafts and universal joints are frequently employed, also, to transmit power at various angles on a particular machine.

In 1946, the American Society of Agricultural Engineers and the Society of Automotive Engineers approved standards establishing the dimensional relationships so that any power-takeoff-driven machine could be used with any make of

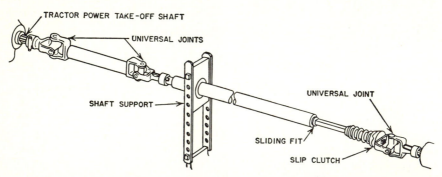

Fig. 4-28 Complete power-takeoff shaft with three universal joints, safety snap clutch, and shaft support. The safety shield has been removed to show all parts of the assembly.

Fig. 4-29 Power-takeoff shaft with universal joints and integral safety shield. (*Rexnord Inc.*)

tractor.[1] The standardized dimensions and relationships of the principal components of the power-takeoff drive are illustrated in Fig. 4-30. Tractors may be equipped with standard power-takeoff drive of either 540 or 1000 r/min; the shaft is $1\frac{3}{8}$ in (3.5 cm) in diameter and has six splines. The 1000 r/min shaft is the same diameter with 21 spline teeth.

Shields are required on the power-takeoff-drive line on all machines at all times to protect the operator. Improvements are being made continuously to make the shielding safer and easier for the operator to keep in place. New standards require that the universal joints, as well as the shafts, be fully shielded, as illustrated in Fig. 4-31.

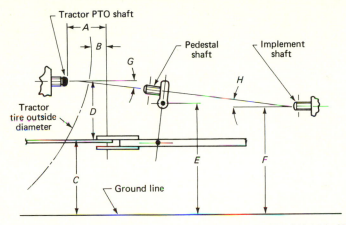

Fig. 4-30 Standard power-takeoff-drive line relationship; (A) 14-in (335.6-mm); (B) 1- to 5-in (25- to 127-mm); (C) 15 ± 2-in (381 ± 50.8-mm); (D) 8-in (203.2-mm); (E) pedestal height should be adjustable for straightest line possible with minimum angles G and H. (ASAE Standard: ASAE S203, S204, and R314, 1973.)

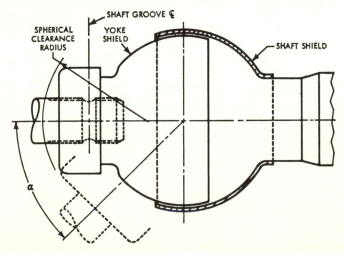

Fig. 4-31 Full shielding of yokes, joints, and line shaft. (ASAE S297T, 1973.)

[1] Power Take-off for Agricultural Tractors, *Agr. Engin. Yearbook*, 1973.

The maximum instantaneous-torque starting values are more important considerations in designing a power-takeoff drive than the average horsepower requirements of the operation. The data in Table 4-2 show the power-takeoff torsional loads for several implements performing different farm operations.

Flexible Shafting

A strong and durable flexible shaft can be used in many cases for the transmission of power in farm power equipment, replacing universal joints and shafts. Figure 4-32 shows an application of flexible shafting for transmitting rotary power to a centrifugal spreader.

The flexible shaft is made up of a number of right- and left-hand layers of helically wound wires wound around a single center wire. Various types of fitting are available for the ends for connecting to the driving and driven members of a machine. The shaft is encased in a flexible metal housing and serves as a guide. The direction of rotation must be specified because of the manner or direction in which the wires are wound.

TYPES OF CLUTCHES

A *clutch* is a gripping device between a power source and a machine or between the working parts in a machine whereby the units can be connected or disconnected. In the operation of farm power equipment, clutches permit the starting of the engine with the machine disconnected. The clutch is then

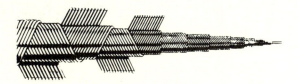

Fig. 4-32   The core of flexible shaft and an application used to drive a centrifugal seeder. (*Stow Manufacturing Company.*)

engaged, and the power is transmitted to the machine by means of shafts, gears, belts, and other devices. Clutches are either positive or friction types.

Positive clutches consist of two parts which have jaws so shaped and placed that when they are brought together, they engage positively as a unit with no slippage (Fig. 4-33). These clutches should be engaged before power is applied to the driving side. They are used when light loads are transmitted at slow speeds, as in drives for planters and grain drills.

Friction Clutches

Friction clutches are used extensively on tractors, trucks, and other equipment such as self-propelled combines, cotton harvesters, and sprayers in which the power is applied gradually to the load. In general, the driver and driven members are coupled through friction surfaces and a suitable mechanism is used for pressing and holding the friction surfaces together. Typical friction clutches used on agricultural machines include disc (Fig. 4-34), cone, belt, and centrifugal. The mechanisms used for engaging and disengaging the frictional surfaces of the clutches include suitable linkages that may be controlled by hand or foot, electromagnets (Fig. 4-35), liquid pressure, and centrifugal force.

(a)   (b)

Fig. 4-33   Positive-type clutches. (*FMC Corporation*.)

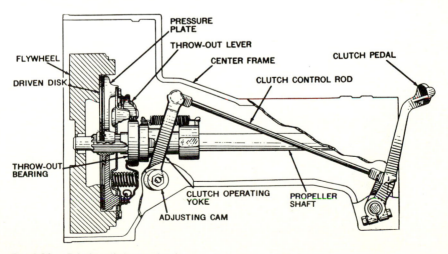

Fig. 4-34   Friction-disc-type clutch.

Table 4-2 Power-Takeoff Torsional Loads

Test no.	Tractor model	Approx. max. tractor, b.hp	Implement make and model	Coupling in P.T.O. drive	Max. starting torque, lb-in		Max. operational torque peaks, lb-in		Avg. torque under normal operating conditions, lb-in	Work being performed
					With normal clutch engagement	With rapid clutch engagement	Average conditions	Near plugged conditions		
1	1	35	Ensilage harvester A	Standard	4900-6400	10,800-15,370	4680-6390	5450-7140	2720	Chopping heavy drilled corn
2	1	35	Ensilage harvester A	Spec. slip	. . . .	8660*	5112-5723	6025-6865	3200	Chopping heavy drilled corn
3	9	40	Ensilage harvester A	Standard	. . . .	11,600	4700-4925	6200-8025	3261	Chopping heavy drilled corn
4	1	35	Ensilage harvester B	Standard	2600-4000*	. . . .	3520-3820	3960-7630*	2390	Chopping heavy drilled corn
5	1	35	Forage harvester C	Standard	. . . .	14,600	3730-7200	6370-7200	2870	Chopping green alfalfa
6	1	35	Forage harvester C	Spec. slip	. . . .	9800-7530*	5230-6700	6100-8700	3270	Chopping green alfalfa
7	1	35	Forage harvester C	Standard	. . . .	12,500-10,900	6060-7460	9500	3600	Chopping green alfalfa
8	1	35	Forage harvester C	Standard	. . . .	21,400	. . . .	. . . .	. . . .	Attempting to start plugged machine
9	1	35	Corn picker D	Standard	1570-1740	3990	822-1031	. . . .	727	Picking corn
10	1	35	Baler E	Standard	. . . .	18,300-20,600	5860-7470	12,100	1140	Baling alfalfa

11	1	35	Baler E	Standard	13,100	··········	6550-8140	11,600-15,000	1545	Recheck of test 10
12	1	35	Baler E	Spec. slip	10,700-12,100*	10,700-12,100*	7250-8920	11,500-13,300*	2250	Baling alfalfa
13	1	35	Baler E	Spec. slip	10,100*	··········	8600-11,100	10,350-12,600	1580	Baling straw
14	9	40	Baler E	Standard	12,250	Standard and universal joints aligned	7749-10,945	10,960-12,095	1938	Baling alfalfa
15	1	35	Baler E	Standard and universal joints aligned	Standard and universal joints aligned	··········	4601-5867		1383	Baling alfalfa
16	1	35	Baler F	Standard	16,500	··········	8600	22,700	····	Baling alfalfa
17	1	35	Baler F	Spec. slip	5000*	5000*	5000*	5000*	····	Baling alfalfa
18	1	35	Combine G	Standard	··········	10,100-16,600*	3760	9380	1890	Combining windrows
19	1	35	Combine G	Spec. slip	··········	··········	7150	7760-9130	1700	Combining windrows
20	1	35	Combine G	Spec. slip	··········	7350-8650*	4160-4200	7470	1600	Straight combining
21	2	25	Hammer mill H	Standard	9030	17,500-20,150	4145	7270	2700	Grinding ear corn
22	1	35	Hammer mill H	Standard	6130	··········	3740	14,900	2140	Grinding ear corn
23	1	35	Hammer mill H	Spec. slip	8230*	8230*	··········	6920	4210	Grinding ear corn
24	4	45	Hammer mill J	Standard	18,150	25,800	7800	13,000	5450	Grinding ear corn

* Safety clutch in P.T.O. line slipped, limiting torsional load to this value.
Source: *Agr. Engin.*, **33**:68, 1953.

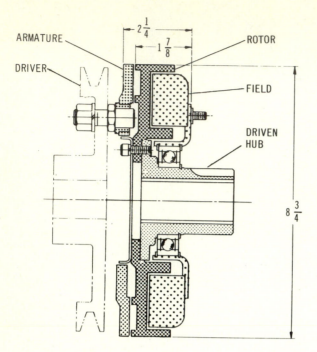

Fig. 4-35   Electromagnetically controlled friction clutch. (*Stearns Electric Corporation.*)

Safety or Slip Clutches

Many machine elements or components are protected from possible damage or overloading by safety clutches. This type of clutch remains engaged at all times but will slip or limit the amount of torque or power that is transmitted through the shaft to the driven member. The two types commonly used on farm machines include the friction type (Fig. 4-36) and the bevel-jaw type (Fig. 4-37).

Overrunning Clutches

Clutches of this type are used to drive machine elements where it is desirable to transmit power in only one direction. The driven members, which are usually large and massive, such as fans on cotton pickers, flywheels on balers, and cutter cylinders on forage harvesters, are allowed to "coast" to a stop without transmitting power back through the driver. High-speed elements normally use roller- or cam-type clutches, as shown in Fig. 4-38, while the pawl-and-ratchet type is used on low-speed shafts (Figs. 4-39 and 4-40).

Another type of clutch is the *belt-tension*. When an idler on a belt is mounted so it can be moved to release the tension around the pulleys, there is not enough friction on either the drive pulley or the driven pulley to transmit power. Thus, the tension applied by the movement of the idler pulley can be made to serve as a clutch to engage and disengage the power.

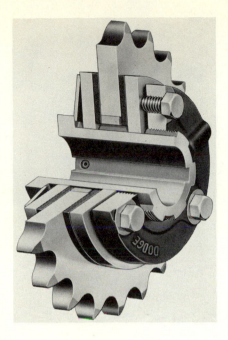

Fig. 4-36 A friction-type slip or safety clutch. (*Dodge Manufacturing Division, Reliance Electric Company.*)

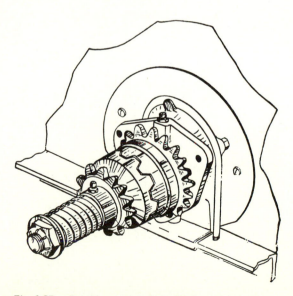

Fig. 4-37 Bevel-jaw-type safety clutch.

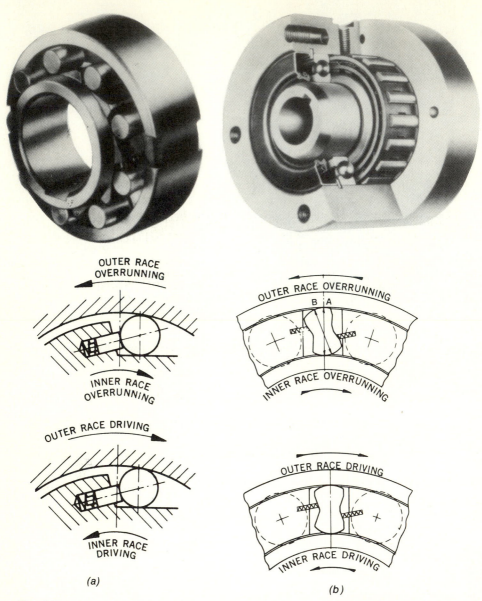

Fig. 4-38 Overrunning clutches: (*a*) roller type; (*b*) cam type. (*Morse Chain, Division of Borg-Warner Corporation.*)

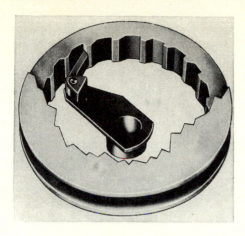

Fig. 4-39 Pawl-and-ratchet over-running clutch.

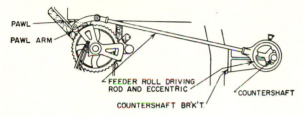

Fig. 4-40   Ratchet-and-pawl drive for manure-spreader apron. Note the eccentric drive for moving the pawl forward and backward.

REFERENCES

Adams, James, Jr.: V-belt Design for Farm Machinery, *Agr. Engin.*, **41**(7):348, 1960.

Confer, L. J.: Standards for V-belt Drives on Farm Equipment, *Agr. Engin.*, **31**(5):237, 1950.

Hansen, Merlin: Load Imposed on Power Take-off Shafts by Farm Machines, *Agr. Engin.*, **33**(2):67-69, 1952.

Kimmich, E. G., and W. Q. Roseler: Variable-speed V-belt Drives for Farm Machines, *Agr. Engin.*, **31**(7):334-336, 1950.

Potgieter, Fred M.: Application of Universal Joints to Farm Machinery, *Agr. Engin.*, **33**(1):21-27, 1952.

QUESTIONS AND PROBLEMS

1   Enumerate the various methods of transmitting power in farm equipment. (*a*) Explain the merits of each method. (*b*) Explain the difference between a direct and a flexible-shaft drive.

2   (*a*) Explain how the sizes of V belts are designated. (*b*) How is the pitch length designated? (*c*) Explain what is meant by a double-sided V belt.

3   The centers of two B-type V-belt sheaves are 24 in (61.0 cm) apart. The drive sheave has a pitch diameter of 14 in (35.6 cm) while the driven sheave has a pitch diameter of 5 in. Find the length of V belt required. Take into consideration the installation take-up allowance.

4   Explain the differences between and the uses of (*a*) single sheave, (*b*) multiple sheave, (*c*) variable-speed sheave.

5   Under what conditions are the following used: (*a*) hook-link chain, (*b*) roller chain?

6   Enumerate the various types of gears and explain their special uses.

7   A tractor is used to operate the horizontal-rotating knives of a stalk cutter-shredder. The tractor power takeoff runs at 540 r/min and drives a pinion gear with 13 teeth. The driven gear has 24 teeth. (*a*) Find the revolutions per minute of the knife blades. (*b*) Find the peripheral travel of the blade ends when cutting a circular pattern 57 in (144.8 cm) in diameter.

8   Explain the functions of the universal joints in a power-takeoff drive. What is the accepted standard size of a power-takeoff shaft for tractors?

9   Explain the differences between the various types of clutches.

10   What are the advantages of an overriding clutch?

Chapter 5

Component Parts
of Machines

The component parts of farm equipment include those parts that are essential to construct a complete high-quality operative machine.

Cam

A *cam* (Fig. 5-1) is a device that produces intermittent motion or a specific motion to a member called the *follower*. When an object is in motion part of the time and at rest between motions, the action is said to be *intermittent*. A cam can best be described as a disc with a lobe on one side (Fig. 5-1). The part that projects is called the *lobe nose*. Anything resting against the cam will be moved only when the nose comes around to it; otherwise, it remains stationary.

Bearings

Bearings in farm equipment are required to hold the various power-transmission parts in position. The proper bearing to use is determined by the amount of wear, the speed at which the shaft is turning, the load it must carry, and the amount of end thrust. Bearings are divided into two general classes: sliding, or plain, and rolling.

Sliding Bearings   Bearings of this type are shown in Fig. 5-2. In plain bearings, the revolving shaft is supported by, and is in direct contact with, a

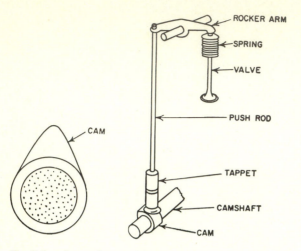

Fig. 5-1   A cam shape at left and the application of a cam at right.

fixed bearing surface. For this reason, friction is high, and the bearing should be lubricated. The bearing metal may be cast iron, babbitt, bronze, or other material.

Rolling Bearings   Bearings of this type have balls or rollers placed between the shaft and the supporting bearing, thus reducing the friction. They are, therefore, called *antifriction bearings*. The lubrication of ball and roller bearings serves to preserve the polished surfaces from corrosion, to act as a cooling agent, and to protect the rubbing surfaces between the rollers, races, and separators. The selection of a lubricant for antifriction bearings is based on the type of bearing housing, the operating temperature, the speed of bearing rotation, and the requirements of the bearing. Some antifriction bearings are packed and sealed, thus requiring no further lubrication for the service life of the bearing. Both ball and roller types of antifriction bearings are used extensively on almost all power-operated farm equipment.

Ball bearings are bearings having one or more rows of small balls placed in a cage or holder (Fig. 5-3). The balls are separated slightly, held in position by a

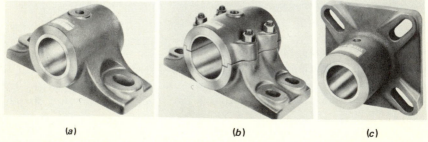

(a)   (b)   (c)

Fig. 5-2   Plain or sliding bearings: (a) solid pillow block; (b) split pillow block; (c) flanged block. (*Bearing Division, FMC Corporation.*)

(a)   (b)

(c)

Fig. 5-3   Ball bearings: (a) single row; (b) double row (*Fafnir Bearing Company Division of Textron*).; (c) thrust bearing (*SKF Industries*).

retainer, and ride in grooves provided in the inner and outer races. Because of the small amount of surface in contact between the balls and the races, friction is very low. Ball bearings are designed to carry (1) radial loads at right angles to the shaft, (2) thrust forces that are parallel to the shaft or tend to shift the shaft out of position, and (3) a combination of radial and thrust loads. There are several types of ball bearings (Fig. 5-3). They are designed to carry the various types of loads listed above. Ball bearings have many applications in all types of farm equipment.

Roller bearings differ from ball bearings in that small cylindrical rollers are substituted for the balls. This gives a much larger bearing surface, such as is necessary for a heavy load. There are also cages to hold the rollers apart, as for the ball bearings. Most roller bearings are designed for only radial loads. Figures 5-4 to 5-9 illustrate several types of roller bearings. Roller bearings may be cylindrical, conical, or tapered, depending on the shape and placement of the rollers. Cylindrical roller bearings can be further divided into (1) the plain roller, (2) the spiral roller (Fig. 5-6), and (3) the needle roller (Fig. 5-7). The plain, straight roller bearing consists of a number of solid-cylinder steel rollers assembled in a cage with an inner and outer race (Fig. 5-4). One further

Fig. 5-4 Cylindrical roller bearing. (*Hyatt Bearing Division, General Motors Corporation.*)

Fig. 5-5   Parts of a plain roller bearing.

classification of plain roller bearings is determined by the manner in which the parts are assembled, as a bearing with a separable inner race, a bearing with a separable outer race, and a bearing with nonseparable parts.

The spiral, or wound, straight, hollow roller bearing is shown in Fig. 5-6. Types are designed to operate with inner and outer races or with no inner race and a split outer race.

The needle-type roller bearing consists of a hardened outer shell containing a number of small-diameter hardened rollers with pointed ends (Fig. 5-7). The needle bearing has the greatest radial load capacity possible for a given housing bore. It can also be used where the bearing space is limited.

Tapered roller bearings are designed to carry radial or thrust loads or a combination of both. The conical rollers are set in the cage at an angle between the inner and outer races, as shown in Fig. 5-8. It is usually necessary to mount tapered roller bearings in pairs in order to balance the radial and thrust loads (Fig. 5-9). Their greatest application is for wheel bearings, but there are many other applications.

Fig. 5-6 The various parts of two applications of a spiral roller bearing. (*Hyatt Bearing Division, General Motors Corporation.*)

Fig. 5-7 The various parts of a needle roller bearing. (*Torrington Roller Bearing Company*.)

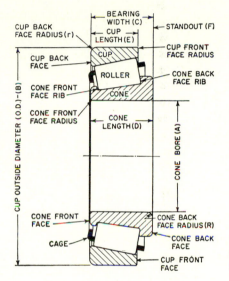

Fig. 5-8 The various parts of a tapered roller bearing. (*Timken Roller Bearing Company*.)

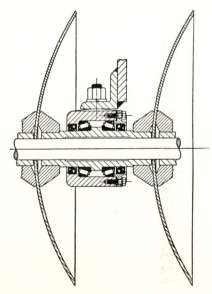

Fig. 5-9 Application of tapered roller bearings on a disk harrow. (*Timken Roller Bearing Company*.)

There are many ways in which bearings may be mounted or housed. Often the bearings may be replaced as a cartridge or unit, such as in a wheel or hub. On most machine components, the bearings are mounted in a special housing, as illustrated in Fig. 5-10. Either roller or ball bearings may be employed in the housings.

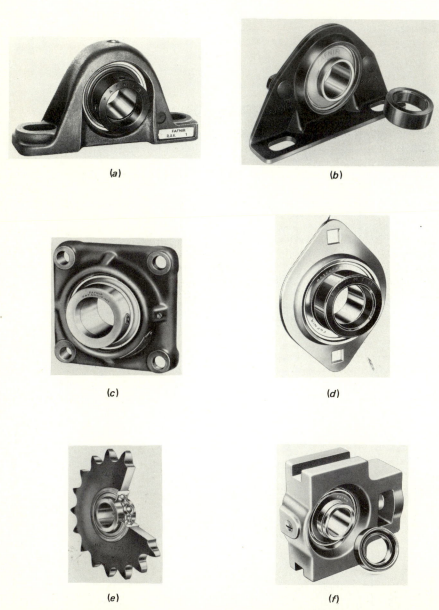

Fig. 5-10   Typical bearing mountings and housings: (a) cast iron pillow block; (b) pressed steel pillow block; (c) 4-bolt cast-iron flange; (d) 2-bolt pressed steel flange; (e) idler; (f) take-up unit. (*Fafnir Bearing Company Division of Textron.*)

Maintenance of Antifriction Bearings   Good maintenance of antifriction bearings is essential to obtain a long, trouble-free service life. All persons who operate and repair farm equipment should obtain a bearing-maintenance handbook from a bearing manufacturer. These handbooks describe and illustrate replacement of bearings; their care, cleaning, and selection; and the use of lubricants. Most antifriction bearings used on farm equipment are sealed to prevent dirt, dust, and moisture from getting into the rollers or balls. These bearings should not be overlubricated to the point of rupturing and damaging the seal.

Bushings   A *bushing* is a replaceable lining for a bearing. It may consist of wood, babbitt, bronze, chilled iron, or other material. Figure 5-11 shows two types of bearing bushings with different types of grooves for distributing the oil along the shaft.

Bolts

The great variety of bolts used in the construction of farm machinery can be classified as follows: machine, carriage, stove, and plow bolts. Most bolts are specified according to length, diameter, and type of thread: N.F. (national fine) or N.C. (national coarse).

Machine bolts are used for holding two pieces of metal together. They have a square or hexagonal head with the stem of the bolt fitting into the head without any change of diameter, as Fig. 5-12.

Carriage bolts (Fig. 5-12), unlike machine bolts, have a rounded or oval-surfaced head with a square shoulder underneath extending out some half an inch, varying according to the size of the bolt.

Plow bolts may have many different kinds of heads, but practically all of them have from one to four shoulder-like points that fit into a groove prepared for them in whatever material they are placed. The undersides of the heads of plow bolts are always countersunk (Fig. 5-12) so that the head can go deep enough into the material to fit flush with the surface. Such bolts are used for holding plowshares.

Stove bolts, as shown in Fig. 5-12, are rather short and are usually less than $\frac{1}{4}$ in (0.6 cm) in diameter. The threads run close to the head, which may be

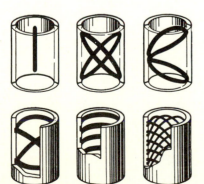

Fig. 5-11   Two types of bearing bushings: (*top*) straight bushings showing types of grooves for oil; (*bottom*) types of grooves for graphited oilless bearing.

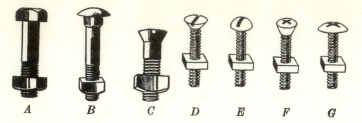

Fig. 5-12 Types of bolts: (A) machine; (B) carriage; (C) plow; (D-G) stove.

either flat or round. Most stove bolts also have a slot cut across the heads so that screwdrivers can be used to prevent them from turning. This type of bolt is used for bolting thin metal together. Some special bolts are shown in Fig. 5-13.

Nuts

The most common types of nuts used on farm machinery are shown in Fig. 5-14. The square nut is used on the cheaper machines, and the hexagon nut is used on the higher-class machines. Castellated nuts are used where vibration is likely to cause the nut to work loose. Wing nuts are used where it is necessary to remove a part frequently. Lock nuts are constructed so that they automatically lock themselves in place (Fig. 5-15).

Screws

Many types of screws are also used in the construction of farm machinery. They may be classified as follows: set, cap, lag, and wood.

Setscrews (Fig. 5-16, 5-17, and 5-18) may have several different shapes for the point. They are so called because they extend through the collar, allowing the point to come in contact with the shaft so that the collar and shaft will be fastened rigidly together and turn as a unit. They are also used in the same way to prevent various parts from moving out of place.

Cap screws (Fig. 5-17) may have square, hexagonal, flat, or button-type heads. Such screws closely resemble a machine bolt with the exception that they do not have a nut on the threaded end; instead, the end passes through whatever it is to hold into a threaded hole which serves as a nut, for example, the cylinder head of an automobile or tractor engine.

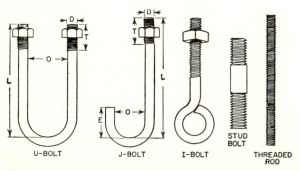

U-BOLT   J-BOLT   I-BOLT   STUD BOLT   THREADED ROD

Fig. 5-13 Types of special bolts.

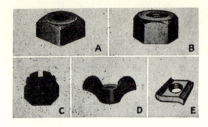

Fig. 5-14 Types of nuts: (*A*) square; (*B*) hexagon; (*C*) castellated; (*D*) wing; (*E*) square lock.

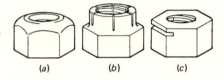

(a)   (b)   (c)

Fig. 5-15 Self-locking nuts: (*a*) soft-collar type; (*b*) slotted type; (*c*) distorted-slot type.

A *lag screw* (Fig. 5-16) has a head like a machine bolt, while the other end is sharp. The threads are coarse and similar to an ordinary wood screw. They are used to attach machinery to floors or beams. The coarse threads, when started, will draw themselves into the wood as the screw is turned with a wrench.

Wood screws, unlike lag screws, are rather small and have slots across the head so that a screwdriver can be used to force them into the wood. (Fig. 5-18).

Washers

Different kinds of washers are used extensively in connection with bolts in farm machinery. They may be used on the end either beneath the head of the bolt or beneath the nut. Washers are of various kinds, as follows: flat malleable-iron, cast-iron, wrought-iron, and spring-lock washers. There is very little difference between malleable- and cast-iron washers, both being rather thick, oftentimes $\frac{1}{2}$ in, and placed where there is a considerable amount of wear. Wrought-iron washers are round discs with holes in the center to allow their being placed under the nut. Conventional spring type lock washers are made of spring steel with one side split from the edge to the center of the hole (Fig. 5-19). The ends

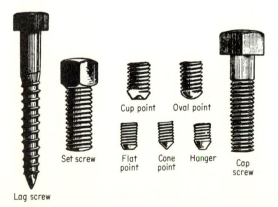

Cup point   Oval point

Set screw   Flat   Cone   Hanger   Cap
             point   point   screw

Lag screw

Fig. 5-16 Types of screw points, lag, and cap screws.

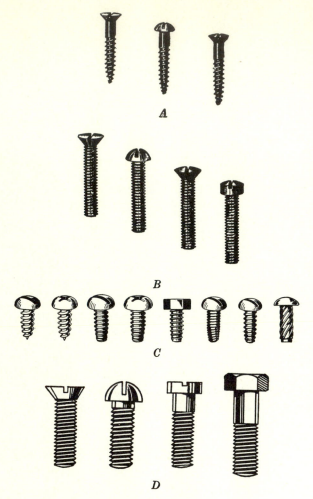

Fig. 5-17   Types of screws: (*A*) wood; (*B*) machine; (*C*) self-threading or tapping; (*D*) cap.

Fig. 5-18   Hollow-head setscrew and wrench.

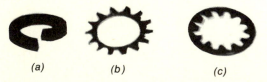

(*a*)   (*b*)   (*c*)

Fig. 5-19   Lock washers: (*a*) spring type; (*b*) external-tooth; (*c*) internal-tooth.

of the split parts are turned in such a manner that they will allow a nut to be turned down easily but resist any effort to turn it off. Other lock washers include internal- and external-tooth types. Quick-repair washers are shown in Fig. 5-20. A plastic liquid material applied to nut and bolt for use as a liquid lock washer is available.

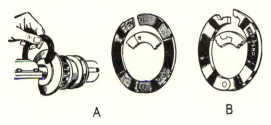

A   B

Fig. 5-20   Quick-repair washers: (*A*) side-latch; (*B*) overlatch.

Fasteners

There are several means that are employed on farm machines to fasten or secure sprockets and pulleys to rotating shafts and to secure other machine components to allow the parts to be assembled or disassembled without the use of special tools and equipment. Keys, as illustrated in Fig. 5-21, are commonly used to fasten pulleys and sprockets to shafts. Pins, such as the hollow steel or tapered pins, are also used. Hollow pins are also used as small shafts in some control linkages. Both pins are designed to fit into drilled holes of smaller diameter than the pin being used to provide a wedge fit. Cotter or split-key pins (Fig. 5-22) are generally used as safety devices to secure nuts on bolts and rotating shafts. The "hair" pin is used to secure parts that are frequently disassembled. Retainer rings, as shown in Fig. 5-23, are used to keep components from sliding on shafts and to fasten other parts where space is limited.

Fig. 5-21   Square and tapered keys.

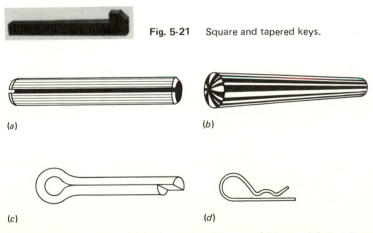

(a)   (b)

(c)   (d)

Fig. 5-22   Pin fasteners: (*a*) hollow-spring; (*b*) taper; (*c*) cotter; (*d*) "hair" pin.

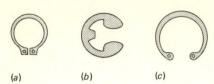

Fig. 5-23   Retaining rings: (a) external;
(a)   (b)   (c)   (b) "E" type external; (c) internal.

Springs

Springs (Fig. 5-24) play an important part in the operation of farm machinery. Extension springs aid in lifting and adjusting heavy implements. Compression and torsion springs facilitate the operation of certain parts of a machine.

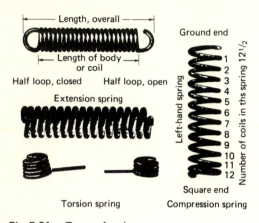

Fig. 5-24   Types of springs.

REFERENCES

Curtis, George W.: Tapered Roller Bearings Practice in Current Farm Machinery Applications, *Agr. Engin.*, **30**(6): 285-293, 1949.

Phelan, Richard M.: *Fundamentals of Machine Design* (McGraw-Hill Book Company, New York, 1970).

Virtue, Byron T.: Application of Needle Bearings in Farm Machines, *Agr. Engin.*, **30**(5):229-232, 1949.

QUESTIONS AND PROBLEMS

1   Explain the function and application of a cam.
2   What is the difference between friction and antifriction bearings?
3   What kind of bearings need bushings?
4   Where are tapered roller bearings used?
5   Enumerate the various types of bolts, nuts, nd screws.
6   What nonmetallic material can be used to lock nuts on bolts?

Chapter 6

Lubricants and Lubrication

Probably the chief cause of machinery wearing out is improper and insufficient lubrication. Much of this trouble can be traced to the lack of maintenance, the poor construction of bearings, and the failure to provide adequate means of conducting the lubricant to the bearing units. Lubrication is needed because of friction.

Friction

Friction is that force which acts between two bodies at their surface of contact to resist the sliding of one body on the other. When one object is being dragged along upon another object, friction between the two tends to stop the one that is being dragged. When one surface rests upon another, there is a tendency for the inequalities of the one to fit into those of the other, causing an interlocking not unlike that produced by putting the cutting edges of two saws together. If such interlocking has occurred, it is possible to move one body over the other only by separating them or by tearing off the interlocking surfaces. No matter how smooth surfaces may be, there are still some elevations and depressions remaining which will permit a small degree of interlocking.

Rolling Friction

When one body rolls upon another, the friction is much less than when one body slides upon another. The resistance in this case is called the *rolling resistance* or *friction.* This can readily be demonstrated by attempting to carry as much upon a sled which has no wheels as upon a wagon or other vehicle mounted on wheels. Most farm implements are now using some type of antifriction bearing in the form of balls or rollers to diminish the amount of friction, which materially increases the efficiency of the machine. Friction in moving parts of machinery causes wasted energy, and therefore it is desirable to reduce it to the smallest possible amount. However, in clutches or to prevent slippage of belts on pulleys, friction is necessary and useful.

Lubrication as a Remedy for Friction

Lubrication tends to reduce friction. The theory of the action of lubrication is that a thin film of the lubricant adheres to the bearing and another to the shaft, completely separating the metal surfaces. Then these films slip one on the other, which reduces the amount of friction. This is because the friction between the films of lubricants is much less than that between the metal parts. A lubricant may act in different ways in reducing the amount of friction: first, by changing the greater resistance of metal to metal to the relatively small resistance of oil over oil; second, by filling up the small depressions in the two frictional surfaces and in this way preventing the so-called *interlocking* (Fig. 6-1).

Forms of Lubricants

Lubricants are available in three forms: fluid oils, semisolids, and solids. Fluid oils are those that flow freely, such as engine oils and gear oils used for lubricating transmissions and gearboxes. Semisolids include the soft greases, such as bearing grease. Solid lubricants consist of graphite and mica. Of these forms, soft greases and oils are most generally used to lubricate farm implements.

Kinds and Sources of Lubricants

All lubricants have three general sources: animal, vegetable, and mineral. Animal oils are lard, tallow, and fish oils. Vegetable oils are cottonseed oil, castor oil, olive oil, and linseed oil. Mineral oils are oils obtained by refining crude petroleum. Of all these, mineral oils are the most universally used on the farm because they can withstand higher temperatures without breaking down.

Manufacture of Lubricants

The diagram in Fig. 6-2 shows the steps in the refining process and the points of extraction where fuels; light, medium, and heavy oils; and the extra-heavy lube stock are obtained.

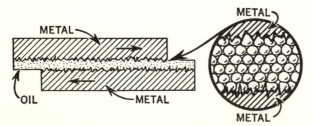

Fig. 6-1   How a lubricant keeps two pieces of metal apart.

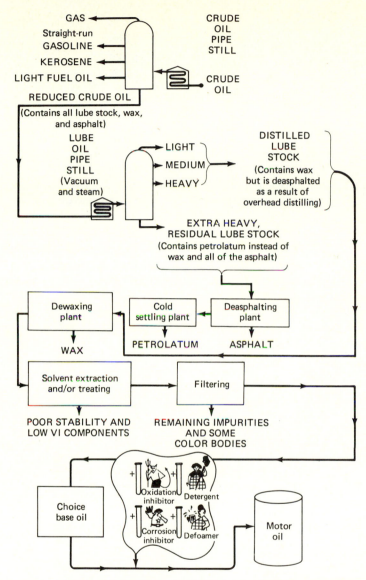

Fig. 6-2 Diagram showing refining process and the points where the oil and greases are obtained. (*Standard Oil Company.*)

ENGINE CRANKCASE OILS

The oil used in the crankcases of most modern engines performs many functions in addition to lubricating the moving parts. Besides reducing friction and wear, oil aids in cooling, reduces shock and noise, helps to seal the cylinders, keeps the parts clean, and prevents rust. To assist the oil in performing these functions, various chemicals or "additives" are used. Some of these include oxidation inhibitors, anticorrosion additives, detergents, and dispersants.

Viscosity

The single most important property of oils is viscosity. *Viscosity* is the internal resistance of a fluid exhibited as one portion or layer of the fluid is moved in relation to the other. In general terms, viscosity is used to describe the resistance of oil to flow (Fig. 6-3). Low-viscosity oils flow easily, whereas high-viscosity oils do not flow readily and are generally referred to as heavy oils. The viscosity is greatly affected by temperature, and oil tends to become thinner when hot and thicker when cold.

Standards have been established by the Society of Automotive Engineers (SAE) by which the viscosity can be measured and designated for engine oils, as shown in Table 6-1. SAE viscosity numbers are widely used and are marked on practically all oil containers; each number indicates a specific range of viscosity at a specific temperature.

As may be noted in Table 6-1, engine oil rated according to viscosity falls into two categories: single-viscosity and multiviscosity. The single-viscosity oils are limited to a narrow range of operating temperatures, whereas the multiviscosity oils can operate under a wide range. For example, a 10W-30 oil has a low-temperature viscosity of a 10W oil and a high-temperature viscosity of a 30 oil. Multiviscosity oils are made possible by an additive called a "viscosity-index improver." The SAE numbers that are followed by "W" are oils that are intended for winter use.

API Service Classification

In addition to selecting the proper viscosity of oil to be used in an engine, it is also important to select the proper type or quality. Because of the wide range of conditions that the many types of engines are operated under, it is necessary to use certain types of oil for each condition. In an effort to assist in clarifying oil specifications and qualities between the manufacturers, the petroleum industry, and the operator, the American Petroleum Institute (API), in cooperation with the American Society of Testing Materials (ASTM) and SAE, has established the API Service Classification System. The three basic factors involved in the system include type of engine, kind of service, and kind and quality of fuel. The classification used until 1971 is shown in Table 6-2. The new API Service Classification which became effective in 1971 is given in Table 6-3. The two systems are compared in Table 6-4.

In selecting the proper oil for an engine, the operator should first consult the operator's manual. The operating conditions that engines are subject to on

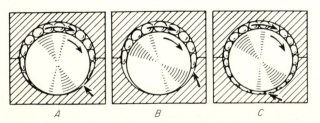

Fig. 6-3   Effects of viscosity in lubricating a bearing.

Table 6-1   SAE Viscosity Numbers for Engine Oils

Single-viscosity oils	Multiviscosity oils
5W	5W-40
10W	10W-30
20W	20W-50
20	
30	
40	
50	

Source: Society of Automotive Engineers.

Table 6-2   API Service Classification of Engine Crankcase Oils Used Prior to 1971

API Service Classifications (1969)	Remarks
Oils for spark-ignition engines	
ML (Motor—Light duty) Discontinued May 1, 1969	
MM (Motor—Moderate duty) Moderate operating conditions but with problems of deposit or bearing corrosion control at high crankcase oil temperatures.	Suitable for light-duty tractor operation. Contains some oxidation and corrosion inhibitor, detergent-dispersant and antiwear additives.
MS (Motor—Severe duty) Special lubrication requirements for deposit, wear or corrosion control under more severe conditions.	Suitable for general to heavy-duty operation at high temperatures. Also ideal for long idling periods. Has greater amounts of oxidation-corrosion inhibitor, detergent-dispenser and antiwear than MM type oils.
Oils for diesel engines	
DG (Diesel—General duty) For operation where there are no severe requirements for wear or deposit control due to fuel, lubricating oil or engine design.	Operating under favorable conditions. Contains additives to give oxidation resistance and detergency. Used with low-sulfur fuels (0.5% or less) within recommended operating conditions.
DM (Diesel—Medium duty) Operating under severe conditions or with fuel with tendency to promote deposits and wear	Satisfactory for most tractors and their operations except where noted under DS. Contains larger quantities of more effective additives than DG oils.
DS (Diesel—Severe duty) Operating under very severe conditions or when using fuel tending to produce excessive wear or deposits. Sometimes called "Caterpillar Series 3" oil or Mil-L-45199B oil.	Highly recommended when operated at high temperatures, intermittent operation at low engine temperatures and when fuel has higher than 0.5% sulfur content.

Source: American Petroleum Institute.

Table 6-3   New API Service Classification of Engine Crankcase Oil, Effective 1971

API service classifications (types)	Remarks
SA—Utility Gasoline and Diesel Engine Service	For engines operated under such mild conditions that the protection provided by additive oils is not required. Not recommended for farm tractors.
SB—Minimum-Duty Gasoline Engine Service	Service typical of engines operated under such mild conditions that only minimum protection afforded by compounding is desired. Oils designed for this service have been used since the 1930s and provide only anti-scuff capability, and resistance to oil oxidation and bearing corrosion.
SC—1964 Gasoline Engine Warranty Service	Service typical of gasoline engines in 1964 through 1967 models of passenger cars and trucks operating under engine manufacturers' warranties in effect during those model years. Oils designed for this service provide control of high- and low-temperature deposits, wear, rust, and corrosion in gasoline engines.
SD—1968 Gasoline Engine Warranty Service	Service typical of gasoline engines in passenger cars and trucks beginning with 1968 models and operating under engine manufacturers' warranties. Oils designed for this service provide more protection from high- and low-temperature engine deposits, wear, rust, and corrosion in gasoline engines than oils which are satisfactory for API Service Classification SC and may be used when API Service Classification SC is recommended.
SE—1972 Gasoline Engine Warranty Maintenance Service	Service typical of gasoline engines in passenger cars and some trucks beginning with 1972 and certain 1971 models operating under engine manufacturers' warranties. Oils designed for this service provide more protection against oil oxidation, high-temperature engine deposits, rust, and corrosion in gasoline engines than oils which are satisfactory for API Engine Service Classifications SD or SC and may be used when either of these classifications is recommended.
CA—Light-Duty Diesel Engine Service	Service typical of diesel engines operated in mild to moderate duty with high quality fuels. Occasionally has included gasoline engines in mild service. Oils designed for this service were widely used in the 1940s and 1950s. These oils provide protection from bearing corrosion and from high-temperature deposits in diesel engines when using fuels of such quality that they impose no unusual requirements for wear and deposit protection.

Table 6-3   New API Service Classification of Engine Crankcase Oil, Effective 1971 (*Continued*)

API service classifications (types)	Remarks
CB—Moderate-Duty Diesel Engine Service	Service typical of diesel engines operated in mild to moderate duty, but with lower-quality fuels which necessitate more protection from wear and deposits. Occasionally has included gasoline engines in mild service. Oils designed for this service were introduced in 1949. Such oils provide necessary protection from bearing corrosion and from high-temperature deposits in diesel engines with higher-sulfur fuels.
CC—Moderate-Duty Diesel & Gasoline Engine Service	Service typical of lightly supercharged diesel engines operated in moderate to severe duty and has included certain heavy-duty gasoline engines. Oils designed for this service were introduced in 1961 and used in many trucks and in industrial and construction equipment and farm tractors. These oils provide protection from high-temperature deposits in lightly supercharged diesels and also from rust, corrosion, and low-temperature deposits in gasoline engines.
CD—Severe-Duty Diesel Engine Service	Service typical of supercharged diesel engines in high-speed, high-output duty requiring highly effective control of wear and deposits. Oils designed for this service were introduced in 1955, and provide protection from bearing corrosion and from high-temperature deposits in supercharged diesel engines when using fuels of a wide quality range.

Source: American Petroleum Institute.

farms and ranches are in many cases quite different from those which automobiles and trucks used on streets and highways must endure. Some equipment manufacturers are supplying oil that has been developed specifically for farm machines.

GEAR OILS

Gear oils are used in standard gear transmissions, final drives, and gearboxes. Because of the amount of power that is developed by modern engines, these lubricants must be capable of withstanding extreme pressures exerted between the teeth of meshing gears. The oil must be capable of withstanding high temperatures and still provide good lubrication for low-temperature operations.

The most important property of gear lubricants is viscosity. Older machines and tractors used extremely heavy oils or greases to lubricate the teeth on the

**Table 6-4   Cross Reference for New and
Old API Service Classifications**

New API designations	1952-70 API designations
SA	ML
SB	MM
SC	MS(1964)
SD	MS(1968)
SE	MS(1972)
CA	DG
CB	DM
CC	DM
CD	DS

Source: American Petroleum Institute.

large, low-speed gears. Many operators still refer to gearbox lubricants as grease. The SAE viscosity numbers for gear oils are given in Table 6-5. The viscosity of the oils, however, is not as heavy as the assigned numbers may indicate. SAE No. 80 gear oil has about the same viscosity as SAE 20 or SAE 30 engine crankcase oil, while SAE 90 gear oil is comparable to SAE 40 or SAE 50 engine oil. The higher numbers are used for gear oils to separate the two types of oils.

Additives are used in gear oils to improve the performance of the oil and to protect the gears and bearings. One of the most important materials used is the extreme-pressure (EP) additive that enables the oil to withstand the high pressures that are experienced between the teeth of gears. High temperatures are usually developed and excessive wear occurs. The additive causes a coating to be developed to protect the metal surfaces and reduce wear. Other additives include rust prevention, antioxidants, and foam inhibitors.

Gear oils are also classed according to the type of service they are adapted for, as shown in Table 6-6. The API gear lubricant number increases as the severity of the service increases. API GL-1 is suitable for low-speed and low-torque gear transmissions. API Service GL-4, 5, and 6 are generally used in tractor transmissions and final drives.

Some transmissions require special oils that do not fall into the above categories. These include machines in which the hydraulic system uses the same

Table 6-5   SAE Classification of Gear Lubricants

SAE viscosity no.	Viscosity range, Saybolt Universal	Consistency must not channel in service at °F
80	100,000 sec at 0° F, max.	Minus 20
90	800 to 1500 sec at 100° F	Zero
140	120 to 200 sec at 210° F	Plus 35

Table 6-6   API Service Classifications for Gear and Transmission Oil

API service classifications	Remarks
GL-1 Regular-Type Gear Lubricant This term designates gear lubricants generally suitable for use in most automotive type transmissions and in most spiral-bevel and worm-gear differentials.	A straight mineral gear oil used where tooth pressure and gear speeds are relatively low. Recommended for a number of farm tractors. May contain antioxidant and antifoam additives.
GL-2 Worm-Type Gear Lubricant	Not used for farm tractors.
GL-3 Mild-Type Extreme Pressure (EP) Gear Lubricant This term designates gear lubricants having load-carrying properties suitable for many automotive transmissions and spiral-bevel differentials under moderately severe conditions of speed and load.	Recommended for a few farm tractors. Usually contains antioxidant and antifoam additives plus additives to reduce friction under heavy loads and provide mild extreme pressure characteristics.
GL-4 Multipurpose-Type Gear Lubricant This term designates lubricants which have properties required to provide satisfactory lubrication of hypoid gears and conventional differentials, including adequate load carrying ability for protection of such gears in sustained high-speed low-torque, and low-speed high-torque conditions in modern high-powered passenger cars and trucks. They are suitable for use in spiral-bevel gears, many transmissions, and for worm gears in some types of service. GL-4 lubricants will generally meet the requirements of Military Specification MIL-L-2105. This rating is now obsolete but some tractor manufacturers still refer to the classification.	Commonly recommended for farm tractor differentials and transmissions. Available from most oil suppliers. Has more load-carrying ability than "Mild Type (EP)". Suitable for most severe applications.
GL-5 Multipurpose-Type Gear Lubricant This term designates lubricants with properties required to meet GL-4 (above) requirements but has higher load-carrying characteristics. GL-5 lubricants will generally meet the requirements of Military Specification MIL-L-2105B	Suitable for more severe applications than GL-4 above.

oil as the transmission and the hydrostatic-type transmission. In all cases the manufacturer's operator's manual should be consulted regarding the correct type of oil to be used for lubricating the gears.

GREASES

The American Society for Testing Materials defines a petroleum lubricating grease as "a semisolid or solid combination of a petroleum product and a soap, or a mixture of soaps, with or without fillers, suitable for certain types of lubrication." In ancient times a grease made of animal fat was probably used to lubricate the axles of chariots. Mineral oil grease was developed soon after the discovery of oil in 1859. Early farm machinery which was drawn by horses had few moving parts, and almost any type of grease would suffice to reduce friction. Modern power-operated machinery has many bearings carrying light to heavy loads. Some parts operate at high speeds and high temperatures and require special high-quality greases. A water-repellent grease is required for water pumps, a soda-base grease of spongy or fibrous texture for wheel bearings and universal joints, a cup grease for distributor shafts, and a soft, tacky grease for chassis joints.

Classes of Greases

There are many types of machines found on a single farm, and a review of the operator's manuals will indicate the many types or grades of greases that are recommended among the makes and models. Too strict adherence to the recommendations among the machines and lubrication points could lead to confusion as well as to the maintenance of a widely assorted stock of greases, such as wheel-bearing grease, water-pump grease, chassis grease, etc. Positive steps have been taken by the manufacturers and the petroleum industry to simplify the problems in this area. The various classes of greases that are currently available are given in Table 6-7. The multipurpose grease has been found to be very effective in meeting the requirements of most lubricating points in farm machinery, thereby minimizing the confusion that might occur as well as reducing the number of different types that must be stocked. Most multipurpose greases will contain additives such as rust inhibitors, antioxidants, and extreme-pressure additives.

Grades for Greases

The National Lubricating Grease Institute (NLGI) has adopted nine grade numbers which are an indication of firmness or hardness of the grease, as shown in Table 6-8. The ASTM-worked penetration at 77°F is shown in relation to the NLGI grade number.

   The penetration test consists of dropping a metal cone of standard size and weight into a worked grease at a specified temperature (usually 77°F) and measuring the penetration of the cone after a specified time interval. The consistency or hardness of greases ranges from No. 000, very soft, to No. 6, a

Table 6-7   Characteristics of Greases

Class	Type	Kind of thickener	Approximate dropping pt.	Characteristics
Lime soap	Cup	Calcium-fat	190° F	Smooth, water-resistant, separates on loss of water content, limited consistency loss on working.
	Complex	Calcium-fat and calcium acetate	Over 500° F	Some or slight fiber, water-resistant, usually some E.P. properties, consistency change on aging.
Soda soap	Fiber or sponge	Soda-fat	375° F	Fibrous, not water-resistant, does not separate at elevated temperatures, variable consistency loss on working.
	Medium or short fiber, smooth	Soda-fat or fatty-acid	375° F	Semismooth to smooth, not water-resistant, does not separated at elevated temperatures, variable consistency loss on working.
	Block or brick	Soda-fat, fatty-acid, and rosin	400° F	Smooth, hard, not water-resistant, does not separate at elevated temperatures.
Lithium soap	Multipurpose	Lithium-fat or fatty-acid	375° F	Smooth, water- and heat-resistant, does not separate at elevated temperatures, limited consistency loss on working.
Nonsoap	High temperature, multipurpose	Inorganic-clay or silica Organic-Arylurea	Usually 500° F or higher	Smooth, water- and heat-resistant, high temperature stability, good mechanical and chemical stability.
Residuum	Pinion or gear compound	Usually none, sometimes lead naphthenate	Rapid thinning when heated	Smooth, black, adhesive, excellent pressure and water resistance.

Source: National Lubricating Grease Institute.

very hard grease. The color of greases varies from light red to pitch black, according to the ingredients used. Most manufacturers recommend either No. 1 or No. 2 grade.

Methods of Applying Greases

Some greases are applied by hand with a paddle, swab, or brush to gears, chains, and wire rope. Slow-moving bearings can be lubricated by placing the grease in cups of the screw-down, spring-loaded, and automatic types.

All well-designed modern farm machinery has high-pressure grease fittings

Table 6-8   National Lubrication Grease Institute (NLGI)
Grades of Greases versus Penetration

NLGI grade, number	ASTM-worked, penetration at 77° F
000	445-475
00	400-430
0	355-385
1	310-340
2	265-295
3	220-250
4	175-205
5	130-160
6	85-115

Source: National Lubrication Grease Institute.

at the points requiring lubrication. The grease is applied through the fitting with high-pressure grease guns (Figs. ɔ-4 to 6-6). The proper lubrication of farm machinery bearings and moving parts is one of the most important factors in care and maintenance. Many manufacturers are installing central lubrication systems called *multi-luber systems.* The system operates by diaphragm pumps activated by vacuum from the engine. On self-propelled machines the system is operated by pressing a push button on the instrument panel. A light on the panel indicates when the lubrication cycle has been completed. Hand-push-plunger multi-luber systems are available for machines not equipped with power units (Fig. 6-7). Companies manufacturing lubricants have available lubrication guides showing all the points on the machine requiring lubrication and the kind of oil or grease recommended for each fitting.

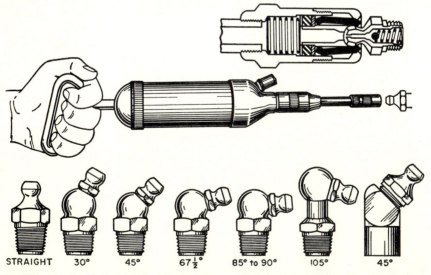

STRAIGHT   30°   45°   67½°   85° to 90°   105°   45°

Fig. 6-4   Hydraulic hand-push grease gun and fittings, showing cross section of nozzle and fitting.

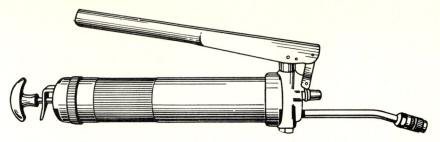

Fig. 6-5   Lever-type grease gun.

Fig. 6-6   Portable hand-tank grease guns and pails: (*a*) interchangeable pail pump with hose; (*b*) installed pail pump on pail; (*c*) interchangeable pail pump for gear oils; (*d*) interchangeable pail hand-gun loader; (*e*) portable grease bucket with pump. (*Alemite and Instrument Division, Stewart Warner Corporation.*)

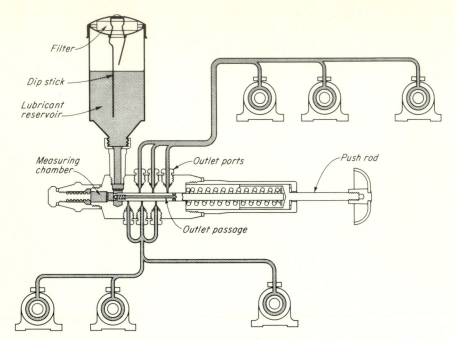

Filter

Dip stick

Lubricant
reservoir

Measuring
chamber

Outlet ports

Push rod

Outlet passage

Fig. 6-7   Arrangement of mutiple feed lines that can be attached to the master hand-gun lubricator shown at the center. (*Lincoln Engineering Co.*)

REFERENCES

Alemite Answers, Alemite Division, Stewart-Warner Corporation, 1945.

API Gear Lube Service Designations, American Petroleum Institute, API Publication 1560, 1966.

American Petroleum Institute, *Tech. Letter* no. 4-397, 1952, and *Publication 1509A,* 1960.

Fuels and Lubricants, American Association for Vocational Instruction, 1973.

Engine Service Classification and Guide to Crankcase Oil Selection, American Petroleum Institute, 1972.

Farm Machinery Lubrication Guides, The Texas Company.

Jones, F. R.: *Farm Gas Engines and Tractors,* McGraw-Hill Book Company, New York, 1952, chap. 23.

Lubricating Greases, *Engin. Bul.* G-208, Standard Oil Company.

Lubrication Guide for Farm Equipment, Socony-Vacuum Oil Co., Inc., 1951 ed.

QUESTIONS AND PROBLEMS

1   Define friction and discuss its advantages and disadvantages.
2   Enumerate the forms, kinds, and sources of lubricants.
3   Explain the API system of service classification of crankcase lubricants for gasoline and spark-ignition engines; name the different kinds of service and give the main features of each service.

4   Explain the API service systems for diesel engines and give the main features of each system.
5   Discuss the SAE grades of engine oils and gear lubricants. What is meant by an *EP* gear lubricant?
6   Explain the meaning of oil viscosity.
7   Give a definition for grease, and list the main types according to use.
8   Explain the NLGI grades of grease.
9   Discuss the various methods of applying greases to farm equipment. What is meant by high-pressure lubrication?
10 Describe a multi-luber system.

Chapter 7

Hydraulic Systems for Farm Equipment

Farm equipment prior to the nineteenth-century era was animal-drawn, guided by hand, and lifted manually. Later, when equipment was mounted on wheels, levers were used to raise and lower the working units. The first mechanical power lift was developed for the trailing-type tractor-drawn plows about 1910 (Fig. 7-1). The tractor power lift was developed about 1930 to raise and lower planters and cultivators mounted on the row-crop tractor. The use of hydraulic power for lifting tractor-mounted equipment was introduced in 1935. Hydraulic power lifts are now used for raising, lowering, and controlling almost all types of field equipment, ranging from the small plow to the platform of a grain combine and the drums of cotton-picking machines. In fact, if it were not for the hydraulic lifts, these heavy units would be extremely difficult to operate.

The extensive use of hydraulic lifts and controls makes it essential that the operator of modern farm equipment have an understanding of the fundamental principles of power-lift hydraulics.

Fundamentals of Hydraulics

There are several branches of hydraulics, but the branch applicable to farm equipment deals with enclosed liquids under pressure. The fundamental law of

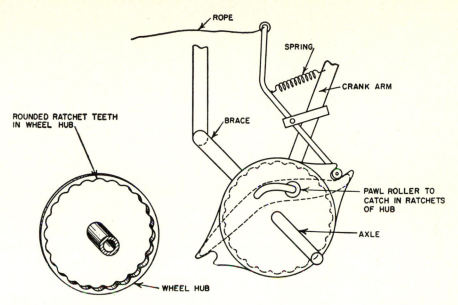

Fig. 7-1 Mechanical power lift for plow.

hydrostatics, or the mechanics of fluids, was defined by Blaise Pascal[1] in 1653 as follows: "Pressure applied to an enclosed fluid is transmitted equally and undiminished in all directions to every part of the fluid and of its restraining surfaces." The application of this law is shown in Figs. 7-2 and 7-3. In Fig. 7-3 is a 1-lb (454-g) weight acting on 1 in^2 (6.45 cm^2) of liquid which is counterbalanced by a weight of 10 lb (4540 g) on 10 in^2 (64.5 cm^2) of liquid. The pressure in the system between the two weights is 1 lb for each in^2 (454 g for

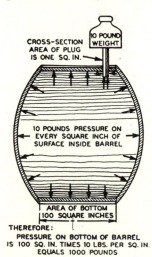

Fig. 7-2 How pressure of a liquid distributes pressure in all directions of its enclosure case. (*International Harvester Company.*)

[1] David Halliday and Robert Resnick, *Fundamentals of Physics*, John Wiley & Sons, Inc., New York, 1974.

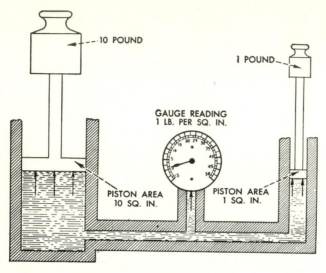

Fig. 7-3 A simple hydraulic system with a 1-lb weight on 1 square inch has the same pressure per square inch as the 10-lb weight on the piston resting on 10 square inches of the liquid. There is also a 1-lb pressure on the gage. (*International Harvester Company*.)

each 6.45 cm^2 or 70.3 g/cm^2) of surface area. Therefore, it can be seen that a force of 1 lb acting on an area of 1 in^2 can be used to lift a weight of 10 lb if the area under the weight is 10 in^2. (In metric units, a force of 454 g acting on 6.45 cm^2 could be used to lift a weight of 4540 g if the weight had an area of 64.5 cm^2). Note, however, that in order to move the weight on the large area that is 10 times the size of the smaller area, the force on the small piston must move through a distance 10 times greater than the distance the larger piston moves.

HYDRAULIC SYSTEM COMPONENTS

A hydraulic system is basically a method of transmitting power from the power source to the machine or component being operated. The same power could be transmitted through other means, such as belts, chains, shafts, or special linkages. The medium that is used to transmit the power in a hydraulic system is the fluid that is contained by the lines between the driver and driven members. The principal advantages of the hydraulic system over other means are that it offers a simple means of transmitting the power to remote sections of the machine and it is easy to convert the rotary motion of the power unit into other forms of motion, such as reciprocating movement produced with cylinders.

A modern hydraulic system will contain many parts or components, since the method of transmitting power is used extensively on machines, such as the self-propelled harvesters, to perform many functions. The two basic components in all hydraulic systems include the pump that converts the power from the engine to fluid power and the actuator, such as a cylinder or motor, that converts the fluid power to the motion and action that are being performed (Fig. 7-4). Other parts of the system include the reservoir, lines, connections, filters, and various types of valves.

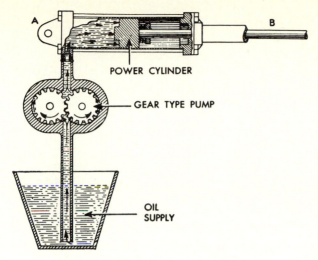

Fig. 7-4   A simple diagram showing how a gear pump pumps oil from a reservoir of oil to a power cylinder. (*International Harvester Company*.)

Pumps

The pump is often referred to as the heart of the system. It is usually driven directly by the engine of the tractor or the self-propelled machine. On some older machines the pump was driven from the power-takeoff shaft, and the main drive clutch had to be engaged before the hydraulic system would operate. Engine-driven pumps operate continually and provide what is commonly referred to as a "live hydraulic system."

The pump is used to create flow of the hydraulic fluid in the system. Pressure in the system is caused by the restriction or obstruction of fluid flow. For example, if there is no load on a hydraulic cylinder, there would be little or no resistance to fluid flow in the cylinder. However, as a load is added to the cylinder, resistance is offered to the flow, and the pumps must be capable of moving or pumping the fluid in the system regardless of the load or the pressure in the system. All pumps used in the hydraulic system are referred to as *positive-displacement* types. This means that the pump is designed to pump the same volume of fluid over a wide range of pressures.

Hydraulic pumps are rated in a number of different ways based upon the amount of fluid the device will pump for a given period of time. Common units to designate the output flow include gallons per minute or liters per minute. The flow rate is determined not only by the speed (r/min) at which the pump is being turned, but also by the physical size of the pump. Another factor that determines the flow rate is the pressure against which the pump is operated. Even on positive-displacement pumps, the discharge or flow rate is reduced slightly as the pressure increases. Therefore, in order to compare pumps or to purchase another pump, the pump capacities must be compared at the same pressure and r/min. Pumps used on farm machinery usually range in size from 10 g/min (38 liters/min) to 50 g/min (190 liters/min).

The pumps may also be rated according to the discharge or volume of fluid that is displaced in a single revolution of the input shaft. This is referred to as "pump displacement," and the units are generally measured in cubic inches, cubic centimeters, gallons, or liters. The displacement of pumps depends upon the internal arrangement of the pumping elements. Pumps may be classed as of a fixed or variable displacement. In a fixed-displacement pump, the amount of fluid that is pumped per revolution is constant. In a variable-displacement pump, the internal arrangement of the pumping elements can be changed during operation to change the displacement so that the discharge rate may be varied from zero to maximum regardless of the pump speed.

There are various types of pumps that are used in hydraulic systems on farm machines and tractors. The vane-type pump, as shown in Fig. 7-5, has a series of sliding vanes that trap the fluid between the rotor and the housing and convey the fluid from the inlet to the outlet of the pump. Vane pumps may be classed as either balanced (Fig. 7-5) or unbalanced. The balanced vane pump has two pumping chambers, one on each side of the rotor. The inlets and outlets for the two chambers are connected together within the pump. The unbalanced vane pump has only one pumping chamber, and the rotor is offset to one side of the pumping chamber it may be constructed either as a fixed- or a variable-displacement type. The variable-displacement pump is designed so that the eccentric housing can be moved in relation to the rotor while in operation.

Gear-type pumps are used extensively on agricultural machines. The types that are often used include the external-gear, Fig. 7-6, and internal-gear pumps as shown in Fig. 7-7. The pumping action of gear-type pumps is obtained by trapping the fluid between the teeth of the gears and the housing. A positive seal between the teeth of the gears prevents the fluid from escaping back to the inlet side. These pumps can usually be driven in either direction without making changes within the pump. They are available usually only as fixed-displacement pumps.

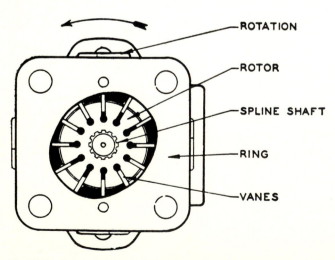

Fig. 7-5   End view of vane-type oil pump with oval ring and rotor with sliding vanes. *(Sperry Vickers, A Division of Sperry Rand.)*

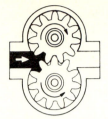  

HYDRAULIC OIL ENTERS   . . . AND IS CARRIED AROUND   . . . AND IS FORCED INTO
INLET SIDE OF PUMP. . .   IN THE TOOTH SPACES. . .   THE OUTLET SIDE OF PUMP.

Fig. 7-6   Operation of external gear pump.

Piston-type pumps, as illustrated in Fig. 7-8, are used in hydraulic systems designed to operate at very high pressures. The pumping action comes from the reciprocating motion of the piston in the cylinder. Fluid is drawn in from the inlet side as the piston is retracted into the cylinder and is forced out as the piston moves outward. Most piston-type hydraulic pumps are equipped with several cylinders to provide a uniform flow of fluid. Piston pumps may be grouped into several categories. One classification is by the arrangement of the cylinders. A radial pump has the pistons perpendicular to the shaft. The most common type used for agricultural applications is the axial pump, in which the pistons operate in a plane parallel to the shaft. The axial pump is further classified as to the method of driving: swashplate and bent-axis. The entire block that contains the cylinder and piston rotates, and the reciprocating motion of the piston is produced by the swashplate, or by the angled shaft in the case of the bent-axis pump. The displacement of the swashplate-type pump can be varied on some pumps by varying the angle of the swashplate and in turn varying the length of stroke of the piston. This type of pump is used extensively in hydrostatic drives to obtain variable speeds.

Motors

Hydraulic motors convert the fluid energy into rotary motion and are frequently used on machine components that are located remotely from the power source.

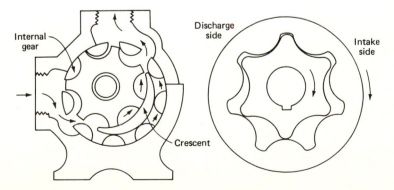

Fig. 7-7   International gear pumps—crescent and gerotor.

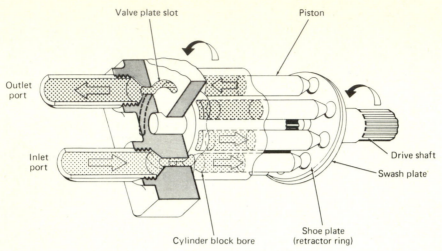

Fig. 7-8   Axial piston pump.

Motors are also used in cases where it is desirable to make frequent changes in the speed and to reverse the driven component. Generally, the motors are similar to hydraulic pumps, and in some cases they may be used as either a pump or a motor. Hydraulic motors are available as gear, vane, and piston. They are rated in much the same way as pumps. The direction of rotation can be changed by reversing the flow of fluid into the motor. The speed is varied by varying the amount of fluid going into the motor. These two features can be accomplished with valves. The amount of torque a motor produces is dependent upon its size and type and the pressure in the system.

Cylinders

The hydraulic cylinder is practically a universal component on farm machines. It is used for raising and lowering cutter bars, plows, disks, and harrows and for controlling the heights and depths of most of these machine components (Fig. 7-9). Cylinders are also used in power-steering systems.

The hydraulic cylinder, as illustrated in Fig. 7-10, is basically a simple device. It consists of a cylindrical tube that is sealed on both ends with a cap. The piston inside the tube is connected to a rod that extends to the outside through an end cap. The piston is equipped with seals to prevent the fluid from leaking past the piston. The movement of the piston is accomplished by forcing fluid into the cylinder.

Cylinders are available as either double-acting or single-acting. A single-acting cylinder has only one hose connection that is made on the piston end of the cylinder. Force may be developed on the rod by the fluid acting against the piston. The piston and rod continue to move as long as fluid continues to fill the cylinder until the rod is fully extended. The rod and piston may be retracted (returned to original position) by either gravity (the weight of the machine component) or a spring. The cylinder can be used to exert force in only one direction.

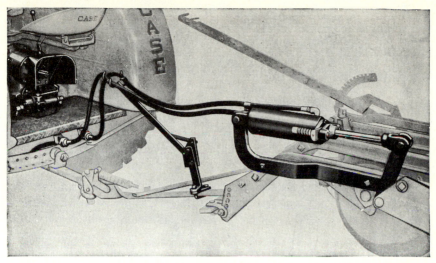

Fig. 7-9 Portable hydraulic-lift cylinder equipped with stop yoke attached to trailing plow so that adjustments can be made for uniform plowing depth. (*J. I. Case Company.*)

A double-acting cylinder is equipped with two fluid connections so that the fluid can be directed to either the rod end or the piston end of the cylinder. A cylinder of this type can exert more force while extending than while retracting. A directional-type control valve is used to direct the flow of hydraulic oil to either end depending upon the desired direction of movement. These cylinders are used extensively for lifting machines and for exerting force downward if required. Some cylinders are equipped with devices that control the travel distance, as shown in Fig. 7-11.

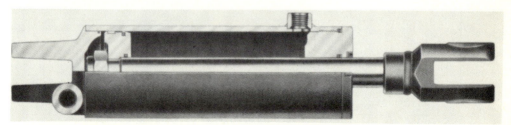

Fig. 7-10 Double-acting hydraulic cylinder. (*Fluid Power Division, Cessna Aircraft Co.*)

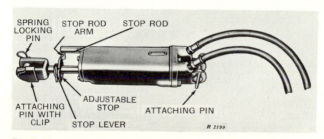

Fig. 7-11 Remote-control hydraulic cylinder. (*Deere & Co.*)

The amount of force that a cylinder can exert is dependent upon the diameter and the system pressure. The forces that can be developed with different sizes of cylinders and pressures are given in Table 7-1. The force developed by a cylinder can be increased for a given pressure by increasing the size of the cylinder, since the pressure of the fluid is acting against a larger surface area. For a given-size cylinder, the force can be increased by increasing the pressure. It should be noted in Table 7-1 that a double-acting cylinder will not exert the same amount of force in both directions. Less force is developed when the cylinder is being retracted (PULL) than when it is being extended (PUSH), because the rod reduces the area of the piston that the fluid acts upon.

The rate of travel of the piston is determined by the rate of fluid flow into the cylinder. Cylinders are available in different lengths and are generally designated by the length of the stroke, that is, by the distance the rod travels from the fully retracted to the fully extended position. The standard stroke for most agricultural cylinders is 8 in (20.3 cm). Typical forces that hydraulic cylinders might be expected to exert are shown for different machines in Tables 7-2 and 7-3.

Valves

There are numerous types of valves that are used in hydraulic systems. They are used basically to control different functions and parameters in the system to make it more flexible and to perform many different functions. Valves are

Table 7-1   Amount of Force Double-acting Cylinders Will Develop at Different Pressures

Size of bore	Piston rod dia.	Pump pressure, lb									
		500 lb/in^2		1000 lb/in^2		1500 lb/in^2		2000 lb/in^2		2500 lb/in^2	
		Push	Pull	Push	Pull	Push	Pull	Push	Pull	Push	Pull
2 in	1$\frac{1}{8}$ in	1570	1074	3141	2148	4710	3221	6282	4296	7851	5369
2 in	1$\frac{1}{4}$ in	1570	957	3141	1914	4710	2872	6282	3828	7851	4786
2$\frac{1}{2}$ in	1$\frac{1}{8}$ in	2454	1957	4909	3915	7363	5872	9818	7830	12272	9787
2$\frac{1}{2}$ in	1$\frac{1}{4}$ in	2454	1841	4909	3682	7363	5527	9818	7364	12272	9204
3 in	1$\frac{1}{8}$ in	3534	3037	7068	6075	10602	9112	14136	12150	17670	15187
3 in	1$\frac{1}{4}$ in	3534	2920	7068	5841	10602	8762	14136	11682	17670	14603
3 in	1$\frac{1}{2}$ in	3534	2651	7068	5302	10602	7952	14136	10604	17670	13254
3 in	2 in	3534	1963	7068	3927	10602	5890	14136	7854	17670	9817
3$\frac{1}{2}$ in	1$\frac{1}{4}$ in	4811	4197	9621	8394	14432	12591	19242	16788	24053	20985
3$\frac{1}{2}$ in	1$\frac{1}{2}$ in	4811	3927	9621	7854	14432	11781	19242	15708	24053	19635
3$\frac{1}{2}$ in	2 in	4811	3240	9621	6480	14432	9719	19242	12960	24053	16199
4 in	1$\frac{1}{4}$ in	6283	5669	12566	11339	18850	17008	25132	22678	31416	28347
4 in	1$\frac{1}{2}$ in	6283	5399	12566	10799	18850	16198	25132	21598	31416	26997
4 in	2 in	6283	4712	12566	9425	18850	14138	25132	18850	31416	23563
5 in	2 in	9817	8247	19635	16494	29450	24742	39270	32988	49085	41236
6 in	2 in	14137	12566	28274	25133	42412	37700	56548	50266	70686	62833

Source: Energy Manufacturing Company.

**Table 7-2   Cylinder Thrust Requirements for
Tandem or Double-Action Disk Harrows**

Width of cut, ft	Maximum cylinder thrust, lb	
	Traveling	Stopped
6	1960	2680
7	3090	2670
7½	3740	3070
8	3080	3210
9½	1900	2400
10	5280	5000
10½	1900	2650
11½	2100	2950
12	2250	4000
14½	8980	11,010

Source: W. H. Worthington and J. W. Seiple, *Agr. Engin.,* **33**(5):273-276, 1952.

generally grouped into three major categories: pressure control, volume control, and directional control.

Pressure control valves may be designed to perform one of the following functions: (1) limiting the maximum pressure, (2) reducing the pressure for a specific part in the system, (3) unloading a pump, or (4) causing operations in a system to occur in a specific order or sequence.

A pressure relief valve (Fig. 7-12) is used to limit the maximum pressure. It is connected in the main line from the pump and has a line running to the reservoir for the excess oil. When the maximum pressure is reached, the pressure of the fluid will open the valve that is held closed by spring force. The spring tension is normally adjustable, and the valve is often referred to as the pressure regulator. The valve should not be changed without using a gage to check the pressure, since the relief valve serves as a safety device.

A pressure-reducing valve, as shown in Fig. 7-13, is used to reduce the pressure from that at which the main system normally operates to a lower pressure that might be required in a secondary system.

An unloading valve is similar to the relief valve in that it will limit the

**Table 7-3   Minimum Hydraulic Cylinder Thrust
for Various Implement-Tractor Groups**

Implement-tractor group	Minimum cylinder thrust, lb	
	Traveling	Stopped
2-plow group	4500	5625
3-plow group	6000	7500
4-plow group	7500	9375
5-plow group	10,000	12,500

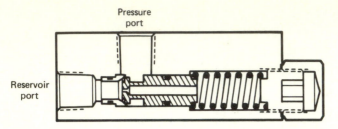

Fig. 7-12   Piston-type pressure relief valve. (*Fluid Controls, Inc.*)

maximum pressure. However, the entire flow from the pump is diverted back to the reservoir, rather than only the excess flow, as with the standard relief valve.

    A sequence valve is used to control the order in which various functions can occur in a system. A typical example of the use of this valve is the delayed-lift system used between the front- and rear-mounted tractor cultivators, as illustrated in Fig. 7-14.

    Volume control valves are used to provide accurate control over the rate of flow of oil in a hydraulic system regardless of fluid pressure. These valves, as shown in Fig. 7-15, are used to control the speed of motors and may be used to control the rate of travel of hydraulic cylinders.

    Directional control valves are designed to control or direct the flow of the oil in the hydraulic system. They vary in type from a simple two-way valve that is used to turn the flow on or off to more complex valves. The controlling mechanism in the valve may be a rotary-disc or spool-type valve, as shown in Fig.

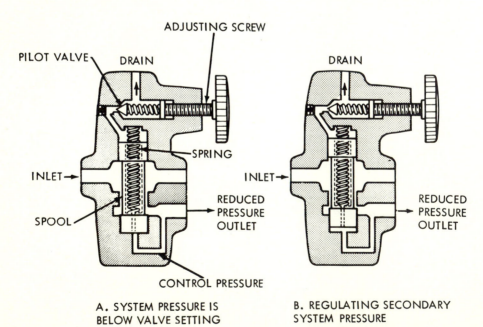

Fig. 7-13   Pressure-reducing valve. (*Sperry Vickers, A Division of Sperry Rand.*)

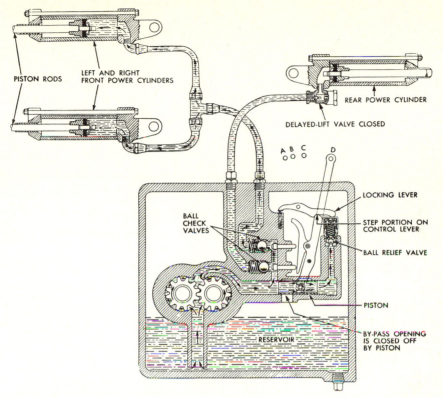

PISTON RODS

LEFT AND RIGHT
FRONT POWER CYLINDERS

REAR POWER CYLINDER

DELAYED-LIFT VALVE CLOSED

A B C
O O O

D

LOCKING LEVER

BALL
CHECK
VALVES

STEP PORTION ON
CONTROL LEVER

BALL RELIEF VALVE

PISTON

RESERVOIR

BY-PASS OPENING
IS CLOSED OFF
BY PISTON

Fig. 7-14 Schematic diagram of a hydraulic-lift system where oil is being pumped into the lifting cylinders. At this stage the pressure is not high enough to open the delayed-lift valve on the rear cylinder. (*International Harvester Company.*)

7-16. The manually operated spool type is the most commonly used valve on farm machines.

Valves are classed according to the number of ports or openings. A two-way valve will have two ports and is generally used to open and close a circuit. This type is not generally used in hydraulic circuits. A three-way valve is often used with a single-acting cylinder. The valve is equipped with a port for the pump, one for the cylinder, and the third port connected to the reservoir.

Four-way valves are used with double-acting cylinders and motors. The four ports include the pressure or pump, the tank or reservoir, and a port for each end of the cylinder or for the inlet and outlet for a motor. These valves have three positions for operation. In the center or neutral position, no fluid will be directed to the actuator. When the valve is moved to the left or right of the center position, the flow to the actuator is directed by the spool. Symbols, as shown in Fig. 7-17, are often helpful in understanding the operation of the valves. In using the symbols and the illustration in Fig. 7-16, it may be noted that when the valve is moved to the left, fluid is permitted to flow from the pump to the cylinder port *A*. It is returned from the other end of the double-acting cylinder through the port *B* to the tank. If the valve is moved to

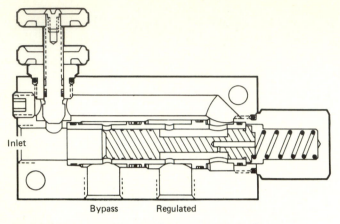

Fig. 7-15 Flow-control valve. (*Fluid Controls, Inc.*)

the right, oil is directed from the pump to the cylinder port *B*, thereby causing the piston to move in the opposite direction. The oil is returned to the tank through cylinder port *A*.

There are three types of spools that are commonly used on hydraulic systems (Fig. 7-17). The basic difference occurs when the valve is in the neutral position. The tandem-center valve is designed to allow the oil to flow freely from the pump to the reservoir in neutral position. The working ports (cylinder ports *A* and *B*), however, are blocked. This type of valve is desirable for use with a

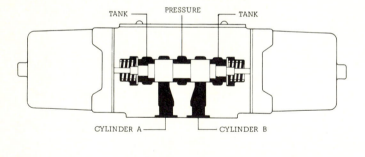

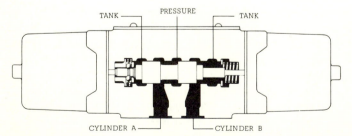

Fig. 7-16 Spool-type, closed-center, four-way directional-control valve: (*top*) neutral position—all parts blocked; (*bottom*) fluid under pressure flows through part *B* to tank. (*Racine Hydraulic Division, Rexnord, Inc.*)

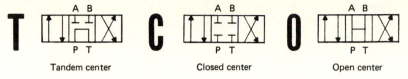

Tandem center   Closed center   Open center

Fig. 7-17   Symbols for directional-control valves.

cylinder to maintain a certain position while the machine is operating. The closed-center valve blocks the flow in all ports in the center position. An unloading valve is usually used with this type of valve. All ports are left open in the open-center valve in the neutral position. This type is often used with motors and with hydraulic cylinders if it is desirable to allow the cylinder to "float" or move freely when it is not being used to raise or lower a machine.

These valves are usually operated manually with a lever. They may be "spring-loaded" to the neutral position, or sometimes detents are used to hold the valves in a working position. Some detent valves are designed to be returned to neutral position automatically when the pressure reaches a certain limit in the system.

Lines and Couplings

The hydraulic fluid or oil is conducted to the various parts in the system by lines made of steel, copper, or synthetic rubber. The lines must be of sufficient size to carry the volume used in the system without causing resistance to flow. They are also designed to withstand the pressure of the fluid. Normally, the lines are rated according to working pressure and bursting pressure. Most hydraulic systems used on agricultural machines have a maximum working pressure of about 2000 lb/in^2 (140.6 kg/cm^2). The bursting pressure of the lines is well above the working pressure.

Flexible hoses, such as illustrated in Fig. 7-18, are used extensively on machines, since they are easy to install and will withstand shock and vibrations. Most hydraulic hoses will contain an inner core that is resistant to oil, reinforcement layers of wire and/or fabric braid, and an outer cover that is also oil-resistant. The strength of the hose is governed by the number of braids and the diameter. Small-diameter lines will withstand higher pressures than larger lines of the same design and construction.

Couplers are used to join hoses or to connect the line to a part in the system. Most hoses on the original machine are equipped with permanent-type couplings, as shown in Fig. 7-19. Care must be taken to be sure that the proper threads are obtained on replacement hoses.

Special tools and equipment are required on the permanent-type couplings, and they cannot be reused. Two types of reusable couplings are shown in Figs. 7-20 and 7-21. No special tools are required to install the reusable fittings, and they are very useful for making field repairs on hydraulic hoses.

Quick-Disconnect Couplers   Quick-disconnect couplers (Fig. 7-22) are used in places where the hydraulic lines are connected or disconnected frequently. Both

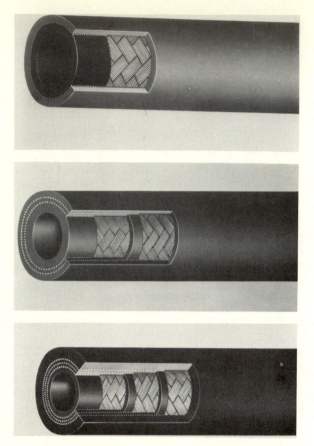

Fig. 7-18 One-wire-, two-wire-, and three-wire-braid hydraulic hose. (*Gates Rubber Company*.)

parts of the couplers are equipped with a check valve that will prevent the loss of fluid when the hoses are disconnected. When the two parts are connected, the check valves are unseated to allow fluid flow. The two parts are held together by a locking device. No tools are required for connecting or disconnecting, but some problems may be experienced on some types if pressure is in the line. This can be overcome by relieving the pressure with the control valve. Care should always be taken to clean the couplers before making connections to prevent dirt and dust from getting into the hydraulic system.

Reservoir or Tank

The primary function of the reservoir is to store the oil or fluid for the system. On many machines, the reservoir is designed to provide some cooling for the system. The filler cap is usually vented to allow for expansion and contraction as the oil is heated and cooled. Baffles are used to prevent the return oil from being used immediately by the system pump. All reservoirs are equipped with some device to indicate the proper oil level.

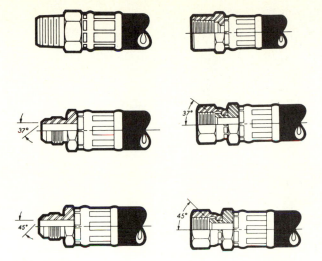

Fig. 7-19 ,Permanent-type hose couplings, including pipe, jic, and sae threads. (*Dayco Corporation.*)

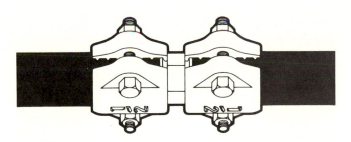

Fig. 7-20   Bolt-type reusable hose coupling. (*Dayco Corporation.*)

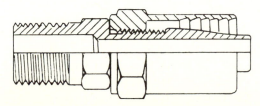

Fig. 7-21   Screen-type reusable coupling. (*Gates Rubber Company.*)

Fig. 7-22 Quick-disconnect coupler with ball check valves. (*Safeway Hydraulics, Inc.*)

Coolers

Because of the amount of heat generated in most of the larger self-propelled machines such as combines, special cooling coils are used. Most systems use a cooler similar to the radiator used for the engine water-cooling system. The coil is often located near the radiator and air is forced through the coil by the engine fan. Some systems, however, may use a water-type cooler that uses the water in the engine cooling system to keep the oil at the proper temperature.

Filters

The filter in a hydraulic system is one of the most important parts as far as determining the useful life of other parts in the system is concerned. In order for the pump to cause fluid to flow against the pressures in most systems, the pump must be constructed with extremely close tolerances. Most other parts are also fitted with precision. Only a small amount of dirt or dust, as well as metal filings generated from normal wear, can ruin a system in a short period of time.

The filters are normally located in the return line and filter the oil before returning it to the tank. Some systems use filters in the pressure line. The filter may be either one of two types: (1) full-flow or (2) bypass (partial-flow). All of the fluid passes through the full-flow type, whereas only a small portion of the fluid goes through the bypass type at one time. Replaceable-type filtering elements are frequently used on machinery hydraulic systems (Fig. 7-23). Reusable metal filter elements may also be used. Because of the importance of the filter to the life of the system, it should be checked and serviced according to the manufacturer's recommendations.

Accumulator

An accumulator is used in some hydraulic systems to "store" oil under pressure to supplement pump delivery and to operate an actuator. It is also used to absorb and cushion surges in a system. The two types that are commonly used are the piston and the bladder (Fig. 7-24). An inert gas is used as the cushioning medium. In the piston accumulator, a floating plunger is used to keep the gas and fluid separated. In the bladder-type accumulator, the gas is contained in the flexible bladder. The gas chambers are generally charged with dry nitrogen to a specified pressure. The fluid-chamber part of the accumulator is designed to receive the quantity of oil that is required to perform specific functions. Therefore the gas pressure plays an important part in its operation.

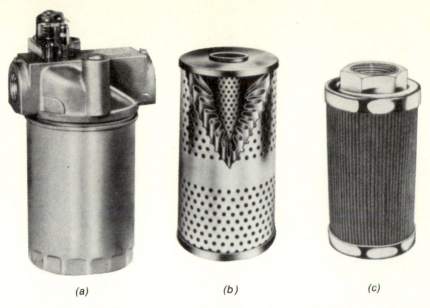

(a)   (b)   (c)

Fig. 7-23 Hydraulic oil filters: (*a*) disposable cartridge element; (*b*) disposable pleated paper; (*c*) reusable steel mesh element. (*Marvel Engineering Company and Schroeder Brothers Corporation.*)

HYDRAULIC SYSTEMS

A hydraulic system contains most of the parts that have been discussed in the previous section, assembled in such a manner as to perform from one to several functions on a machine. For example, on a cotton picker (Fig. 7-25) the same hydraulic pump is used to supply fluid to both the lift cylinders and the basket cylinders. Other components on the same machine that may be operated hydraulically included steering and the basket compactor. The same pump is used to operate the various "circuits" through the use of the directional control valves.

There are two basic types of systems that are used on farm machines. These are referred to as the *open-center* system and the *closed-center* system. The basic difference between the two is the type of directional control valves that are used.

The *open-center* system may use either a tandem-center or an open-center control valve. Both types of valves provide "free" flow of the fluid from the pump to the tank when they are in the center or neutral position. Therefore, when no machine component is being actuated by the hydraulic system, there will be little or no pressure in the system. A fixed-displacement-type pump is used, which means that the volume of oil flowing through the system is constant and the pressure will vary with the type of load that is placed on the cylinders or motors.

The *closed-center* system uses closed-center control valves. Recall that in the neutral position the pump and tank ports are blocked on this type of valve.

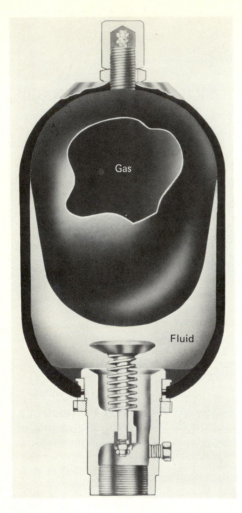

Fig. 7-24 Bladder-type accumulator.
(*Greer Hydraulics, Inc.*)

Therefore some means must be provided to take care of the flow of oil from the pump. There are two methods that are employed to "unload" the pump when all valves are in the neutral position. One method is to use a variable-displacement pump. The pump automatically moves to a neutral position, and no fluid is pumped when a specified pressure between the valves and the pump is reached. Pressure is held constant by the pump. When a valve is operated, the pump moves from the neutral position and forces oil through the system being operated.

The second type of closed-center system uses a fixed-displacement pump with an unloading valve and an accumulator. When the valves are placed in neutral, the pump charges the accumulator with a small volume of oil. The unloading valve, which is located between the pump and the accumulator, opens automatically and diverts the pump flow back to the tank. A check valve between the accumulator and the unloading valve prevents the loss of pressure and fluid stored by the accumulator. When a control valve is actuated, stored

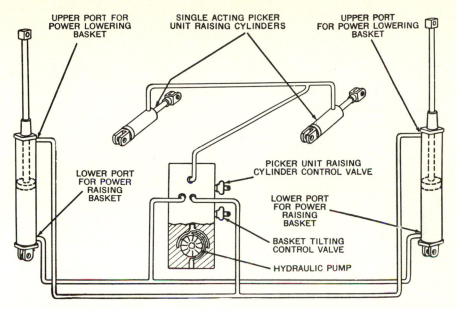

Fig. 7-25 Schematic diagram of hydraulic-lift system for cotton picker, showing hydraulic pump, oil reservoir, a system of passages, and valves to direct the flow of oil to raise the cotton-picker drums and tilt the cotton basket. (*Deere & Co.*)

fluid under pressure will flow into the circuit being operated. The unloading valve will close automatically when the pressure in the system drops to a certain level, and the pump flow will be diverted into the system. Fluid in the working part of the system is held at a constant pressure in both types of closed-center systems; however, the flow rate varies with the demand of the circuits that are operated.

Hydrostatic Drives

The hydrostatic drive is basically a system of transmitting the power from the engine to the drive wheels of the machine with fluid under pressure. A pump is driven by the engine, which forces fluid to a motor that drives the power train attached to the drive wheels. This type of drive is used extensively on harvesters such as combines, cotton pickers and strippers, and others where it is desirable to change the travel speed without having to "change gears" as with a standard gear-type transmission.

Most hydrostatic drives use a variable-displacement pump that forces fluid to either a fixed- or a variable-displacement motor. The most common type of pump and motor used is the axial-piston type. The displacement is varied on both units by changing the angle of the swashplate. The direction of flow can also be reversed by the swashplate so that the machine travel direction can be reversed when required. The speed of the machine can be changed while in motion to any desired rate of travel from zero to the maximum limit. This is accomplished by varying the rate of flow of the fluid between the pump and motor. Most hydrostatic drives operate as closed systems in that no other parts on the machine are operated by the system. The system contains its own oil, separate from the regular hydraulic system.

Hydraulic Fluids

There are many types of oils or fluids that are used in hydraulic systems. Since the fluid used is the "lifeblood" of the system, it is imperative that the proper fluid be used. Most of the problems in hydraulic systems are often related to the use of improper types of fluids or contaminants in the fluid. The fluid must perform several functions other than transmitting power under high pressures. Lubrication of the moving parts, including the pump, valves, and actuators, is of equal importance. The fluid must protect the system from rust and corrosion, resist foaming, and maintain proper viscosity over a wide range of temperatures.

Practically all hydraulic systems used on agricultural machines are based on petroleum oils. There are other types of fluids that are made for specific purposes and systems. These include special fire-resistant fluids that are generally classed as water-based or synthetic fluids. The latter fluids should be used only in systems for which they are specified.

Hydraulic systems on tractors and machinery may be designed to use engine crankcase oils, gear oils, or special hydraulic oils. These oils should not be interchanged or mixed in the hydraulic system. The manufacturer's recommendations should always be followed when replacing or changing the oil in the hydraulic system. The viscosity of hydraulic oils is determined in the same manner as that of engine oils but is usually described in units of Saybolt Seconds Universal (SSU) rather than SAE numbers. Special hydraulic oils will usually contain antifoaming, antiwear, antioxidants, and other additions to improve the performance of the oil in the system.

CARE OF HYDRAULIC SYSTEMS

1   Keeping the oil and the system clean is the most important single factor in the service and performance of hydraulic systems. The importance of this maintenance has been discussed as it relates to various parts in the system.
2   The system should never be allowed to run low on oil. The quantity of the oil in the system is usually easily determined with a dip stick or level indicator in the reservoir. If the system is allowed to run without oil, air will get into the system through the pump. This is usually called "cavitation" and is indicated by a buzzing or rattling sound. The machine should be stopped at once and oil should be added.
3   Use oils with the proper viscosity. Heavy oils, especially during cold-weather operation, will cause slow or sluggish operation of the system parts. Pump cavitation may also occur with the use of heavy oils. The use of oils that are too light may cause wear due to lack of lubrication.
4   The oil should be drained and replaced periodically, according to recommendations. It should be drained while the system is warm. Use clean containers to replace the new oil of the recommended type.

5   The hydraulic filter should be serviced by replacing or cleaning, depending upon the type of element used, at the recommended intervals.

6   All leaks should be repaired as quickly as is practical, not only to prevent the loss of oil, but also as a matter of safety and machine cleanliness. The slippery surfaces left by the oil may cause a serious accident if a person steps into the area. The petroleum-based hydraulic oils will burn. Therefore, leaks close to the engine or exhaust systems should be repaired immediately.

REFERENCES

Application of Hydraulic Remote Control Cylinders to Agricultural Tractors and Trailing-Type Agricultural Implements, *Agricultural Engineering Yearbook*, ASAE Standard: ASAE S201.4, 1973.

Application of Remote Hydraulic Motors to Agricultural Tractors and Trailing-Type Agricultural Implements, *Agricultural Engineering Yearbook*, ASAE Standard: ASAE S316T, 1973.

Barrett, R. D.: Determination of Loads Required to Lift Hydraulically Operated Implements, *Agr. Engin.*, **30**(10):479-480, 1949.

Fluid Power Handbook and Directory, Hydraulics and Pneumatics, 1972/1973.

Hydraulic Systems for Industrial Machines, Mobil Oil Corporation, 1970.

Peterson, W. A., T. L. Hanna, and J. A. Weber: Efficiency Comparison of Axial-Piston Pumps, *Transactions of ASAE*, **14**(1), 1971.

Selection and Care of Hydraulic Fluids, Abex Corporation, 1966.

Service and operator's manuals.

Worinton, Wayne H., and J. Waldo Seiple: Hydraulic Capacity Requirements for Control of Farm Implements, *Agr. Engin.*, **33**(5): 273-276, 1952.

QUESTIONS AND PROBLEMS

1   Give the fundamental law of enclosed fluids.

2   List at least five advantages of using hydraulic systems to transmit power on farm machines.

3   What are five types of pumps that may be used in hydraulic systems?

4   What is the difference between a fixed-displacement pump and a variable-displacement pump?

5   List the principal parts that would be required in a complete hydraulic system to perform a single function such as lifting the header on a combine.

6   List the types of oils and fluids that have been used in hydraulic systems.

7   What are the differences between a closed-center and an open-center system?

8   If a load of 6000 lb (2727.3 kg) had to be lifted by a hydraulic system that had a maximum pressure of 1000 lb/in^2 (70.3 kg/cm^2), what size cylinder would be required?

9   List the types of fittings that can be used on hydraulic hoses.

Chapter 8

Rubber Tires for Farm Equipment

Most modern farm tractors, self-propelled machines, and implements are equipped with rubber tires. Although there are problems and disadvantages with rubber tires, the advantages over steel wheels and other materials greatly outweigh the problems. When used on farm machines, such as combines, sprayers, trailers, planters, and other trailing implements, rubber tires reduce power requirements, decrease fuel consumption, permit higher speeds, and reduce vibration, noise, and dust. The investment in tires used on farm equipment may amount to several thousand dollars on a single farm. There are many types and sizes available for specific purposes and machines. Those who operate and service the equipment should know how to select and care for these tires so they will give long service life.

Kinds of Rubber Tires

In general, rubber tires used on farm equipment will fall into three categories, namely, traction, steering, and implement tires (Table 8-1).

Traction Tires These are used on self-propelled machines such as combines, cotton pickers and strippers, and forage harvesters. The function of these tires is to propel the machine through the field and support the weight of

Table 8-1   Agricultural Tire Code Designations

Tire description	Code marking
Traction tires (rear, tractor)	
Regular tread	R-1
Deep tread (cane and rice)	R-2
Shallow tread	R-3
Industrial	R-4
Steering tires (front, tractor)	
Single-rib tread	F-1
Triple-rib tread	F-2
Industrial tread	F-3
Implement tires	
Rib tread	I-1
Moderate traction	I-2
Traction tread	I-3
Bevel tread (plow tail wheel)	I-4
Smooth tread	I-6
Garden tractors	
Regular tread (bar type)	G-1
Intermediate	G-2
Shallow tread	G-3

the machine and the crop being harvested on machines like the cotton picker and the combine with baskets or hoppers. Various types of traction tires are illustrated in Fig. 8-1. It is difficult to design and specify a single tire-tread design for traction wheels that would be suitable for all conditions. In soil types such as loose sand or freshly plowed loam soil, smooth- or shallow-tread tires would be desirable for self-propelled machines. For machines that must operate in wet, soft fields and slippery conditions, tires with deep lugs, such as the rice and cane tires, are more desirable. Since most machines are not used in a single soil type or field condition, most tire manufacturers provide a regular or general-purpose tread design that will operate satifactorily under a wide range of conditions.

Steering Tires   They are used on the wheels used in steering the self-propelled machines. They also must support part of the weight and provide good flotation. These tires are equipped with deep ribs to reduce skidding while turning. Different rib designs are shown in Fig. 8-2.

Implement Tires   These are designed to support the weight of the implement and provide the least amount of rolling resistance. The most common type that is used is the rib-tread, as shown in Fig. 8-3. The ribs aid the wheel in rolling straight forward and reduce side slip. Where moderate traction is required on implements for operating the machine, such as ground-driven manure spreaders, tires with traction lugs are used (Fig. 8-3). Special flotation-type tires

Fig. 8-1 Tires for traction or drive wheels: (*a*) regular tread; (*b*) shallow tread; (*c*) industrial, intermediate tread; (*d*) cane and rice, deep tread. (*Goodyear Tire and Rubber Company.*)

Fig. 8-2 Single-rib and triple-rib tires for steering wheels. (*Armstrong Rubber Company.*)

(a)   (b)   (c)

Fig. 8-3 Implement tires: (*a*) rib; (*b*) smooth; (*c*) traction. (*Goodyear Tire and Rubber Company.*)

(Fig. 8-4) are often used on self-propelled machines such as sprayers and fertilizer applicators, so that these operations can be performed under a wide range of field conditions.

Agricultural Tire Code

The Tire and Rim Association and Rubber Manufacturers Association have developed and approved a code numbering system for tires, as shown in Table 8-1. The *F*, *R*, and *G* series are traction-type tires, while the *I* series are implement tires. The *F* series are used on front wheels of tractors and steering wheels of self-propelled machines. This code designation is stamped on the sidewall of all tires just under the size and ply rating.

Ply Rating of Tires

The Tire and Rim Association and the Rubber Manufacturers Association have defined ply rating as follows: "The term 'ply rating' is used to identify a given type of tire with its maximum recommended load, when used in a special service. It is an index of tire strength and does not necessarily represent the number of cord plies in the tire." The ply rating for agricultural tires ranges from two to twelve, depending upon the type of service. Small, lightweight tractors require only two-ply tires, while large, heavy tractors carrying mounted equipment require eight- to ten-ply tires.

Tire Sizes

The size of tires used on tractors and implements is designated by cross-sectional diameter and the diameter of the rim. The first size designation, such as 13.6-38, refers to the cross-sectional diameter in inches, and the second number is the rim diameter in inches. In an effort to aid manufacturers and users of machines,

Fig. 8-4   Flotation-type tires on a self-propelled fertilizer applicator. (*Ag-Chem Equipment Company.*)

recommendations have been established by the American Society of Agricultural Engineers for the purpose of providing selection tables of tires for application to machines of future design.

Inflation Pressures

Recommended inflation pressures for various loads and sizes for tractor steering and implement tires are given in Tables 8-2, 8-3, and 8-4. Proper inflation is one of the most important factors in the satisfactory performance and maintenance of tractor and implement tires. Improper tire pressure is a key contributor to tire failure. Underinflation (pressure too low) causes the side wall to flex abnormally and will result in breaks and separations in the cord body. Overinflation (pressure too high) makes the tire body rigid and reduces its resistance to impact, thus making it more susceptible to failure breaks.

Tractor tires operate most of the time in the field, where the lugs can penetrate the soil, making it possible for all portions of the tire to come in contact with the soil (Fig. 8-5). The machines, however, are moved along the highway in many cases. Hard surfaces that do not allow the lug to penetrate cause unnatural flexing on the wall and excessive wear on the lugs. It is advisable to

Table 8-2 **Tire Inflation Pressures (lb/in^2) at Various Loads for Traction-Type Tires**

Tire size	Tire type	Ply rating	Tire loads at various cold inflation pressures—maximum speed 20 mi/h								
			12	14	16	18	20	22	24	26	
8.3-24	R-1, R-3	4	970	1060	1150	1230	1310	1380(4)			
9.5-16	R-1, R-3	4,8	910	1000	1080	1150	1230(4)	1300	1370	1430	1840 @ 40(8)
9.5-24	R-1, R-3	4,6	1210	1330	1430	1540	1630(4)	1720	1810	1900	2070 @ 30(6)
11.2-24	R-1	4,6	1470	1610	1740	1860(4)	1980	2090	2200	2310(6)	
11.2-28	R-1	4,6	1570	1710	1850	1980(4)	2110	2230	2350	2460(6)	
11.2-34	R-1	4	1710	1870	2030	2170(4)					
11.2-36	R-1	4	1760	1930	2090	2230(4)					
11.2-38	R-1	4,6	1820	1990	2150	2300(4)	2440	2580	2720	2850(6)	
12.4-16	R-3	6,8	1350	1480	1600	1710	1820	1920	2020(6)	2120	2390 @ 32(8)
12.4-24	R-1	4,6,8	1760	1920	2080(4)	2230	2370	2510	2640(6)	2760	3120 @ 32(8)
12.4-26	R-1	4	1820	1990	2150(4)						
12.4-28	R-1	4,6	1880	2050	2220(4)	2380	2530	2670	2810(6)		
12.4-36	R-1	4,6	2110	2310	2500(4)	2680	2850	3010	3170(6)		
12.4-38	R-1	4,6	2170	2380	2570(4)	2750	2930	3100	3260(6)		
13.6-24	R-1	4,6,8		2270(4)	2460	2630	2800	2960(6)	3110	3620	3410 @ 28(8)
13.6-26	R-1	4,6		2350(4)	2540	2720	2890	3060(6)			
13.6-28	R-1, R-3	4,6	*2210	2420(4)	2620	2810	2980	3160(6)			
13.6-38	R-1, R-2	4,6		2810(4)	3030	3250	3460	3660(6)			
13.9-36	R-1	4,6		2730(4)	2950	3160	3360(6)				
14.9-24	R-1, R-3	4,6	*2470	2700(4)	2920	3130	3330(6)				
14.9-24	R-4	6,8		2700	2920		3330(6)	3520	3700	3880(8)	
14.9-26	R-1	4,6,8		2790(4)	3020	3230	3440(6)	3640	3830	4010(8)	
14.9-28	R-1, R-4	4,6,8		2880(4)	3120	3340	3550(6)	3760	3950	4140(8)	
14.9-30	R-1	6		2980	3220	3450	3670(6)				
14.9-38	R-1, R-2	4,6		3340(4)	3610	3870	4120(6)		4320		
15.5-38	R-1, R-2	6,8		3150	3410	3650	3880(1)	4110		4530(8)	

Table 8-2 Tire Inflation Pressures (lb/in²) at Various Loads for Traction-Type Tires (Continued)

Tire size	Tire type	Ply rating	Tire loads at various cold inflation pressures—maximum speed 20 mi/h								
			12	14	16	18	20	22	24	26	
16.9-24	R-1, R-3, R-4	6,8	*3000	*3280	3550	3800(6)	4040	4270	4490(8)		
16.9-26	R-1, R-3	6,8	*3090	*3390	3660	3920(6)	4170	4410	4640(8)		
16.9-28	R-1	6,8			3780	4050(6)	4310	4560	4790(8)		
16.9-30	R-1	6,8			3900	4180(6)	4450	4700	4950(8)		
16.9-34	R-1	6,8			4140	4440(6)	4720	4990	5250(8)		
16.9-38	R-1	6,8			4380	4690(6)	4990	5280	5550(8)		
18.4-16.1	R-1, R-3	6,8,10	*2370	*2660	2810(6)	3010	3200(8)	3380	3560	3730(10)	
18.4-26	R-1, R-2, R3	6,8,10	*3700	*4050	4380(6)	4700	4990(8)	5280	5560	5820(10)	
18.4-28	R-1, R-4	6,8,10			4530(6)	4850	5160(8)	5450	5740	6010(10)	
18.4-30	R-1, R-2, R-3	6,8,10	*3950	*4320	4670(6)	5000	5320(8)	5630	5920	6210(10)	
18.4-34	R-1, R-2	6,8,10			4960(6)	5310	5650(8)	5970	628ᵒ	6580(10)	
18.4-38	R-1, R-2	6,8,10,12			5240(6)	5620	5980(8)	6320	6650	6970(10)	7870 @ 32(12)
20.8-34	R-4	8,10			6000	6430(8)	6840	7230(10)			
20.8-38	R-1, R-2	8,10,12	*5300	*5800	6350	6800(8)	7240	7650(10)	8050	8440	8810 @ 28(12)
23.1-26	R-1, R-2, R-3	8,10,12			6280(8)	6720	7150(10)				
23.1-30	R-1, R-2, R-4	8,10,12			6690(8)	7160	7650(10)				
23.1-34	R-1, R-2	8,10,12			7100(8)	7600	8090(10)				
24.5-32	R-1, R-2	10,12			7630	8170	8690(10)				
28.1-26	R-1, R-2	10,12			7270	7790(10)	8280(12)				
30.5-32	R-1	12,16			9110	9760	10,380(12)	10,970	11,550	12,100(16)	

* Loads at these inflation pressures are for R-3 type when in agricultural service only.
Note: Numbers in parentheses denote ply rating for which boldface loads and inflations are maximum for all speeds up to 20 mi/h.
Source: Tire and Rim Association.

Table 8-3   Tire Inflation Pressures (lb/in^2) at Various Loads for Steering Wheels

Tire size	Tire loads at various cold inflation pressures											
	20	24	28	32	36	40	44	48	52	56	60	64
4.00-12	330	370[2]	400	430	470	490	520	550[4]				
4.00-15	390	430	480	510	550	590	620	650[4]				
4.00-19 (4-19)	470	520	570	620	670	710	750	790[4]				
5.00-15	540	600	660	710	760	810[4]	850	900	940	980	1020	1060
5.50-16	660	740	810	870	940[4]	1000	1050	1110	1160[6]			
6.00-12	630	700	760	830	890	940	1000	1050[6]				
6.00-14	680	760	830	900	960	1030	1080	1140[6]				
6.00-16	760	840	920	1000[4]	1070	1140	1200	1260[6]	1320	1380	1440	1490[8]
6.50-16	850	950	1040	1130[4]	1210	1280	1360[6]					
7.50-10	820	910	1000	1080	1160	1230[6]	1300	1370	1440	1500	1560	1620[10]
7.50-16	1100	1220	1340[4]	1450	1550	1650[6]	1740	1830	1920[8]			
7.50-18	1190	1330	1450[4]	1570	1680	1790[6]						
7.50-20	1280	1430	1560	1690	1810	1930[6]						
9.00-10	1100	1230[4]	1340	1450	1550	1650	1750[8]	1840	1930	2010[10]		
9.50-20	1850	2060	2250	2440[6]	2610	2770[8]						
9.50-24	2100	2340	2560	2760[6]								
10.00-16	1750	1950	2130[6]	2310	2470	2630[8]						
11.00-12	1640	1820	1990	2160	2310	2460	2600	2730	2870	2990[12]		
11.00-16	2070	2300	2520[6]	2720	2920[8]	3100	3280	3450	3610	3780[12]		

Note: Superscript numbers denote ply rating for which accompanying load and inflation are maximum.
Source: Tire and Rim Association.

Table 8-4   Tire Inflation Pressures (lb/in^2) at Various Loads for Rib-Type Implement Tires

Tire size	Ply rating	Tire loads at various cold inflation pressures—maximum speed 20 mi/h								
		20	24	28	32	36	40	44	48	
4.00-9	4	360	400	440	480	510	540(4)			
4.00-12	2,4	450(2)	500	540	590	630	670(4)			
5.00-15	4	730	810	890	960(4)					
5.50-16	4	900	1000	1090	1180(4)					
5.90-15	4	860	950	1040	1130(4)					
6.00-16	4	1020	1140	1240(4)						
6.40-15	4,6	960	1060	1160(4)	1260	1350	1440	1520(6)		
6.50-16	4,6	1150	1280(4)	1410	1520	1630	1730(6)			
6.70-15	4,6	1070	1190	1300(4)	1400	1500	1600(6)			
7.50-16	4,6,8,10	1480	1650(4)	1810	1950(6)	2090	2230	2350(8)	2480	2710 @ 56(10)
7.50-18	6	1540	1720	1880	2030(6)					
7.50-20	4,6	1590	1770(4)	1940	2100(6)					
7.60-15	4,6,8	1250	1390(4)	1530	1650	1770(6)	1880	1990	2090(8)	
9.00-16	8,10	1960	2180	2380	2580	2760	2930(8)	3100	3270(10)	

Source: Tire and Rim Association.

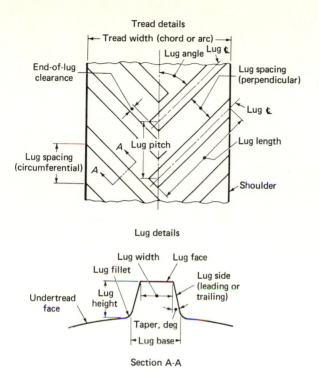

Fig. 8-5 Tractor tire lug and tread diagram. (*Agricultural Engineering Yearbook, 1973, ASAE, R 296.1.*)

increase the air pressure to maximum recommendation if the machine is driven on the highway for extended periods, to reduce the wear on the lugs.

It has been reported in studies of the effects of tire pressure on rolling resistance that lower pressures will reduce the rolling resistance and the amount of power required to move a machine in loose, plowed soil. This is accounted for by the fact that the tire surface conforms to the soil surface more easily at the lower pressures. Rolling resistance increases, however, with low pressures on hard surfaces. Care should be exercised that the sidewalls are not damaged when tires are operated at low pressures.

The tire pressure on tractors should be increased when plowing if one of the wheels operates in the furrow. This is required because of the shift in weight to the furrow wheel.

Agricultural tractor and implement tires are designed for a maximum speed of 20 mi/h (32.0 km/h). Tires used on farm trailers should be of the automotive or truck type, which are designed for high speeds.

Rims for Tractor and Implement Tires

Rims now in use are usually of the drop-center type with a shallow or deep well. The trend is toward the use of wide-base rims with a shallow drop center (Fig. 8-6). The wide-base rim on a trailing implement allows for better lateral stability

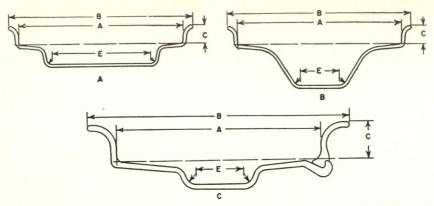

Fig. 8-6 Types of wide-base rims for tires: (*A*) wide-base drop-center with shallow well; (*B*) wide-base drop-center with deep well; (*C*) wide-base semi-drop-center with split side ring.

and provides a better-shaped tire section to carry loads. A cross section of a wide-base semi-drop-center rim is shown in Fig. 8-6. There is a removable split-side ring, which makes it easier to remove and mount heavy tires

Life of Agricultural Pneumatic Tires

There are many factors that affect the life of tires used on tractors and implements, such as (1) type of use, (2) type of farming and crop, (3) abrasive wear, (4) cuts and chipping, (5) punctures and blowouts, (6) exposure to weather, (7) improper inflation, (8) annual use, and (9) general care.

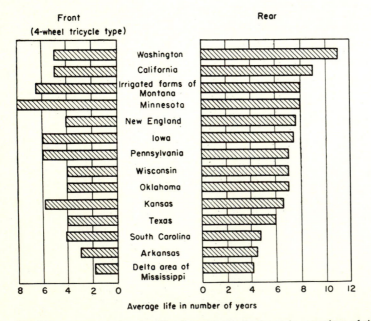

Fig. 8-7 Bar graph showing life of tractor tires in the various sections of the United States.

The bar graph in Fig. 8-7 shows the average life of front and rear tractor tires for various sections of the United States. The difference in the life of tractor tires in the different areas is greatly influenced by many of the factors enumerated above. High speeds, overloads, and underinflation will increase tire temperatures, damage the tire, and shorten its service life.

REFERENCES

Agricultural Engineering Yearbook, St. Joseph, Mich., 1973.
Data Book, The Firestone Tire and Rubber Company, Akron, Ohio.
Farm Tire Handbook, Goodyear Tire and Rubber Company, Inc., Akron, Ohio.
Gill, William R., and G. E. Vandenberg: Soil Dynamics in Tillage and Traction, *Agricultural Handbook* No. 316, USDA, 1967.
McKibben, E. G., and J. B. Davidson: Effect of Inflation Pressures on Rolling Resistance of Pneumatic Tires, *Agr. Engin.*, 21:25-26, 1940.
Reed, I. F., C. A. Reaves, and J. W. Shields: Comparative Performance of Farm Tractor Tires Weighted with Liquid and with Wheel Weights, *Agr. Engin.*, 34(6):391-395, 1953.
———and J. W. Shields: Effect of Lug Height and Rim Width on the Performance of Farm Tractor Tires, *SAE Journal*, 58(12):40-41, 1950.
Rubber-Tired Equipment for Farm Machinery, *Ohio Agr. Expt. Sta. Bul.* 556, 1935.

QUESTIONS AND PROBLEMS

1   Enumerate and discuss the advantages and disadvantages of pneumatic tires on farm equipment.
2   Explain the differences in traction and implement tires and the agricultural tire code system.
3   Discuss the various factors that affect the life of agricultural pneumatic tires.
4   Discuss the various factors that affect the traction of tires.
5   Under what conditions will an underinflated tire result in less rolling resistance?
6   What are the general differences between the types of tractor tires?

Primary Tillage Equipment

Tillage is the preparation of the soil for planting and the process of keeping it loose and free from weeds during the growth of crops. The primary objectives and fundamental purposes of tillage are divided into three phases: (1) to prepare a suitable seedbed, (2) to destroy competitive weeds, and (3) to improve the physical condition of the soil. The equipment used by the farmer to break and loosen the soil for a depth of 6 to 36 in (15.2 to 91.4 cm) is called *primary tillage equipment*. It includes the moldboard, disk, rotary, chisel, and subsoil plows.

MOLDBOARD PLOWS

The moldboard plow is adapted to the breaking of many types of soils and is well suited for turning under and covering crop residues.

Development of the Moldboard Plow

Recorded history in the form of hieroglyphs and cuneiform characters shows that the ancients had a type of plow thousands of years B.C. It is recorded that about 900 B.C. Elisha was found "plowing with twelve yoke of oxen before him."[1]

[1] Kings xix: 19.

The Roman plow, which was improved by the Dutch, was imported into England about 1730. The Essex plow of about 1756 had an iron moldboard. The Norfolk wheel plow of 1721 had a cast-iron share and an iron rounded moldboard. A curved moldboard made its appearance in 1760 on the Suffolk swing plow. The Rotherham plow was improved by James Small, who wrote a book on plow design in 1784. The close of the eighteenth century saw the change in England from the wooden plow to the iron plow.

In America, Thomas Jefferson and Daniel Webster were among the first to advance improvements of the plow. Charles Newbold of Burlington, New Jersey, secured the first patent on a cast-iron plow in 1797. Farmers rejected this iron plow because they thought it poisoned the soil. Jethro Wood developed a moldboard in 1814 of such curvature as to turn the soil in even furrows. The first steel plow was made from three sections of an old handsaw by John Lane about 1833. He also secured in 1868 a patent for soft-center steel, which is used at the present time in making moldboards for plows. In 1837, John Deere at Grand Detour, Illinois, made a steel plow (share and moldboard in one piece) from an old sawmill saw. Ten years later he established a factory at Moline, Illinois.

James Oliver was granted a patent in 1868 for hardening cast iron, which was known as chilled iron.

In 1856, M. Furley patented a single-bottom sulky or wheel plow which permitted the operator to ride. In 1864, F. S. Davenport patented a riding two-bottom horse-drawn gang plow. Three- and four-bottom gang plows often required 10 to 12 horses to pull them.

The large 10- to 15-bottom plows were pulled by steam tractors in the 1890s and by the large, slow, cumbersome gasoline-engine tractors from about 1900 to 1910. The early two- to five-bottom trailing-tractor plows were equipped with hand-lever lifts. In the early 1920s, mechanical power lifts were developed. They were used until the hydraulic lift was introduced in the 1930s. The integral-tractor-mounted unit assembly and unit-lifted plow or the three-point hitch were developed in the early 1940s by Ferguson. This type of plow is now standard on small and average-sized farms. The trend is toward quick-coupler units.

The Moldboard Plow Bottom

The part of the plow that actually breaks the soil is called the *bottom* or *base* (Fig. 9-1). It is composed of those parts necessary for the rigid structure required to lift, turn, and invert the soil. The parts which form the moldboard plow bottom are the *share*, the *landside*, and the *moldboard*. These three parts are bolted to an irregular-shaped piece of metal called the *frog*. The *beam* can also be attached to the frog (Fig. 9-1).

When a bottom is used to turn the soil, it cuts a trench called a *furrow* (Fig. 9-2). The ribbon of soil cut, lifted, and thrown to the side is called the *furrow slice*. When plowing is started in the middle of a strip, a furrow is plowed across the field, then the tractor and plow are turned about, and on the return trip a furrow slice is lapped over the first slice. This leaves a slightly higher ridge than

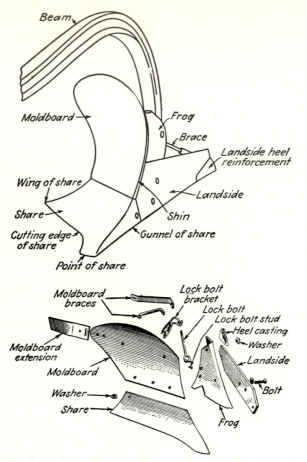

Fig. 9-1   Assembled and exploded view of moldboard plow bottom and parts.

the second, third, and other slices. This raised ridge is called a *back furrow* (Fig. 9-2). When two strips of land are finished, the last furrows cut out leave a trench about twice the width of one bottom. This open trench is called a *dead furrow* (Fig. 9-2). The unbroken side of the furrow is called the *furrow wall* (Fig. 9-2). When the land is broken by continuous lapping of furrows, the land is said to be *flat broken*. If the land is broken in alternate back furrows and dead furrows, it is called *bedded* or *listed* land.

The Moldboard   The moldboard is that part of the plow just back of the share. It receives the furrow slice from the share and turns it. When the action of the plow bottom upon the soil is considered, the moldboard is the most important part of the plow because it is upon the moldboard that the furrow slice is broken, crushed, and pulverized (Fig. 9-1). On some moldboards an extension is provided to turn the soil over more gradually and completely (Fig. 9-3). Different soils require different-shaped moldboards to give the same degree of pulverization. For this reason moldboards are divided into several classes, namely, *stubble, general-purpose, blackland, breaker*, and *high-speed* (Figs. 9-3

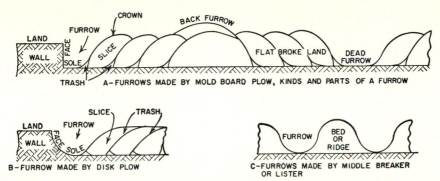

Fig. 9-2   Kinds of furrows made by different types of plows.

and 9-4). In classifying moldboards, it should be kept in mind that there are hundreds of shapes of each class. Such diversity of shapes has resulted in an attempt by manufacturers to make a plow that will do good work in all types of soils, but one that will work successfully everywhere is yet to be made. A special shape called the *blackland* bottom (Fig. 9-4) is used extensively in Texas and in other localities where the soil does not scour (leave the surface clean and polished) well.

Fig. 9-3   Three commonly used types of moldboard plow bottoms: (*top*) stubbles; (*middle*) general-purpose; (*bottom*) breaker. (*International Harvester Company*.)

Fig. 9-4 Three special types of mold-board plow bottoms: (*top*) high-speed; (*middle*) deep-tillage; (*bottom*) blackland (*International Harvester Company.*)

The general-purpose moldboard is a combination of the sod and stubble types and can be used easily for sod or stubble land. It has less curvature than the stubble moldboard. Hence, it is called a *general-purpose plow*.

The stubble type of moldboard (Fig. 9-3) is broader and bent more abruptly along the top edge. This causes the furrow slice to be thrown over quickly, pulverizing it much better than the other types of molds. This type is best suited to work in soil that has been cultivated from year to year, called *stubble soil* because of the fact that the stubble of plants from the previous crop is still on the land. Unlike the sod plow, the furrow slices lap upon one another.

A breaker bottom is designed to work in sod land and in land that has remained idle for a number of years.

The high-speed bottom (Fig. 9-4) has slightly less curve to the upper section of the moldboard than does the general-purpose moldboard. It is designed to throw the furrow slice just far enough to lap upon the previous furrow slice.

Slat moldboards (Fig. 9-5) are often used where the soil is sticky and does not scour on solid moldboards.

Generally, there are three materials used in the manufacture of moldboards,

Fig. 9-5   Slat-type moldboard plow bottom. (*J. I. Case Company*.)

namely, *soft-center steel, crucible steel,* and *chilled cast iron*. Soft-center-steel moldboards are the best to use under most conditions because the majority of soils will scour better on this type of material. For the Middle West, the steel plow seems to give satisfaction in most cases. Because of their wear-resistant qualities owing to the hardness of the material of which they are made, chilled plows are better for sandy, gritty, and gravel soils. Chilled plows are adaptable to all parts of the South, where there is much sandy land.

The Share   The share of a moldboard plow (Fig. 9-1) provides the cutting edge. The principal parts of the share are the *point*, the *wing*, the *cutting edge*, and the *gunnel* (Fig. 9-1). The kinds of shares are the *regular with gunnel, two-piece,* and *straight*. The latter two are designed so that, when they become worn and dull, it is more economical to replace them with new ones than to try to resharpen them (Fig. 9-6). Chilled-cast-iron shares can be resharpened by grinding.

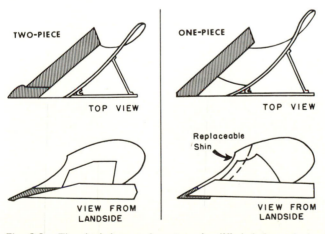

Fig. 9-6   The shaded areas show two simplified designs of plowshares: (*left*) two-piece; (*right*) one-piece. (*U.S. Dept. Agr. Farmers' Bul. 2172, 1961.*)

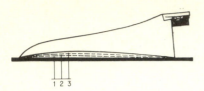

Fig. 9-7   Three degrees of share suction (*1*) regular-suck: 3/16 in, (3.8 mm), for light soil easy to penetrate; (*2*) deep-suck: 5/16 in (7.9 mm), for ordinary soil that is dry and hard; (*3*) double-deep-suck: 3/8 in (8.5 mm), for stiff soils, gravel land, and other soils where penetration is difficult.

Figure 9-7 illustrates three degrees of suction: regular-suck, deep-suck, and double-deep-suck. The amount of suction is around $\frac{3}{16}$, $\frac{5}{16}$, and $\frac{3}{8}$ in (4.8, 7.9, and 9.5 mm), respectively.

Vertical or *down suction* is the bend downward of the point of the share to make the plow penetrate the soil to the proper depth when the plow is pulled forward. The amount of suction will vary from $\frac{1}{8}$ to $\frac{3}{16}$ in (3.2 to 7.9 mm), depending on the style of the plow and the soil it was made to work in.

Horizontal or *land suction* is the amount the point of the share is bent out of line with the landside (Fig. 9-8). The object of this suction is to make the plow take the proper amount of furrow width.

Landside   The *landside* is that part of the plow bottom which slides along the face of the furrow wall. It helps to counteract the side pressure exerted by the furrow slice on the moldboard. It also helps to steady the plow while it is being operated. The *shin* (Fig. 9-1) is the cutting edge of the moldboard, just above the landside.

Size of the Plow   The *size* of a moldboard plow is its width in inches. This is determined by measuring the distance from the wing to the landside. The rule is held perpendicular to the landside. Tractor plow sizes are 10, 12, 14, 16, and 18 in (25.4, 30.5, 35.6, 40.6, and 45.7 cm). Special brush plows may be as large as 18 and 20 in (45.7 and 50.8 cm).

Types of Tractor Moldboard Plows

In general, tractor moldboard plows may be grouped into three types: *trailing*, *semimounted*, and *integral-mounted*.

Trailing Moldboard Plows   The *trailing*, or *pull-type*, tractor plow is a complete unit in itself, supported by two or three wheels; when hitched to the drawbar of the tractor, it trails behind the tractor.

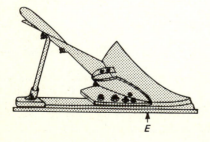

E

Fig. 9-8   Position of straitedge to measure the horizontal or land suction at *E*.

Regular Trailing Plows   Trailing moldboard plows are built in sizes ranging from one to five bottoms. The size of the bottoms may range from 12 to 18 in (30.5 to 45.7 cm), but the most common size is 14 in (35.6 cm).

The two-, three-, four-, five-, and six-bottom moldboard plows can be obtained with hydraulic lifts (Fig. 9-9). The smaller-sized trailing plows may have the rear furrow wheel either rigidly attached or arranged so that the wheel will castor, or a long landside may be substituted for the wheel. Generally, trailing plows are provided with A-frame-type hitches. Some trailing plows have an overhead curved beam of I-shaped steel, while others have a fabricated trussed standard-type beam, which, it is claimed, gives a higher clearance. Most of the multiple-bottom plows are designed so that one bottom can be removed, thereby reducing the size. Wheels are usually equipped with rubber tires.

The heel of the landside on the rear bottom of trailing plows should be adjusted to run from $\frac{1}{2}$ to $\frac{3}{4}$ in (12.7 to 19.0 mm) above the furrow sole and out from the furrow wall (Fig. 9-10). The rear furrow wheel carries almost one-third of the plow's weight and load. If a multiple-bottom plow is throwing unequal furrows and is difficult to adjust, it should be checked for sprung beams by measuring from share point to beam. Fig. 9-9 shows a plow equipped with hydraulic trips to clear obstacles.

Two-Way Trailing Plows   This plow is named *two-way* because it has both right- and left-hand bottoms and, therefore, throws the furrow slices both to the right and to the left of the operator as the plow is reversed when the tractor is

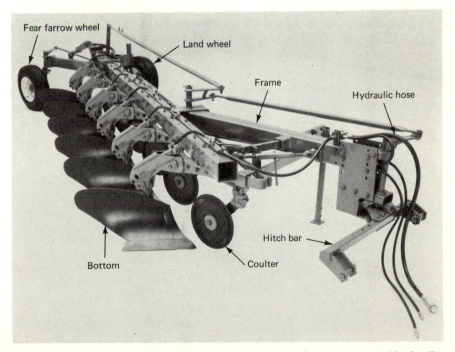

Fig. 9-9   Five-bottom trailing moldboard plow equipped with remote-control hydraulic-cylinder wheel lifts and hydraulic cylinders to reset bottoms. (*Deere & Co.*)

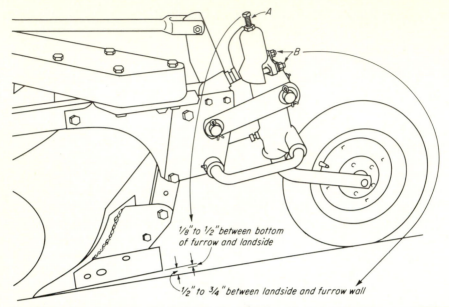

A

B

1/8"to 1/2"between bottom
of furrow and landside

1/2" to 3/4" between landside and furrow wall

Fig. 9-10 Showing range of adjustment for vertical and horizontal suction on a trailing moldboard plow. (*J. I. Case Company.*)

turned at the turn row. The furrow slices, therefore are thrown in one direction in relation to the field.

The two-way plow is used in plowing irrigated lands and where it is desirable to break the land without leaving a dead furrow, such as hillsides, terraced fields, and irregular-shaped fields.

Semimounted Moldboard Plows   This type of plow has the front end directly connected to and supported by the tractor (Fig. 9-11). The rear end of

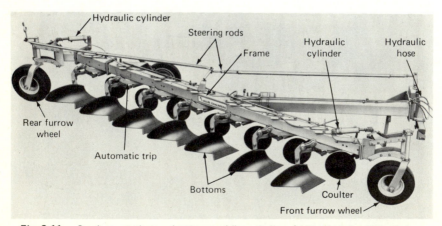

Hydraulic cylinder

Steering rods

Frame

Hydraulic cylinder

Hydraulic hose

Rear furrow wheel

Automatic trip

Bottoms

Coulter

Front furrow wheel

Fig. 9-11   Semimounted seven-bottom moldboard plow. (*Allis-Chalmers Mfg. Co.*)

the plow is supported by a furrow and a land wheel. The raising and lowering of the rear part of the plow on the furrow wheel may be accomplished either by a mechanical linkage or by a remote-controlled hydraulic cylinder. The front end of the plow is raised and lowered by the tractor hydraulic linkage system. Usually this type of plow is attached to the tractor by a quick-coupling mechanism.

Integral-mounted Moldboard Plows   This type of plow is also called a *direct-connected, tractor-mounted,* or *tractor-carried* plow.

Regular Mounted Plows   The integral-mounted plow is really a tractor attachment, as it depends upon the tractor for its power lift and upon the power of the tractor engine for its general operation. The entire weight of the plow is carried on the tractor when lifted (Fig. 9-12). The depth of plowing is controlled in some cases hydraulically, in others by levers and gage wheels. The number of bottoms ranges from one to five, depending upon the size of the tractor. The smallest tractor of 8 to 10 drawbar hp may be equipped with a single 12-in (30.5-cm) bottom. The medium-sized tractor carries two bottoms, while the large-sized tractor can carry up to five bottoms.

The weight and length of the tractor ahead of the drive wheels determine the weight and length of the plow that can be mounted and lifted behind the drive wheels. In some cases extra weight must be added to the front of the tractor to counterbalance the plow and prevent interference with steering.

The integral-mounted plows are unit assemblies, and the unit is mounted and attached to the tractor by specially designed hookups for quick attaching and detaching. The larger plows usually have a linkage-hitch arrangement for lifting and leveling the plow. All mounted plows are lifted by hydraulic lifts. As the plows are tractor attachments, no wheels are required except for a small landside wheel at the rear.

Fig. 9-12   Integral-mounted four-bottom moldboard plow.

Two-Way-mounted Moldboard Plows The two-way-mounted plow performs the same functions as the trailing two-way plow. A different arrangement is provided to change from right- to left-hand bottoms. This is accomplished by rotating the unit 90° for some plows and 180° for others (Fig. 9-13). One company describes this change as a *vice versa* action. The action may be either mechanically or hydraulically controlled.

Integral-mounted Middlebreakers The middlebreaker is known by different names in different sections. In the South, it is often called a *middlebuster* or *bedder*, but where crops are planted in the furrow, it is called a *lister*. The bottom is really right- and left-hand moldboard plows joined together. In most cases the share or point is one piece with wings to fit against each moldboard. Shares may have the wings and the central point separated into three pieces that fit closely. The shapes of the moldboards range from the stubble to the blackland types (Fig. 9-14). In some areas for certain jobs sweeps are substituted for the moldboard bottoms (Fig. 9-14).

The *middlebreaker* is usually attached to a tool bar (Fig. 9-15) and is easily detached. In most makes, the removal of two bolts is all that is necessary. The beams also can be moved sidewise along the tool bar for different row spacings. When plowing in soil where there may be hidden obstructions such as roots or stones, a spring or hydraulic trip is commonly used (Fig. 9-9).

When the land has been bedded or listed for some time and rains and winds have partially filled the furrows with soil, it is desirable to clean out the furrows with the middlebreaker bottom, throwing some soil up on the edges of the bed and thereby reshaping it. This operation is called *hipping* by farmers of the Mississippi Delta area.

There are several attachments for middlebreaker bottoms, such as gage wheels, rolling colters, root cutters, and planting equipment.

Design of Moldboard Plows To design a plow which will perform satisfactorily under all soil conditions is a problem that has never been entirely

Fig. 9-13 Four-bottom integral-mounted two-way moldboard plow. (*Deere & Co.*)

Fig. 9-14 Three types of middle-breaker bottoms: (*top*) general-purpose; (*middle*) blackland; (*bottom*) middle-breaker sweep. (*International Harvester Company.*)

solved, yet more work has been done on perfecting the plow bottom than on any other agricultural implement. Upon its performance depends the quality of the seedbed the farmer can prepare, which in turn will influence the germination of the seed, the growth of the plant, and the yield that will be obtained in the end. Therefore, the farmer should strive to do a good job of plowing. Good plowing consists of turning and setting the soil into even, clean furrows of roundish conformation.

Fig. 9-15   Integral-mounted eight-row bedder-lister. (*Deere & Co.*)

The main points to consider are the following:

1   The top of the furrow may be slightly ridged.
2   The soil must be pulverized thorougly from the top to the bottom of the furrow.
3   Each furrow must be perfectly straight from end to end on level land.
4   All back furrows must be slightly raised, and all trash completely covered.
5   The outline of the furrows must be in a point without break or depression.
6   All trash must be buried completely in the lower right-hand corner of the furrow.
7   Furrows must be thoroughly uniform.
8   The depth of all the furrows must be the same, continuing in uniform depth.
9   The dead furrows must be free of all trash.
10   Unbroken strips must not be left between furrows in contour plowing.

These are the standards to consider when plows are used and, of course, do not apply where it is desirable to leave the crop residue on the soil surface. The main thing to consider in plowing is that the land should be completely broken and that the soil be thoroughly pulverized (Fig. 9-16).

The whole bottom is essential for good plowing, the share cutting and slightly lifting the furrow slice, the landside controlling and steadying the plow, while the moldbord completes the lifting, pulverizing, and inverting of the furrow slice. It is upon the moldboard that the main part of successful plowing depends. The curvature and length of the moldboard determine the degree of pulverization given the furrow slice. Russian scientists have made an extensive mathematical study of the shape of the moldboard plow bottom and its effect on the furrow slice.

Forces That Act on the Plow   Lindgren and Zimmerman analyze the many forces that act upon the plow bottom as follows:

First. The principal vertical forces:
    a   Those that are due to the weight of the plow.
    b   Those that are due to the downward pressure exerted during the lifting of the soil.
    c   The lifting component due to the hitch being above the point of resistance.

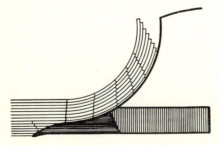

Fig. 9-16   Diagram showing the pulverizing or breaking of the furrow slice as it travels over the moldboard. Shorter and deeper curves cause more and finer pulverizing. The longer and shallower curves cause less and coarser pulverizing. (*Allis-Chalmers Mfg. Co.*)

d   That force developed when the plow is dull and worn, which has an upward component, the result of the sloping undersurface of the share.

Second. The principal horizontal cross-furrow forces:
a   The cross component caused by the friction of the soil on the moldboard.
b   The cross component caused by transferring the soil sidewise the width of the furrow.
c   The cross component due to the cutting and wedging of the sloping share edge in operation.
d   The component of the line of draft.
e   Such cross component as may result from the rear-furrow-wheel reactions in multiple outfits where used.

Third. The principal longitudinal forces acting lengthwise of the furrow:
a   The soil resistance to cutting.
b   The friction between the furrow wall and the landside.
c   The friction due to the weight and pressure upon the bottom of the plow, according to the setting or condition of the cutting wedge.
d   The component of friction of the earth sliding over the moldboard.[1]

For equilibrium we have the sum of the draft produced by the motive power.

Thus it can be seen that the moldboard, which is a modified warped surface, as analyzed by White,[2] will have a great deal to do with the proper functioning of the plow, depending upon its width, curvature, and length. Moldboards which have a greater curvature, being bluffer, will naturally give a better pulverizing action upon the furrow slice because of their pinching, crushing action (Fig. 9-16).

Clyde[3] explains and illustrates the relative size of the forces acting on moldboard plows by the length of arrows shown in Fig. 9-17A and B. The soil resistance on the share and moldboard is shown by the line RH. The position and angle of RH will vary with the condition of the soil. At any point on RH, the two unequal forces L and S can be substituted (Fig. 9-17B). RH makes an angle of $13\frac{1}{2}°$ with the direction of travel, the side force S being 24 percent of L. If the soil were perfectly uniform, theoretically the plow could be pulled with a force equal but opposite to RH and no side support would be needed.

If the pull is straight ahead, as shown in Fig. 9-17B, the plow will need side support from the furrow wall. This support causes frictional drag, which is shown by the angle of Q. For side balance, the side component of Q must equal S, the side component of the soil force acting to the left. If a parallel-sided figure is constructed on RH and Q, the single force PX will balance them. PX is a little more than L in Fig. 9-17A, because PX also overcomes the landside friction. Note that PX passes through H, the junction of RH and Q. The point H is

[1] Lindgren and Zimmerman, *Amer. Soc. Agr. Engin. Trans.,* **15**:150, 1921.
[2] E. A. White, *Jour. Agr. Res.,* **12**(4):149-182, 1918.
[3] A. W. Clyde, Mechanics of Farm Machinery, *Farm Impl. News,* Jan. 6 to Mar. 16, 1944; Force Measurement Applied to Tillage Tools, *Amer. Soc. Agr. Engin. Trans.,* **4**:153, 1961.

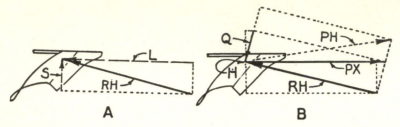

Fig. 9-17   Diagrams showing horizontal forces that act on a moldboard plow and their resultant. (*A. W. Clyde, University of Pennsylvania.*)

commonly called the *center of resistance* of a plow bottom. If the pull is at an angle *PH*, it must go through *H*. This angled pull forces the plow harder against the furrow wall. In Fig. 9-17*B*, the angle of *PH* is such that its side component is nearly 80 percent of *S*, and it is in addition to *S*.

As a well-shaped share lifts the furrow slice, the vertical forces appear to be downward. A rolling colter operating above the share point must be forced into the soil, and therefore the forces are upward. The degree of the upward forces exerted by the colter is influenced by the hardness of the soil. This is shown in Fig. 9-18.

Center of Resistance or Center of Load of a Plow Bottom   The above discussion shows that a point where all the horizontal and vertical forces meet is considered to be the *center of resistance* or *load* (Fig. 9-19). It cannot always be determined just exactly where this point will be on a plow bottom, but it will usually come within the range of the following dimensions for a 14-in (35.6-cm) bottom: vertical forces will be in equilibrium 2 to $2\frac{1}{2}$ in (5.1 to 6.4 cm) up from the floor; the horizontal forces 2 to 3 in (5.1 to 7.6 cm) to the right of the shin; the longitudinal forces 12 to 15 in (30.5 to 38.1 cm) back from the point of the share. Briefly, we can say that the center of resistance of any moldboard plow bottom will be on the surface about where the share and moldboard intersect and to the right of the shin (Fig. 9-19). If two or more bottoms are used, the center of resistance will be the average of all the centers. The style of bottom as to shape, type of share, and moldboard will influence the point where all the various forces acting on the bottom will be in equilibrium.

Influence of Friction on Design   After all the above principles are taken into consideration, they resolve themselves into one general principle of plow design that must be considered in every type of plow, whether it is stubble, general-purpose, or sod. That principle is that friction will be the greatest at the point of the share and gradually decrease backward to the end of the moldboard. This can be seen readily on any plow bottom after considerable use. The greatest

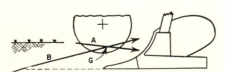

Fig. 9-18 The line *A* represents upward forces for fairly easy plowing, while line *B* is for hard plowing. The point *G* is the average soil resistance, which is slightly below the soil surface. (*A. W. Clyde, University of Pennsylvania.*)

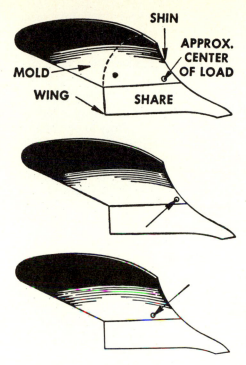

Fig. 9-19 Each of the circles on the bottoms shows the approximate point where the various forces act upon the plow in equilibrium for different soil types: (*top*) average soils; (*middle*) light soils; (*bottom*) tough soils. (*International Harvester Company.*)

amount of wear is found to be at the point; this gradually decreases backward to the tip of the wing of the moldboard. This is why the stubble moldboard, which has a greater amount of curvature, gives better pulverization to the furrow slice. It is also seen that this type of moldboard will pick up the soil quicker and turn it over harder than any other type. That makes this type of plow more adaptable to plowing the loams and the sandy-loam soils. The general-purpose moldboard has a smaller amount of curvature than the stubble; it is in this class that the blackland type of plow falls, because its curvature is not so pronounced as that of the stubble moldboard.

Influence of Speed on Design   In the last few years, there has been much agitation regarding the designing of plows for high speeds. It is not as difficult to design a plow for high speeds as it is to obtain pulverization. The bottom designed for high speeds must have gradual curves, which approach closely those of the sod type of plow. It can be seen readily that it is not necessary to have the moldboard as wide in this case in order to lift and invert the furrow slice. The higher velocity will carry the soil up over the moldboard, throwing it farther to the side. Much difficulty is likely to result from plows for high speed which must incorporate a plow bottom of long slopes. They may scour well while going at a high rate of speed, but when the speed drops to 2 or 3 mi/h (3.2 to 4.8 km/h), will they continue to scour at this speed? Will they do the same type of work as at the higher speed?

Type of Soil   Another important factor influencing plow design is the type of soil. In fact, if it were not for the soil factors, designing of plows would be a comparatively simple matter. Brown[1] says: "The types of soil, from sands, through the loams to the clays, are affected differently by the same plow bottom, and since the prime object of plowing is to put the soil in the proper condition of tilth for the successful growing of crops, it follows that there must be a variety of plow shapes."

Methods Used to Aid Scouring of Moldboard Plows   A plow is said to *scour* when the soil sheds clean from the moldboard and leaves a polished surface. Plow designers have used various materials in an effort to obtain good scouring qualities. Materials such as steel, iron, glass, aluminum, plaster of paris, and hog hide have been used. It was found that plaster of paris and hog hide gave better results than steel, iron, glass, and other materials. Other attempts have been made with special types of bottoms having holes through which water could flow and keep the surface of the moldboard wet. Fairly good results have been obtained by heating the moldboard by the use of the tractor exhaust flowing into an enclosed space underneath the moldboard. Slatted moldboards of iron and wood have been used to aid scouring. Late experiments with plastics such as nylon and Teflon fused to the surface of the moldboard show good scouring qualities and reduce the draft 15 to 25 percent. The author could find no reference as to where high-frequency electrical vibrations had been tested as a scouring aid. In general, the shape and construction of the steel moldboard have been relied on more than anything else to obtain scouring qualities. The low incline and slight curvature of the blackland bottom shown in Fig. 9-4 have the best scouring qualities of the various solid moldboard shapes available.

The Design of the Plow Framework   To support the plow bottom or bottoms, there must be a framework consisting of beams, braces, crank arms to raise and lower the plow, arrangements for hitching or mounting the plow to the tractor, and power lifts, either mechanical or hydraulic. The designer must consider the size and strength of the steel members and fit them together to withstand severe stresses, both longitudinally and horizontally.

Accessories for Moldboard Plows   The moldboard plow bottom is a working unit in itself and is used extensively without accessories. There are a number of devices that can be used as aids and attachments to assist the bottom in doing a good job of plowing. The list of accessories includes gage wheels, colters, jointers, and trash-covering aids of hooks and wires.

Where the soil is soft, gage wheels are required if the plow is to maintain a uniform depth. They can be attached to the beam ahead of the bottom or to the side of the beam.

A colter is used to cut the furrow slice from the land and leave a clean wall. It also cuts through trash so the plow can cover it better.

The *rolling colter* (Fig. 9-20) is a round, flat steel disc which has been sharpened on the edge and suspended on a shank and yoke from the beam. The edge of the colter may be made smooth, rippled or fluted, or notched. (Fig. 9-20). It is so constructed that it can be adjusted up and down for depth and

[1] Theo Brown, *Trans. of the ASAE,* **19**:24, 1925.

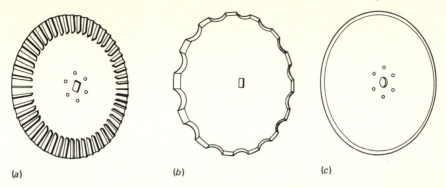

(a)   (b)   (c)

Fig. 9-20   Different types of colters: (a) fluted or rippled; (b) notched; (c) plain or smooth.

sidewise for width of cut. The proper setting of a rolling colter is shown in Fig. 9-21. This type of colter is used more than any of the others because it will leave a smooth furrow face and will also cut trash much better than the other types.

The *jointer* is a small, irregular-shaped piece of metal having a shape similar to an ordinary plow bottom (Fig. 9-22). It is a miniature plow. Its purpose is to turn over a small, ribbonlike furrow slice directly in front of the main plow bottom. This small furrow slice is cut from the left and upper side of the furrow slice and is inverted, so that all trash on top of the soil is completely turned under and buried in the right-hand corner of the furrow.

The jointer is used not only by itself but also in combination with the rolling colter (Fig. 9-23). This gives a *combination rolling colter and jointer*. The rolling colter cuts the main furrow slice and all trash vertically from the furrow wall, and the jointer turns its miniature furrow slice as when working alone. The advantage of the combination rolling colter and jointer is that the rolling colter cuts all trash and allows the jointer to turn its furrow slice without any trash hanging around the shank. The performance of the plow at high speeds is greatly aided by the rolling colter and jointer.

The regular *concave rolling-disk colter* (Fig. 9-24) is a relatively new innovation in colter design. The concave blade cuts and turns a shallow furrow slice and thus does the work of both a jointer and a rolling-disk colter. This type of colter is not recommended for hard and stony conditions.

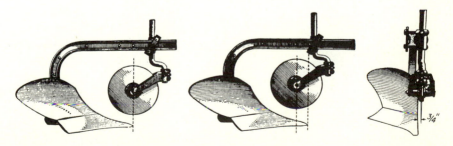

Fig. 9-21   Views of rolling colter showing proper setting in relation to the plow bottom. (*International Harvester Company.*)

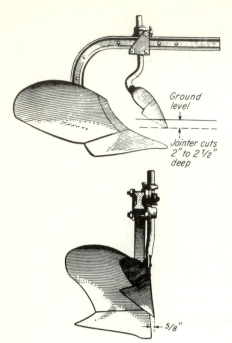

Ground level

Jointer cuts 2" to 2 1/2" deep

⊢ 5/8"

Fig. 9-22 Proper setting for a jointer. (*International Harvester Company*.)

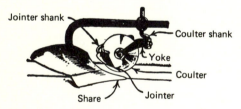

Jointer shank

Coulter shank

Yoke

Coulter

Share   Jointer

Fig. 9-23 Combination rolling colter and jointer, showing how the hub of the colter is set over the point of the share.

Fig. 9-24 Concave disk colter. (*Allis-Chalmers Mfg. Co.*)

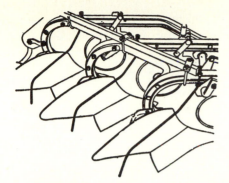

Fig. 9-25 Properly attached and shaped weed hooks are an aid in plowing under tall green vegetation. (*Iowa State Col. Bul. p. 95.*)

The common type of weed hook consists of a rod attached to the beam and extending out to the front and side of the plow bottom (Fig. 9-25). The object is to bend the weeds over in such a manner that they will be completely buried in the bottom of the furrow. Smooth wires 10 to 12 ft (3.0 to 3.6 m) long, fastened to the colter shank and to the axle of the furrow wheel, are of much value in covering trash and tall plant growth (Fig. 9-26). The wires should be rigidly fastened to reduce swinging and dragging under the bottoms when turns are made with the plow lifted.

Draft of Moldboard Plows   The draft of a moldboard plow is affected by many factors, such as the type and shape of bottom, more especially the moldboard; the sharpness of the share; the over-all adjustment of the plow; the depth and width of furrow; and the many soil types and characteristics of the soil. The speed at which the plow is operated is an important factor affecting draft of plows. When all these factors are considered, there will be a wide range in the draft of any type or shape of bottom from field to field, depending on the type and condition of the soil. Thus, the draft may range from 5 to 12 lb (2.2 to 5.4 kg) pull per in^2 of furrow section.[1] A 14-in (35.6-cm) bottom plowing 8 in (20.3 cm) deep will turn a furrow section of 112 in^2. Assuming the draft is 8 lb (3.6 kg) per in^2, then the draft for the bottom will be 896 lb (406 kg).

The draft of any plow can be determined by an instrument called a *dynamometer*, which registers the pull or draft of the plow over a measured

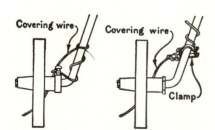

Fig. 9-26 Method of fastening covering wires to plow axle. (*U.S. Dept. Agr. Farmers' Bul. 1690.*)

[1] American Society of Agricultural Engineers Data: Crop Machines Use; I. Draft and Power Requirements of Crop Machines, *Agr. Engin. Yearbook,* 1954, p. 69.

distance. Then, knowing the rate of travel, the horsepower as well as the average draft per unit of the cross section of the furrow slice can be determined.

$$Hp = \frac{\text{force x distance traveled in ft/min}}{33,000}$$

The dynamometer measures the total draft required for the plow. It should be kept in mind that the weight of the plow alone requires power to move it. Collins[1] states: "The draft of the plow on the ground is 18 percent; draft due to turning furrow slice, 34 percent; draft due to cutting slice, 48 percent."

Thus it is seen that practically 50 percent of the total draft of the plow is used in cutting the furrow slice. A test was run to determine the effect of dull shares and sharp shares upon the draft of the plow. In a test on sandy-loam soil, the difference in draft of a sharp share was almost negligible. In a field of bluegrass sod, there was a difference of 14 percent in favour of the sharp share. In soil that is soft and mellow, the sharpness of the share will not matter so much; but if there are many roots, or if the soil is comparatively hard or lacks moisture, a sharp share is to be advocated.

Effect of Speed on Draft   Where high performance is desired in plowing with tractor power, speed is highly important.

Results of the tests made in California by Davidson, Fletcher, and Collins[2] were as follows: "In clay loam speed 1 mi/h (1.6 km/h)–draft, 100 percent. Speed 2 mi/h (3.2 km/h)–draft 100 to 114 percent. Speed 3 mi/h (4.8 km/h)–draft 128 percent. Speed 4 mi/h (6.4 km/h)–draft 142 percent." Tests in Iowa black-loam soil gave the following results: "Speed 1 mi/h (1.6 km/h)–draft 100 percent. Speed 2 mi/h (3.2 km/h)–draft 117 percent. Speed 4 mi/h (6.4 km/h)–draft 126 percent."

The conclusions were that an increase of the field speed of a plow with a general-purpose moldboard from 2 to 3 mi/h (3.2 to 4.8 km/h) resulted in an increase of draft from 8 to 12 percent, varying with the soil. Doubling the speed will result in an increase in draft from 16 to 25 percent. The amount of work accomplished is increased from 50 to 100 percent, respectively. It is to be remembered that practically 50 percent of this task of plowing is cutting the furrow slice. The conclusions reached by Collins in his tests in Iowa in 1920 were that the increase in draft, due to speed, is applied to that part of the total which is required for turning and pulverizing. This varies with the speed from less than one-third to about one-half the total draft of the plow within a range of 2 to 4 mi/h (3.2 to 6.4 km/h).

Studies made in Ohio by Ashley, Reed, and Glaves[3] indicated that with two bottoms the average increase in draft due to increased speeds was 1.17 lb (0.53 kg)/in^2 of furrow slice for each mi/h increase in speed.

[1] E. V. Collins, *Amer. Soc. Agr. Engin. Trans.*, 14:39, 1920.
[2] J. B. Davidson, L. F. Fletcher, and E. V. Collins, *Amer. Soc. Agr. Engin. Trans.*, 13:69, 1920.
[3] W. M. Ashley, I. F. Reed, and A. H. Glaves, Progress Report on Draft of Plows Used for Corn Borer Control, *U.S. Dept. Agr. BAE*, 1932.

Table 9-1   Miles and Meters Traveled in mi/h and km/h

mi/h	km/h	ft/min	m/min	mi/h	km/h	ft/min	m/min
1.00	1.6	88	26.8	3.75	6.0	330	100.6
1.25	2.0	110	33.5	4.00	6.4	352	107.3
1.50	2.4	132	40.2	4.25	6.8	374	114.0
1.75	2.8	154	46.9	4.50	7.2	396	120.7
2.00	3.2	176	53.6	4.75	7.6	418	127.4
2.25	3.6	198	60.4	5.00	8.0	440	134.1
2.50	4.0	220	67.1	5.25	8.4	462	140.7
2.75	4.4	242	73.8	5.50	8.8	484	148.5
3.00	4.8	264	80.5	5.75	9.2	508	154.8
3.25	5.2	286	87.2	6.00	9.6	528	160.0
3.50	5.6	308	93.9				

Effect of Grade on Draft   When a tractor is on a grade, its effective drawbar pull is lessened 1 percent for each percent of grade. For example, the weight of the tractor ready for work with an operator and a three-bottom plow is approximately 7600 lb (3450 kg). To negotiate a 10 percent grade with this outfit would require an additional power equivalent to a pull at the drawbar of 760 lb (344.3 kg).

DISK PLOWS

The disk plow was probably developed about 1890. Models were listed in implement catalogs by 1895. One of the earliest patents for a disk plow was secured by M. A. and I. M. Cravath, Bloomington, Illinois. J. K. Underwood, D. H. Lane, and M. T. Hancock made improvements on the disk plow and made it practical. Since 1900, the development of the disk plow has followed trends similar to that of the moldboard plow.

The disk[1] plow was brought out in an effort to reduce friction by making a rolling bottom instead of a bottom that would slide along the furrow. It cannot be said with authority that, after the extra weight is incorporated into the plow, it will have any less draft than the moldboard type. The results of the disk-plow usage, however, show that it is adapted to conditions where the moldboard will not work, such as the following:

1   Sticky, waxy, gumbo, nonscouring soils and soils having a hardpan or plow sole
2   Dry, hard ground that cannot be penetrated with a moldboard plow
3   Rough, stony, and rooty ground, where the disk will ride over the rocks
4   Peaty and leaf-mold soils where the moldboard plow will not turn the slice
5   Deep plowing

[1] As used in this text, *disk* means a round, concave piece of metal while *disc* means a round, flat piece of metal.

This type of plow is used in the South and North and very extensively in the Southwest and the semihumid regions of the Middle West. It is of special value in Texas because of the large areas of soil having a close texture which will not scour on the average moldboard plow.

Tractor Disk Plows

Tractor disk plows are divided into trailing, semimounted, and integral-mounted types.

Trailing Disk Plows   The trailing disk plows can be divided into two types, *regular* and *one-way*.

Regular Trailing Disk Plows   As the name indicates, these plows are pulled behind the tractor (Fig. 9-27). The plow is a unit in itself and is attached to the tractor by a hitch that can be adjusted both vertically and horizontally. In the older models, the frame is arranged to the side and below the top of the disk bottoms. This is called a *side-frame* plow. Newer models of trailing disk plows have a high frame above the disk. This plow is called an *overhead* disk plow. Longer standards are required to attach the disk bottoms to the overhead frame than to the side frame. The high frame gives sufficient clearance for plowing in trashy fields. Arrangements are provided for adjusting the vertical angle of the disk, which influences penetration.

The trailing disk plow is provided with three wheels: two furrow wheels and a land wheel. The front furrow wheel is at the front end of the frame and is connected to the hitch to aid in guiding and turning the plow. The rear furrow wheel is usually allowed to swivel on left-hand turns but is limited in its movement to the right so it will hold the plow in proper position when plowing. Both furrow wheels are inclined to hold the plow in position (Fig. 9-27). The land wheel is usually located toward the rear of the plow but slightly forward of the rear furrow wheel. The power lift is always a part of the land wheel. When plowing hard soils, additional weights on the wheels will help to hold and steady the plow. If desired, the wheels can be equipped with rubber tires. Levers and

Fig. 9-27   Four-bottom trailing disk plow equipped with hydraulic remote-control cylinder to lift plow. (*International Harvester Company.*)

screw cranks provide a means of adjusting for depth and for leveling the plow. Hydraulic remote-control lifts can be obtained for trailing disk plows. Heavy-duty disk plows can be obtained for deep plowing and for plowing in heavy soils.

Tiller or One-Way Trailing Disk Plows As shown in Fig. 9-28, the one-way disk plow is a combination of the principles of the regular disk plow and the disk harrow and is often termed a *wheatland, cylinder, harrow,* or *tiller plow.*

The one-way tiller plow was developed in the Great Plains area about 1927. It was designed primarily as a one-way disk harrow. As its use spread, farmers began to adopt it for shallow plowing. Improvements have made the tool into a popular and widely used plow. The speed of the plow should not be over 4 mi/h (6.4 km/h). High-speed operation increases the power requirements, causes too much pulverizing of the surface soil, and does not leave trash on the surface to prevent wind erosion.

The one-way disk plow has the frame, wheel arrangement, and depth-adjusting devices of the regular disk plow, but the disk bottoms are assembled on a single shaft and turn as a unit similarly to a gang of disks on a disk harrow (Fig. 9-29). The disks may be spaced 8 or 10 in (20.3 or 25.4 cm) apart. The number per plow may vary from 2 to 35. The size will range from 20 to 26 in (50.8 to 66.1 cm). When ten disks are spaced 8 in (20.3 cm) apart, the approximate cut will be 6 ft (1.8 m), but when disks are spaced 10 in (25.4 cm) apart, the cut will

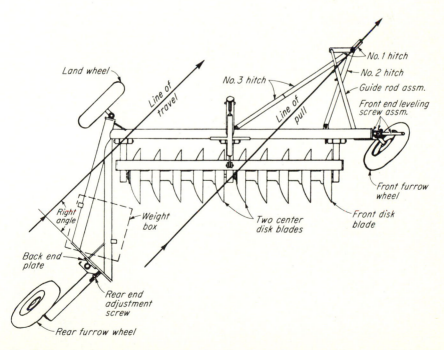

Fig. 9-28 Overhead view of one-way plow showing arrangement of frame and disk gangs. (*Schafer Plow Company.*)

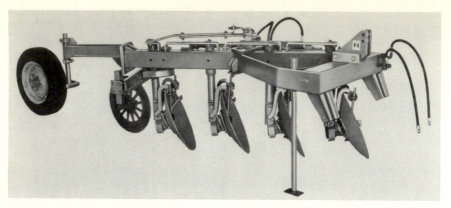

Fig. 9-29   Semimounted four-bottom disk plow. (*International Harvester Company*.)

be approximately $7\frac{1}{2}$ ft (2.9 m). A plow with 35 disks will cut a strip about 20 ft (6.0 m) wide.

Special features and uses of the one-way plow include (1) a seedbox attachment (Fig. 9-30) which permits the seeding of small grains and grass, (2) removable sections of the gangs to reduce the size, (3) disks mounted eccentrically for the forming of pits and basins to conserve water and prevent wind erosion, and (4) the use of the one-way plow for the building of broad-based terraces.

Semimounted Disk Plows   The semimounted disk plow is also called a *direct-connected* plow. The front of this plow is connected to and mounted on the tractor, thus eliminating the front furrow wheel and the land wheel (Fig. 9-31). A furrow wheel supports the rear end. This close-coupled plow is compact

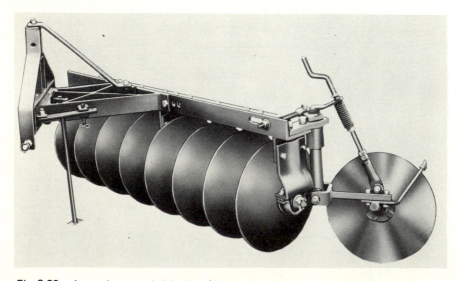

Fig. 9-30   Integral-mounted disk plow. (*Towner Manufacturing Company*.)

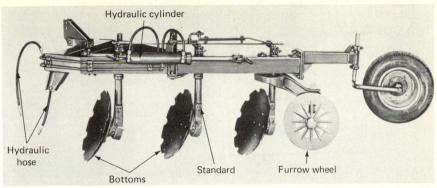

Fig. 9-31   Side view of two-way or reversible four-bottom disk plow. (*International Harvester Company.*)

and easy to handle. It is easy to maneuver because short turns can be made, enabling the operator to work close to fences. It can also be backed into corners. The rear wheel is automatically controlled from the steering mechanism at the front of the tractor. These plows are built in three sizes according to the number of bottoms. A hydraulic lift raises the front of the plow high enough so it can be turned and transported easily. Depth of plowing is adjusted by the lever at the rear. Wheel weights and scrapers for the disks can be obtained.

Five-bottom semimounted two-way or reversible disk plows are now available.

Integral-mounted Disk Plows   The integral mounting of disk plows on the rear of a tractor so they could be lifted or picked up with hydraulic lifts was first thought to be impractical. This was true as long as the heavy weights were retained on the rear furrow wheel. Designers found that integral-mounted disk plows would give good performance if the weight was moved from the rear furrow wheel forward onto the frame. A rear furrow wheel on an integral-mounted disk plow may serve to counteract side pressures, hold the plow in alignment, and act as a gage wheel for plowing depth. The depth in some makes is controlled by adjusting the hydraulic lift.

Figure 9-32 shows a two-way or reversible disk plow integrally mounted. The two bottoms are reversed by a lever or hydraulic arrangement that permits the bottoms to swing sidewise on a quadrant located on the beam above the front bottom (Fig. 9-32).

Special integral-mounted gangs of disks for bedding land and for barring off stubble or crop rows are shown in Fig. 9-33. The disks on a gang may all be the same size, or they may vary in size. The gangs may be adjustable for pitch and gather an may also be reversed.

Integral-mounted border-making disks are available. The number of disks per gang varies from one to four. The diameter of the disk ranges from 20 in (50.8 cm) where three and four disks are used to 28 in (71.1 cm) where a single disk is used.

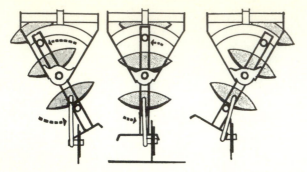

Fig. 9-32   Most reversible disk plows are reversed by moving the front of the plow to the right or left or center.

Design of Disk Plows

The disk-plow bottom is a perfectly round, concave disk of heat-hardened steel, sharpened on the edge to aid in the penetration of the soil. The size of a disk-plow bottom is the diameter of the disk and ranges from 20 to 38 in (50.8 to 96.5 cm). The average thickness of the steel for disk-plow bottoms is $\frac{3}{16}$ in (4.8 mm) for the smaller sizes and may be as much as $\frac{3}{8}$ in (9.5 mm) for the larger sizes. The amount of concavity varies with both the different diameters and the same diameter, as shown in Table 9-2.

Angling of Disk-Plow Bottoms   The disk bottom is attached to a standard which may extend downward from the heavy steel beam. The disk rotates on a chilled-iron ball or roller bearing. The standard is adjustable to give variable degrees of vertical and horizontal angle to the disk bottom.

The disk plow can be made to penetrate more easily when the disk is set more in a vertical position (Fig. 9-34). The flatter it sits, the less tendency there

Fig. 9-33   Four-row integral-mounted disk bedder. (*Deere & Co.*)

Table 9-2   Size, Concavity, and Radius of the Average Disk Plow

Size		Concavity		Size		Concavity	
in	cm	in	cm	in	cm	in	cm
20	50.8	$2\frac{7}{8}$	7.3	26	66.1	$4\frac{1}{2}$	11.4
23	58.4	$3\frac{1}{2}$	8.9	28	71.1	$4\frac{1}{4}$	10.8
24	61.0	$3\frac{3}{8}$	9.2	28	71.1	$5\frac{3}{8}$	14.3
24	61.0	$3\frac{11}{16}$	9.4	32	81.3	$4\frac{1}{4}$	10.8
26	66.1	$3\frac{3}{4}$	9.5	32	81.3	$6\frac{1}{2}$	16.5
26	66.1	4	10.1	38	96.5	$6\frac{1}{2}$	16.5

Source: R. C. Ingersoll, The Development of the Disk Plow, *Agr. Engin.,* 7(5):172, 1926.

will be for it to penetrate. Further to enable the disk plow to take the soil properly, weight is added to the frame and wheels to force the plow into the ground. There is one great difference between moldboard and disk plows: the moldboard plow is pulled into the ground by the suction of the plow, while the disk is forced into the ground by added weight or force and by the suction of the disk due to the angle at which it is set.

The horizontal angle of the disk influences the width of the furrow slice and the tendency to roll. Disks set more nearly perpendicular to the direction of travel cut wider furrows and do not turn so freely as those set more nearly parallel to the furrow. When the disk plow is pulled forward, the disk will turn as a result of the action of the furrow slice upon it. The top of the disk is revolving to the tractor operator's left. The furrow slice, then, is cut by the left edge of the disk, brought under and up to the right, and then thrown out to one side. The furrow slice is pulverized to some extent when carried over the concave surface of the disk.

The Center of Resistance   The center of resistance on disk plows is closer to the furrow wall than on moldboard plows. Its location is to the left and below the center of the disk. The exact point varies with the vertical and horizontal angles, the depth, and the amount of concavity of the blade.

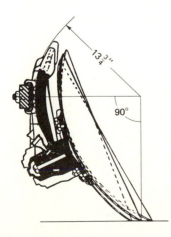

Fig. 9-34 Vertical angle of disk bottom can be easily changed.

Accessories for Disk Plows

Disk-plow bottoms should be equipped with scrapers which can be adjusted to work from the center to the edge of the disk. With the aid of the scraper, it is possible to get greater pulverization of the furrow slice. It is also possible to invert the furrow slice and cover trash much better. Weights aid in forcing the disks into the soil and hold the wheel in the furrow to prevent the disks riding out of the soil when plowing extra-dry and hard soil. Most types of disk plows can be obtained with hydraulic lifts. Levers or screw cranks aid in leveling the plow and adjusting for depth.

Draft of Disk Plows

The disk plow is slightly lighter in draft than the moldboard when plowing under similar conditions and turning the same volume of soil. The type of soil is the greatest external factor to consider in the draft of any plow. In very hard ground, it is often necessary to add weight to the wheels to force the plow into the soil. Of course, the added weight will create more draft.

Factors incorporated in the plow are very important. The bearings of the disk-plow blade also affect the draft. According to tests conducted by Hardy[1] a plain cone bearing will pull 23 percent heavier than a ball or roller bearing.

The type of scraper used to clean the disk will also affect the draft. Hardy's tests indicate that the revolving type gave slightly less draft than the spade type. Draft of disk plows is, of course, affected by the depth and width of the cut per bottom and for the complete plow.

ROTARY PLOWS

Rotary plows are discussed separately from moldboard and disk plows because they are of an entirely different design. They are neither moldboard nor disk plows. Shawl[2] states that the rotary plow was invented about 90 years ago. The rotary plow has been used in Europe for many years, but the American farmer has only recently become interested in this type of plow. The reason for this lack of interest was the high cost and the large power requirements. In general, rotary plows may be divided into three types: the pull auxiliary-engine, the pull power-takeoff-driven, and the self-propelled garden type.

Pull Auxiliary-Engine Rotary Plow

This rotary plow is pulled forward by a tractor but has the cutting knives driven by an auxiliary engine mounted on the frame of the plow. This type of plow is made in 4-, 5-, and 6-ft (1.28-, 1.52-, and 1.82-m) sizes and requires 60 to 90 hp. The L-shaped cutting knives are mounted on a horizontal power-driven shaft. Tractor-propelled, auxiliary-engine-driven rotary cultivators are also available.

---

[1] E. A. Hardy, University of Saskatchewan, Saskatoon, Saskatchewan, Canada.
[2] R. I. Shawl, Rotary Plowing as a Means of Seedbed Preparation, *Farm Impl. News*, 67(6):50-53, 1946.

Tractor-mounted Power-Takeoff-driven Rotary Plow

The rotary plow shown in Fig. 9-35 not only is pulled forward by the tractor but also has the cutting knives driven by the tractor. This type is usually 3 to 4 ft (0.91 to 1.28 m) wide and requires 10 to 15 hp for each foot of width. The cutting knives or tines are generally mounted on a horizontal power-driven shaft which operates at about 300 r/min. The knives on some machines are provided with a shock-cushioned friction clutch that prevents the knives from breaking when they come in contact with a rock or solid obstacle.

CHISEL AND SUBSURFACE PLOWS

The Chisel Plow

The chisel-type plow is a tool with a rigid curved or straight shank with relatively narrow shovel points. It may be termed a heavy-duty deep cultivator. The soil is stirred more or less in place (Fig. 9-36). The standard or shank is constructed of nickel-alloy heat-treated spring steel that is given a long, gradual curve flatwise to permit a slight spring action. The standards are arranged on heavy frames in two or three staggered rows to permit trash to pass between them without choking. Most chisel plows are provided with coil cushion springs in conjunction with the

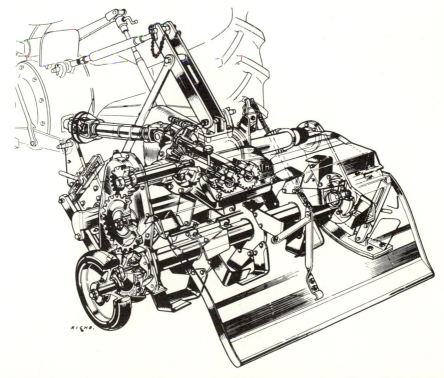

Fig. 9-35 Skeleton view of power-takeoff-driven rotary plow. (*Howard Rotavator Company*.)

Fig. 9-36   Front view of a trailing hydraulically lifted chisel plow with spring-loaded shanks.

clamps. This permits the ground tool to swing back and up and to unhook the point (Fig. 9-36). The furrows may be as close as 12 in (30.5 cm) or as wide apart as 2 to 3 ft (61.2 to 91.4 cm). The depth of plowing may be as shallow as desired or as deep as 18 in (45.7 cm) or more.

Some types of chisel plows are available in sizes ranging from 5- to 45-ft (1.5- to 13.7-m) widths. On some of the larger plows the outer sections can be folded up on the center section. This aids in transporting the tool. Pneumatic-tired wheels control the depth or serve for transportation. Hydraulic lifts are provided for all sizes of plows.

As the soil is broken by stirring, it is not inverted and pulverized to the extent that moldboard and disk plows crush the soil. Therefore, the chisel plow is often used to loosen hard, dry soils before the regular plow is used. The chiseling and stirring operation does not throw enough soil to cover trash completely. Hence, the chisel-type plow is used for stubble-mulch or subsurface tillage practices. This type of plow is also useful in breaking up hard layers of soil just below the regular plowing depth. This layer of soil is called by several names, such as *hardpan* and *plow sole*. Special plow shovels, lister shovels, sweep and knife assemblies (Fig. 9-37), and seeding attachments make it possible to do many jobs with the chisel-type plow.

Subsoilers

Subsoil plows are built heavier than the chisel plows, since they are used to penetrate the soil to depths of from 20 to 36 in (50.8 to 91.4 cm). Tractors of 60 to 85 hp may be required to pull a single standard ripping through a hard soil at a depth of 3 ft (91.4 cm). The standard on subsoil plows is usually long and narrow with a heavy, wedgelike point (Fig. 9-38). One standard is generally used for the deeper depth, but two or more can be used for shallower operations. Large hydraulic cylinders are provided for lifting the plows. In some types of poorly drained soils, a mole ball, shaped somewhat like a torpedo, is drawn

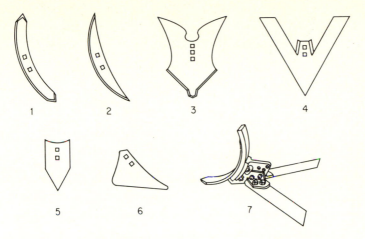

Fig. 9-37   Types of points that can be used on chisel standards: (*1*) chisel; (*2*) spike; (*3*) furrower; (*4*) sweep; (*5*) shovel; (*6*) drill shoe; (*7*) combination chisel and sweep.

through the soil behind a subsoil standard. The mole ball leaves a tunnel which serves as a drainage channel for water. Three types of subsoil standards are shown in Fig. 9-39.

Subsoil plows are available in both trailing and mounted units. Subsoil standards can also be attached to regular tractor tool bars.

Generally, the subsoiler plow is a heavy-duty tool designed to operate below the normal depth of tillage and to loosen the soil by lifting or displacement.

Fig. 9-38   Trailing subsoil plow with hydraulic lift. (*Atlas Scraper and Engineering Company.*)

Subsoiler shape affects draft

1 and 2 require about 25% less pull than 3

Fig. 9-39 Three types of subsoiling standards. (*Cornell Univ., Agron., Mimeo, 61-2.*)

RATE OF PLOWING

One 14-in (35.6-cm) plow bottom pulled at the rate of $2\frac{1}{2}$ mi/h (4.0 km/h) will plow approximately $\frac{11}{32}$, or 0.3, acre (0.12 hectare) in 1 hour. Some time, however, must be allowed for turning, which will depend on the shape and size of the field and how it is laid out. For example, with a two-plow outfit in a field 80 rods long, where lands of average width are struck out and the turning is done on headlands, about 6 percent of the time is spent in turning at the ends. Table 9-3 shows the acreage plowed with plows of different width when drawn at different speeds.

A four-plow outfit, of course, will accomplish about twice as much as the two if both of them are run at the same speed, and a six-plow outfit twice as much as the three-plow outfit. One acre contains 43,560 ft^2 (2.48 hectares) or 160 rd^2. A 14-in (35.6-cm) furrow 1 mi (1.6 km) long equals 6,160 ft^2 (70.6 m^2).

The number of acres plowed per day depends upon the width of strip plowed, the rate of travel (speed), and the length of time operated.

This can be expressed in the following formula, which allows about 17.5 percent of the operating time for loss in turning:

Acres per 10-h day = W x mi/h

where W = width of strip plowed
mi/h = miles per hour or rate of travel

Example: How long will it take to plow a 40-acre field with three 14-in-bottom plows pulled by a tractor at 3 mi/h?

Three 14-in bottoms cut a strip $3\frac{1}{2}$ ft wide.

W x mi/h = $3\frac{1}{2}$ x 3 = 10.5 acres plowed in a 10-h day
40 ÷ 10.5 = 3.8, or almost 4 days required to plow the 40 acres

GIANT PLOWS

In some areas where good land has been covered by blow sand or flood deposits, giant plows are used to bring up good soil from depths of 2 to 6 ft (61.0 cm to

Table 9-3 Chart Showing Acres Covered per Hour with Different Widths of Implement at Various Speeds

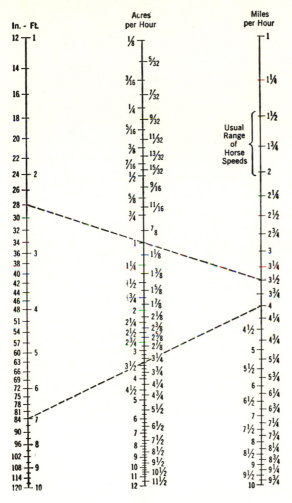

1.8 m) and lay it on top of the sand. Giant plows are available in both moldboard and disk types (Fig. 9-40).

HITCHES FOR PLOWS

A plow or implement may be well designed and built of high-grade materials, but unless it is properly hitched to the prime mover, or the power unit, it cannot give the highest possible performance. The hitch may be simple, consisting of only one or two parts, or it may consist of a multiplicity of bars, braces, angles, linkages, and levers arranged to absorb certain vertical and horizontal forces. The problem is to get all the pulling forces of the power and the resisting forces of

Fig. 9-40 Giant disk plow with large disk for deep plowing. (*Towner Manufacturing Company.*)

the load in equilibrium or nearly so, both vertically and horizontally. Even though most moldboard plows now available are either semimounted or integral-mounted, the trailing hitch forces on the plow are discussed.

Moldboard Trailing-Plow Hitches

The perfect hitch for a trailing plow or any other trailing tool would be to have the center of the pulled load directly behind the center of the power unit. This condition is rarely obtainable because of the varying widths of the different sizes of tractors and the different widths and sizes of the plows or pulled units.

    Vertical Principles   Above, under "Plow Design," the various forces that act on a single plow bottom were discussed. Under "Hitches for Plows" these principles must be expanded and applied to plowing units from two to six or more bottoms.

    Clyde[1] states that "the location of the combined backward and vertical resistance is important for deciding where the pull should be applied." Under average soil conditions, he found that the vertical center of resistance of a

[1] A. W. Clyde, Mechanics of Farm Machinery, *Farm Impl. News,* January-March 1944.

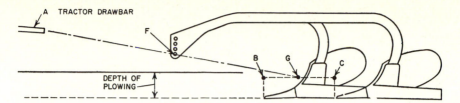

Fig. 9-41   The vertical center of resistance for a two-bottom moldboard plow is located above the share. The vertical line of pull should be a straight line from the center of load *G* through *F* to *A*. (*A. W. Clyde, University of Pennsylvania.*)

moldboard plow is a little below the surface and *above the share point.*[1] Figure 9-41 shows the points of resistance for a two-bottom plow at *B* and *C,* while the center of load and point of pull for the two bottoms are at *G.* The tractor drawbar is shown at *A.* The vertical line of pull should be a straight line between *A* and *G.* Therefore, the beam must be extended forward and provided with hitch plates that extend downward to the line *AG* at *F.* A string stretched from the tractor drawbar *A* to the center of resistance *G* will show where the hitch bar of the plow should be placed.

If the hitch at *F* (Fig. 9-41) is above the line of pull, the bottom will tend to run on its nose, causing excessive wear on the share and undue strain on the axle and wheel bearings. Should the hitch at *F* be below the line of pull, there will be a lifting action on the front of the plow and the bottoms will tend to run shallow, particularly in hard ground.,

The Vertical Line of Draft   From the above description of vertical forces it is seen that the vertical line of draft or line of pull is a straight line from the center of the load through the point where the drawbar is attached to the plow to the point where the drawbar is attached to the drawbar of the tractor (Figs. 9-41 and 9-42).

Horizontal Principles   The balancing of the horizontal forces for various sizes of tractors and plows is the most important and the most difficult of the hitching principles. The objective in the horizontal hitch is to determine the

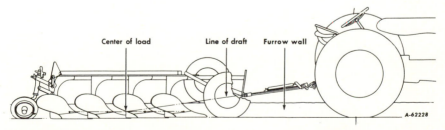

Fig. 9-42   The correct line of draft for a four-bottom trailing moldboard plow. (*International Harvester Company.*)

[1] Most literature shows the vertical center of resistance located at the same point as the horizontal center of resistance.

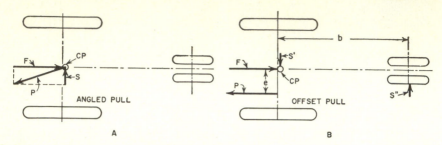

Fig. 9-43   Forces acting on a tractor when the center of the load is to the right of the center of pull. The pull *P* is at an angle in *A*, while in *B* the pull is offset and straight back. (*A. W. Clyde, University of Pennsylvania.*)

center of pull, or center of power, and the center of the load, or the center of the trailing unit, and arrange the hitch so the center of load is as nearly as possible directly behind the center of the power unit (Fig. 9-43).

The Horizontal Line of Draft   The horizontal line of draft should be a straight line from the center of resistance or load to the point where the plow drawbar is attached to the tractor drawbar. As shown in Fig. 9-44 the line of draft in the two-hitch arrangement does not follow the plow drawbar. This is

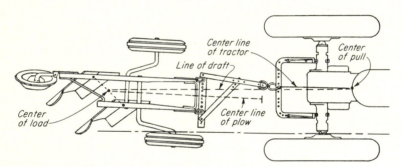

Sidedraft is all on the plow in this illustration. Strain is on the rear axle and wheel— landsides will wear rapidly, load is increased and plowing will be of poor quality.

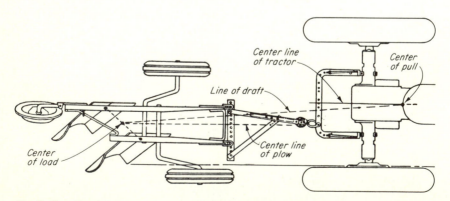

Fig. 9-44   Two-hitch arrangements for a two-bottom trailing moldboard plow drawn by a row-crop tractor. (*International Harvester Company.*)

because of the difference in the widths of the tractor and the plow. The drawbar of the plow is held at an angle in an effort to bring the center of the load directly behind the center of the power or tractor and prevent side draft.

Center of Power   The center of power is often described as the *true point of hitch* or *center of pull*. Whatever the term used, the point referred to is the center of the power, which is chiefly horizontal, but the vertical forces must also be considered (Fig. 9-44). On a tractor, this is the point where the front end of the drawbar is attached, which is always the middle of the tractor halfway between the wheels. In most tractors, the rear end of the drawbar can be swung sidewise to compromise the hitch on the tool. Such a drawbar is termed a *swinging drawbar*. It is useful when an angle pull is required.

Center of Load or Resistance   The factors involved in finding the center of resistance of a single moldboard plow bottom were discussed in detail in connection with the design of a moldboard bottom. It should be recalled that the center of resistance for a single bottom is located on the face of the bottom to the right of the shin, a distance equal to about one-fourth the width of the plow or width of cut (Figs. 9-17 and 9-19). When two 14-in (35.6-cm) bottoms are used together, the center of load for the plow will be approximately halfway between the center of resistance of the two bottoms, or one-half the width of the second bottom, which is 7 in (17.8 cm) added to the $10\frac{1}{2}$ in (26.7 cm) for the first bottom. The center of the load for the two bottoms is $17\frac{1}{2}$ in (44.4 cm) to the left of the wing of the first bottom. Thus, if three 14-in bottoms are used, the center of load will be $24\frac{1}{2}$ in (62.2 cm) from the wing of the first bottom. For each 14-in bottom added, the center of load moves 7 in (17.8 cm) to the left.

Side Draft or Side Pull   If the center of the load is directly behind the center of the pull or power, there should be little or no side draft or side pull (Fig. 9-43*A*). If, however, the center of the load is a few inches to the right of the center of the power, there will be side draft or pull (Fig. 9-43*B*). When a swinging drawbar (Fig. 9-45) is used, the pull will be through the angle *P*. Part of the side pull of the load tends to pull the tractor sideways. In this arrangement

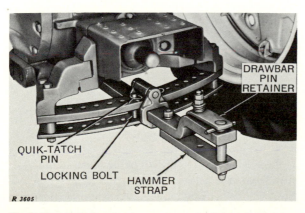

R 3605

Fig. 9-45   Swinging drawbar on a tractor. (*Deere & Co.*)

the center of pull *CP* is located at a point where the side forces *S* and the forward pull of the tractor *F* meet. Where the line of pull is through *P*, practically all the side force is carried by the rear tractor wheels.

If the point of hitch is moved to the side on the tractor drawbar, as in Fig. 9-43*B*, the line of pull is straight back but offset instead of angled as in Fig. 9-43*A*. In this case, the offset pull tends to rotate the tractor clockwise. That is, the rear of the tractor is being forced to the left while the front of the tractor is being forced to the right. The principles of the horizontal hitch are shown in Fig. 9-44.

Hitching of Semimounted Plows

As described above and shown in Fig. 9-11, the semimounted moldboard plow is connected directly to the tractor. The weight of the front part of the plow is on the tractor hitch. The weight of the rear part of the plow is supported by a rear furrow wheel when in operation and when raised. The plow can be attached to the tractor by either a two-point connection of a three-point linkage arrangement. Some makes of semimounted plows are equipped with a land wheel which aids in keeping the plow level. Where there is no land wheel, all the depth-control and leveling adjustments are provided in the hitch. The instructions in the operator's manual should be followed in making adjustments.

Hitching of Integral-mounted Plows

The principles of hitching the trailing plow will also apply to the mounted plow. The hitching, however, is not left to the plowman because the design engineer necessarily had to establish the points of hitch and build the plow to suit a particular size and make of tractor. The plowman, therefore, has only to adjust the spacing of the rear tractor wheels to give the correct width of cut for the front bottom. Each of the tractor wheels should be adjusted an equal distance from the center of the tractor to keep the center of pull in the correct relation to the line of pull.

The designer of mounted plows, however, must consider the problems of vertical forces. A well-designed, depth-regulating device must be provided. Trailing plows were built heavy to aid the penetration of the soil. Mounted plows must be built of strong but lighter materials to avoid overloading the tractor. The design engineer must have a good knowledge of mechanics and also a good conception of the reactions to be expected from soil types. Some mounted plows pull from a single point, but most designers use a three-point linkage (Fig 9-46). The plow is lifted by hydraulic power. In some plows the depth is adjusted by hydraulic controls.

The plow is equipped with a crossbar which extends across the front of the plow. The two lower hitch links on the tractor are connected to the center of load of the plow.

The center of resistance of a single disk is to the left of the center of the disk. In operation, it will be slightly below the surface of the furrow slice. When two disks are used, the center of load will be halfway between the center of resistance for each of the two disks (Fig. 9-47). As the number of disks is increased, the

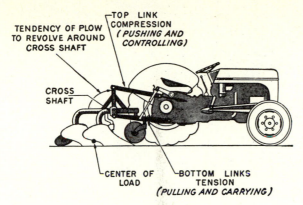

Fig. 9-46 A three-point hitch linkage that utilizes the resistance and weight of the soil to increase rear-wheel traction. (*Massey-Ferguson Ltd.*)

center of load will be the average for all the disks, or the center of cut as measured on the ground.

Most mounted two-way or reversible moldboard plows are attached to the tractor by means of the three-point hitch. The adjustments for lifting, leveling, and depth control are the same as for the regular moldboard plow.

Disk-Plow Hitching

The principles of hitching trailing moldboard plows are also applicable in hitching trailing disk plows. However, different hitch adjustments are necessary to obtain correct vertical and horizontal hitching. The vertical hitch is quite simple, because the hitch bar can be adjusted up and down on the plow so there will be a straight line from the tractor drawbar through the plow connection to the center of load of the plow (Fig. 9-47).

When a trailing disk plow is hitched, the center of load should be as nearly as possible directly behind the center of pull or the power unit. For a narrow-tread tractor, place tractor and plow in position for hitching and stretch a string from the center of pull on tractor drawbar to the center of load on the plow (Fig. 9-41). Then adjust the drawbar or hitch bar of the plow over the string from tractor drawbar to the plow clevis. If the tractor has a wide tread, stretch the string from the center of load of the plow to the tractor at a point about 3 in to the right of the center of pull. Then adjust the draw rod over the string. The front furrow wheel is adjusted by the guide arm so that it runs straight ahead or with a slight toe-out.

It is sometimes desirable to change the total width of cut of a disk plow. *To change to a narrower cut, the hitch on front of the plow is moved to the left.* This causes the front of the plow to move to the right and the rear to swing to the left (facing the front of the plow), causing the disks to fall more in line or to *trail*. Since the front furrow wheel is held in position by a guide arm (Fig. 9-48) connected to the hitch bar, it is necessary to adjust this arm whenever the hitch bar is moved either to the right or to the left. *To change to a wider cut, the hitch bar is moved to the right on the plow.* This will cause the rear of the plow to

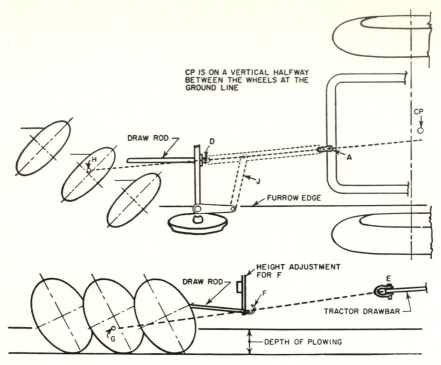

Fig. 9-47   Horizontal and vertical principles of hitching a disk plow. (*Pa. Agr. Ext. Cir. 259.*)

swing to the right so that each of the disks cuts more. The front of the plow is held in position by giving the front furrow wheel a lead toward the furrow wall.

The hitching of the semimounted and the integral-mounted disk plows is similar to that of the moldboard semimounted and integral-mounted plows.

Hitching the One-Way Disk Plow

The hitch arrangement for a one-way disk plow is slightly different from that of the regular trailing disk plow, but the same principles of horizontal and vertical forces are involved. The gang of disks is set at about a 45 to 50° angle to the direction of travel, while the beam of the reguar disk plow is set at an angle of 30 to 35°. The center of load for the entire plow will be farther from the first furrow for the larger plows. Therefore, a horizontal clevis bar extends across the front of the plow to the left and beyond the line of pull (Fig. 9-49). Near the outer end of the horizontal clevis bar is a drawbar that extends back to points about midway and to the rear of the frame or gangs. These bars keep the disks from trailing.

Two types of hitches are used on one-way plows, the figure-A type and the figure-four type (Figs 9-49 and 9-50). The free-swing end of the figure-A type should be approximately in line with the line of pull for the plow. For the figure-four, the long hitch bar should be in line with the center of pull.

The center of load for a one-way plow having small disks with a shallow

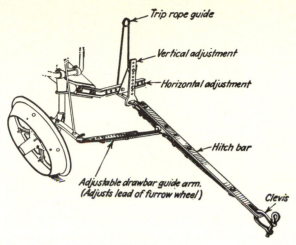

Fig. 9-48 A disk-plow hitch.

dish is approximately one-half the width of cut. Larger disks with a larger dish have the center of load farther toward the plowed ground but never closer to the furrow than one-third the width of the cut. The vertical center of load is at the same point as the horizontal center of load but about one-half the plowing depth.

It is important that the hitch bar from the plow clevis to the tractor drawbar be in line with the vertical line of pull. A wide range of vertical adjustment is provided on one-way plows. If the tractor drawbar is too high or the hitch bar too high on the plow, there will be a downward pull on the plow. This will throw excessive weight on the front furrow wheel and the frame will be tipped forward, which will reduce the weight on the rear furrow wheel. The

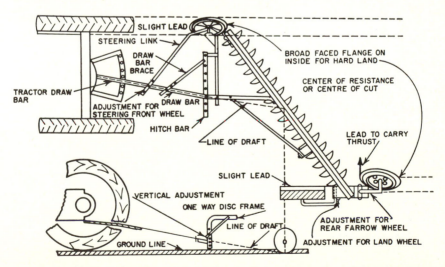

Fig. 9-49 Horizontal and vertical hitches for a one-way plow. (*Kansas Engineering Extension Department.*)

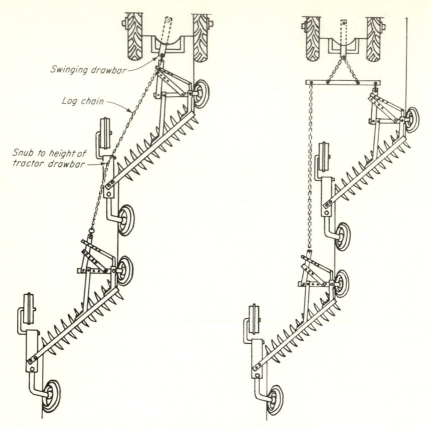

Swinging drawbar

Log chain

Snub to height of
tractor drawbar

Fig. 9-50 Two types of drawbar hitches used on one-way plows, and two methods of hitching two one-way plows in tandem. (*Kansas Engineering Extension Department.*)

reduction in weight on the rear furrow wheel will tend to let it climb out of the furrow, and the plow will swing out of line. On some one-way plows, it may be necessary to add an extension to the vertical-adjusting bar to get the hitch bar at the plow low enough to prevent pulling down on the front end. The one-way plow generally operates best with a swinging tractor draw bar, as it permits shorter turns.

Multiple-Unit and Tandem Hitches   Many farmers who have large tractors often prefer to use two or three medium-sized one-way plows rather than one large unit. Multiple-unit hitches for one-way plows are shown in Fig. 9-51.

Tandem hitching is used where one or more units performing different operations are trailed one behind the other. A plow may be followed by a disk harrow or a spike-tooth harrow. A tandem disk harrow is often used behind power-operated stalk cutter-shredders.

Offset hitches are used for trailing a clod crusher to the side of a plow or a trailer beside a corn picker.

Multiple-width hitches are used when several units of the same kind are

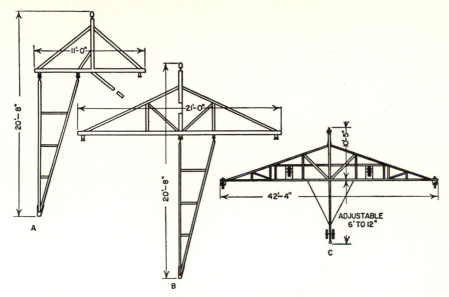

Fig. 9-51 Three multiple-implement hitches: (*A*) hitch frame for two implements; (*B*) frame for three implements; (*C*) frame for three, four, five, or six implements. (*Caterpillar Tractor Co.*)

used to cover wide swaths, such as harrows, grain drills, and planters (Fig. 9-51). On some of the larger farms, the swath harrowed or seeded may be as wide as 40 ft (12.2 m).

MAINTENANCE AND CARE

1   Check plow for proper alignment of bottoms.
2   Check the hitch arrangement of the moldboard plow for proper performance.
3   Keep the shares on moldboard plows sharp.
4   Check the automatic trip for moldboard bottoms.
5   On disk plows, check hitch arrangement for proper performance of the plow.
6   Store in off-season in a shed, on concrete slab, planks, or blocks.
7   Remove wheels with rubber tires and store in shed.
8   Tag wheels for plow and location on plow.
9   Apply rust-resistant grease or heavy oil to bottoms.
10   Make list of repairs needed, if any, and place order for replacements.
11   Cover hydraulic hose ends with plastic, tied securely.
12   Cover with plastic or canvas sheet, tied securely.

REFERENCES

American Society of Agricultural Engineers Data: Crop Machines Use; I. Draft and Power Requirements of Crop Machines, *Agr. Engin. Yearbook*, p. 69, 1954.

ASAE Standards: Three-point Three-link Attachment for Hitching Implements to Agricultural Wheel Tractor, *Agr. Engin. Yearbook,* 8th ed., p 75, 1961.

Better Plowing with Moldboard Plows, International Harvester Educational Series 4.

Browning, G. M.: Principles of Soil Physics in Relation to Tillage, *Agr. Engin.,* **31**(7):341-344, 1950.

Clyde, A. W.: Mechanics of Farm Machinery, series of articles in *Farm Impl. News,* Jan. 6 to Mar. 16, 1944.

——: Cushion Hitch Developments, *Agr. Engin.,* **30**(4):169-171, 1949.

Crop Yields as Related to Depth of Plowing, *S. Dak. Agr. Expt. Sta. Bul.* 369, 1943.

Farming Practices, Allis-Chalmers Manufacturing Company, Milwaukee, 1947.

Ferguson, I. M., and H. H. Ramsour: *One-way Disk Plow Adjustment,* Kansas Engineering Extension Department Farm Machinery Series 4, 1950.

Field, C. G.: Do You Really Need to Plow? *Country Gent.,* **123**(3):35, 1953.

Goryachkin, V. P.: Collected works in three volumes, vol. II. Translated from Russian. Published for the U.S. Dept. of Agr. and the National Science Foundation, 1972.

Heith, D. C.: The Kinematics of Tractor Hitches, *Agr. Engin.,* **33**(6):343-346, 1952.

Hull, Dale O.: Three-Point-Hitch Plow Adjustment, *Iowa Agr. Ext. Ser. Pham.* 219, 1955.

——: Tractor Plow Adjustment and Operation, *Iowa State Col. Ext. Serv. Bul.* P95, 1949.

Hume, A. N.: Depth of Plowing and Crop Yields, *S. Dak. Agr. Expt. Sta. Bul.* 344, 1940.

James, Paul E., and Dale E. Wilkins: Deep Plow: An Engineering Appraisal, *Trans. of the ASAE,* **5**(3):420, 1972.

Kaufman, L. C., and D. S. Totten: Development of an Inverting Moldboard Plow, *Trans. of the ASAE,* **15**(1):55, 1972.

Lah, Radheg: Measurement of Forces on Mounted Implements, *Amer. Soc. Agr. Engin. Trans.,* **2**(1):109, 1959.

Martin, E. B., and B. T. Stephanson: One-way Disc Maintenance and Operation, *Univ. of Alberta and Alberta Dept. Agr. Pub.* 6, 1948.

McCreery, W. F., and M. L. Nichols: The Geometry of Disks and Soil Relationships, *Agr. Engin.,* **37**(12):808-812, 1956.

Mohsein, Nuri, et al.: Wear Tests of Plowshare Materials, *Agr. Engin.,* **37**(12):816, 1956.

Nichols, M. L., I. F. Reed, and C. A. Reaves: Soil Reaction to Plow Share Design, *Agr. Engin.,* **39**(6):336-339, 1958.

——, and C. A. Reaves: Soil Reaction to Subsoiling Equipment, *Agr. Engin.,* **39**(6):340-343, 1958.

Promersberger, W. J., and S. L. Vogel: Multiple Hitches for Large Tractors, *N. Dak. Agr. Ext. Ser. Cir.* AE-45, 1953.

Seim, Dick: Big Yields without Plowing, *Farm Journal,* **97**(8):18, 1973.

Tanquary, E. W., and A. W. Clyde: New Principles in Tractor Hitch Design, *Agr. Engin.,* **38**(2):88-93, 1957.

Technical Features of Tillage Tools, *Pa. Agr. Expt. Sta. Bul.* 465, part 2, 1951.

Vogel, S. L.: Hitching Moldboard Plows, *N. Dak. Agr. Ext. Ser. Cir.,* AE-56, 1957.

QUESTIONS AND PROBLEMS

1   Define primary tillage and list the types of equipment used.
2   Draw sketches of the various kinds of furrows and name their parts.
3   Name the types of moldboard plow bottoms and explain their differences and uses.
4   Discuss the functions, advantages, and disadvantages of (a) two-way plows and (b) middlebreaker plows.
5   Discuss the advantages and disadvantages of trailing and integral-mounted plows.
6   Discuss the various factors involved in the design of moldboard plow bottoms.
7   Determine the acres plowed per hour when a tractor is operating at 3.5 mi/h (8.8 km/h) and is pulling four 14-in (35.6-cm) moldboard bottoms at a depth of 5.5 in (14.0 cm). What will be the total draft in pounds pull, if there is a draft of 8 lb/in^2? How many acres can be plowed in 10 hours when the field efficiency is 78 percent?
8   Discuss the conditions under which it may be advantageous to use a disk plow.
9   Explain the differences in design of the one-way and the regular disk plows.
10 Discuss the use of rotary and chisel plows.
11 Explain the vertical and horizontal principles involved in hitching a trailing moldboard plow equipped with four 14-in (35.6-cm) bottoms to a tractor on which the inside walls of the tires are 67 in (1.7 m) apart.
12 Discuss the principles of the three-point hitch for integral-mounted plows.
13 Explain why the center of resistance is not always a fixed point on a moldboard plow bottom.
14 Define vertical and horizontal lines of draft.
15 Define side draft.

Chapter 10

Secondary Tillage Equipment

The term *secondary tillage* as used in this discussion means stirring the soil at comparatively shallow depths. In many cases secondary tillage follows the deeper primary-tillage operation. It is possible to use some of the primary-tillage tools to do secondary-tillage operations. For example, the one-way plow and certain types of chisel plows can be adjusted and equipped with attachments to till the soil at shallow depths.

The general objectives of secondary tillage are stated as follows:

1   To improve the seedbed by greater pulverization of the soil
2   To conserve moisture by summer-fallow operations to kill weeds and reduce evaporation
3   To cut up crop residue and cover crops and mix vegetable matter with the topsoil
4   To break up clods, firm the topsoil, and put it in better tilth for seeding and germination of seeds
5   To destroy weeds on fallow lands

There are many types of machines that can be used for secondary tillage. They are the various types of harrows, rollers, and pulverizers, and tools for mulching and fallowing.

HARROWS

A *harrow* is an implement used to level the ground and crush the clods, to stir the soil, and to prevent and destroy weeds. Under some conditions, harrows can be used to cover seeds. There are three principal kinds of harrows, namely, the disk, the spike-tooth, and the spring-tooth.

Research by Rogin[1] revealed that "an early form of the rotary harrow, known as the Nishwitz harrow, aroused considerable interest soon after the Civil War." This harrow consisted of slightly concave disks mounted on separate standards to an A-shaped frame. The disks on each leg of the frame threw the soil outward. "It was apparently not until the late seventies, when the LaDow and the Randall disk harrows appeared, that this type of implement attained considerable vogue." Disk harrows were largely made in blacksmith shops up until about 1880, when the Keystone Manufacturing Company of Sterling, Illinois, started factory production. The cutaway disk was introduced in the early nineties.

The spring-tooth harrow was patented in 1877, and two factories started making such harrows in 1878. Homemade types of peg-tooth or straight-tooth square and triangular harrows were mentioned as being used as early as 1790.

Disk Harrows

Uses   The disk harrow is adapted to a wide variety of uses in many types of farming practices and management. Some of these uses are enumerated as follows:

1   It is used before plowing to cut up vegetable matter that may be on the surface, such as cornstalks, cotton stalks, and weeds, and to pulverize the top of the soil to such an extent that the furrow slices will make better connection with the bottom of the furrow soles, preventing air spaces when slices are turned. Disk harrows are frequently used in combination with crop-residue shredders to mix the shredded material into the soil.

2   It is used after plowing to pulverize the soil and put it in better tilth for the reception of the seed. Oftentimes, land plowed in the fall will need disking in the spring. This will save replowing and put the soil in the best possible condition for spring seeding.

3   It puts all plowed ground in condition for spring planting.

4   It is used for the cultivation of crops.

5   It is used for summer fallowing.

6   When seeds are sown broadcast, it is used to cover them.

Types   Disk harrows are available in sizes suitable for any size tractor. There are many types of disk harrows, but they can be divided into two general classes, trailing and mounted. Figs. 10-1 and 10-2 show several gang arrangements.

---

[1] Leo Rogin, *The Introduction of Farm Machinery*, University of California Press, Berkeley, Calif., 1931.

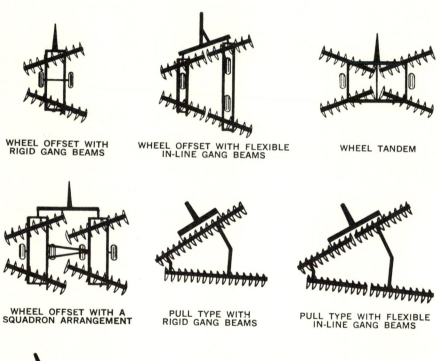

Single-action   Double-action or tandem   Offset

Fig. 10-1   Three types of disk harrows according to gang arrangement.

WHEEL OFFSET WITH
RIGID GANG BEAMS

WHEEL OFFSET WITH FLEXIBLE
IN-LINE GANG BEAMS

WHEEL TANDEM

WHEEL OFFSET WITH A
SQUADRON ARRANGEMENT

PULL TYPE WITH
RIGID GANG BEAMS

PULL TYPE WITH FLEXIBLE
IN-LINE GANG BEAMS

PULL TYPE
SQUADRON

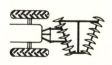

TRACTOR MOUNTED
OFFSET

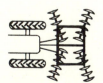

TRACTOR MOUNTED
TANDEM

Fig. 10-2   Various arrangements of disk harrow gangs. (*Towner Manufacturing Company.*)

Trailing Disk Harrows   Trailing disk harrows are hitched to the drawbar of the tractor and pulled by the tractor.

Single-action trailing disk harrows consist of two gangs of disks placed end to end which throw the soil in opposite directions (Fig. 10-1). The cutting width for single-action disk harrows may range from 4 to 20 ft (1.2 to 6.0 m). Sizes wider than 12 ft (3.6 m) have end sections that fold over on the main harrow so the harrow can pass through gates (Fig. 10-3). The cutting width may exceed 20 ft (6.0 m).

The *double-action* disk harrow is often called a *tandem harrow* because a set of two gangs follows behind the front gangs and is arranged so that the disks on the front gangs throw the soil in one direction (usually outward) and the disks on the rear gangs throw the soil in the opposite direction (Figs. 10-1 and 10-2). Generally, the trailing-type disk harrow cannot be lifted off the ground for turns, but the harrow shown in Fig. 10-4 is equipped with transport wheels and a remote-control hydraulic lift. The wheels may also serve as gage wheels. The sizes in width of cut may range from 5 to 25 ft (1.5 to 7.6 m).

The *offset* disk harrow is given this name because the harrow can be operated in offset positions in relation to the tractor (Fig. 10-5). A change in the hitch can cause the harrow to run to either the right or left of the tractor. Thus, it is possible to operate a harrow under limbs, near trees in an orchard, while the tractor runs out beyond the limbs. The offset disk harrow is becoming popular for straight-field disking, because the tandem-arranged gangs of disks give a double disking and leave the soil smoother than is usually obtained with a four-gang double-action harrow. Offset harrows are available in sizes ranging from $4\frac{1}{2}$ to 30 ft (1.3 to 9.1 m). Rubber-tired transport wheels (Fig. 10-6) are available for many makes. These carrier wheels are used to transport the harrow to and from fields and across grass waterways and may serve as depth gages.

Special heavy-duty disk harrows are built heavy enough to chop up brush on pasture lands and excessively heavy crop residues in sugar-cane fields. They are

Fig. 10-3   Double-action or tandem disk harrow equipped with folding wings or outriggers for extra width. (*Deere & Co.*)

Fig. 10-4   Tandem disk harrow with lifting wheels. (*Massey-Ferguson, Ltd.*)

suitable for deeper harrowing of heavy soils. One company lists a heavy-duty disk harrow as follows: width of cut 13 ft (3.9 m), 24 disks 28 in (71.1 cm) in diameter, total weight 8900 lb, requiring a tractor having 110 to 130 drawbar hp.

Heavy-duty harrows are usually equipped with cutaway or notched-edge disk blades (Fig. 10-7). They are available in both single- and double-action types.

Mounted Disk Harrows   This type of disk harrow is designed to be used with tractors equipped with three-point hitch and hydraulic lift systems.

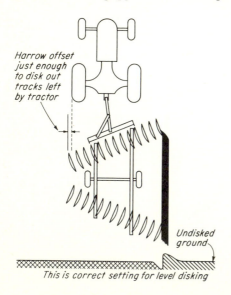

Harrow offset just enough to disk out tracks left by tractor

Undisked ground

This is correct setting for level disking

Fig. 10-5   Correct hitch for an offset disk harrow. (*J. I. Case Company.*)

Fig. 10-6 Trailing offset pull-type disk harrow available in various widths. (*International Harvester Company*.)

Different manufacturers have given the mounted disk harrow different names, such as *direct-connected, pickup*, and *lift-type*. As the mounted disk harrow is generally a unit assembly with fixed-angle construction, it can be lifted with hydraulic power lifts, backed into corners, and used close to fences and along ditches or borders. The hydraulic control permits adjustment of the disks to desired depths (Fig. 10-8).

Most of the regular two-gang single-action mounted disk harrows are of the heavy-duty type.

The conventional four-gang (double-action) tandem power-lifted disk harrow has a fixed-angle construction (Fig. 10-8). It is available in widths ranging from 5 to 11 ft (1.5 to 3.3 m). There is sufficient flexibility in the hitch not to interfere with the steering of the tractor.

The *offset* double-action mounted disk harrow consists of two gangs of disks arranged in a fixed-angle frame so that one gang runs behind the other (Fig. 10-6). The rear gang may be set to run directly behind the front gang, giving a double disking to the soil, or it can be adjusted sidewise to run in an offset position to the right or to the left as desired. Adjusting brackets are available for leveling the harrow sidewise and for increasing or decreasing pressure on either side of the harrow. The harrow can also be adjusted to regulate the penetration of the rear gang. Mounted offset harrows are available in widths up to 21 ft (6.3 m).

Component Parts of Disk Harrows   A disk harrow consists of a number of units or component parts, such as the disks, disk gangs, frame, standards, bearings, bumpers, scrapers, weight boxes, and leveling devices.

Disks   Round, smooth-edged disks (Fig. 10-6) are used on most disk harrows. Special harrows are equipped with disks having a *cutaway, notched*, or

Fig. 10-7   Heavy-duty offset disk harrow equipped with cutaway or notched-edge disk blades. (*Towner Manufacturing Company*.)

Fig. 10-8   Tandem disk harrow available in width sizes from $25\frac{1}{2}$ to 30 ft (7.8 to 9.1 m). (*J. I. Case Company, Cat. A76072*.)

scalloped edge (Fig. 10-7). Where there is much residue to be cut, cutaway disks are recommended for the front gangs and round disks for the rear gangs. Disk blades for harrows range from 16 to 28 in (40.6 to 71.1 cm) in diameter. The 18- to 24-in (45.7- to 61.0-cm) sizes are popular for regular farm use. Heavy-duty disk harrows are equipped with disks ranging from 26 to 28 in (66.1 to 71.1 cm) in diameter. Disks for harrows are made of high-grade heat-treated steel.

Disk Gangs   Gangs for disk harrows consist of three to thirteen disk blades assembled on a long *gang bolt*, or *arbor bolt*, which is generally square. The spacing between disks ranges from 6 to 9 in (15.2 to 22.9 cm) for light-duty harrows and from 10 to $12\frac{1}{2}$ in (25.4 to 31.8 cm) for heavy-duty harrows. The disk blades are held an equal distance apart by a spool-shaped casting.

Harrow Frame   Each gang of disks has a strong rectangular or tubular frame supported above the gang by standards that rest on axle bearings. The gang frames of a double-action harrow are connected by a linkage arrangement that, in most cases, permits adjustment of the angle of the gangs to obtain varying degrees of soil penetration. The pull or hitch bar is attached to the frames of the two front gangs. The unit-lift or pickup harrows may have a rigid frame construction to hold all gangs in a fixed position.

Bearings   Lightweight disk harrows generally have two bearings per gang, while heavy-brush and bog harrows may have several bearings per gang. These bearings consist of a specially designed spacer spool around which is bolted a malleable cast-iron housing. The standards for the frame and the pull bars are attached to the bearing housing. A wood bushing can be used between the spool and the housing. The spool on many harrows is made of white, chilled cast iron and serves as the moving part of the bearing. Some harrows are equipped with antifriction bearings. These may have pressure lubrication fittings or may be seal-packed for their lifetime.

Bumpers   Where two gangs of disks are set to throw the soil outward, as shown in Fig. 10-1, the sidewise forces acting on the disks as they are drawn forward tend to force the gangs toward each other. A half-moon-shaped cast-iron plate weighing several pounds is placed on the convex side of the disk at the end of the gang. Rather than absorb all the sidewise pressures in the frame, standards, and bearings, the gangs are allowed to bump together against the bumper plates. Bumper plates can be placed on the outer ends of the rear gangs of double-action disk harrows to prevent the head of the gang bolt hanging and snagging on posts and trees. The weight of the bumpers also aids in holding the gang level for uniform penetration from end to end of the gang.

Scrapers   Scrapers are placed on the disk harrow to clean and remove soil that may stick to the concave side of the disk blades.

Weight Boxes   A boxlike framework is often provided on the frame so that weights can be placed on the harrow gangs. In some cases, specially shaped iron weights can be attached to the harrow frame. Where the gangs are long, the weight should be added near the bumper end of the gang to offset the tendency for that end of the gang to be forced upward by the sidewise forces acting on the gang.

Leveling for Even Penetration   Clyde[1] analyzes the forces acting on a disk-harrow gang as follows:

[Figure 10-9] shows how forces S and V act on a gang as viewed from the rear. S is balanced by an equal side thrust, S', applied either at the bumper or at the bearings. It is evident that the pair of forces, S and S', tends to lift the bumper end and to force the other end down. To prevent that, the net weight on the gang must be offset from V. In practice this is done by applying most of the frame's weight near the bumper end. Sometimes the frame must borrow weight from the concave end of the gang (pull up on it) in order to hold down hard enough on the bumper end. Thus, there is a natural tendency for the bumper end of the gang to run shallow. Uneven penetration causes uneven wear of the disks.

In offset disk harrows the basic principle is that the side force against the front gang is opposed by the side force of the rear gang, these forces being out of line. To balance them, the pulling forces must be out of line with the backward forces on the gangs. Another method of getting the same results is to combine the backward and side forces into one force, RH, on each gang, as in [Figure 10-10]. The pull balances these two, hence they are placed with their arrows at their junction H, making the parallelogram, and get the diagram PX, which is the pull needed. H may be called the center of resistance in the same sense as used with the plow. The numbers in [Figure 10-10] are shown for 22-inch (55.9-cm) disks and the angles are chosen so that the S of one gang equals the S of the other. This requires that the rear gang be angled more than the front, because the rear has less net weight acting on its disks.

Leveling devices to give even penetration of gangs on double-action disk harrows may consist of *tension springs, turnbuckles, snubber blocks*, and *hold-down bars.*

Level-Action Disks   A recent development in the design of disk harrows permits level disking. A natural line of draft hitching places the entire weight of the harrow on the disk blades. The front and rear gangs cut the same depth, so that the soil thrown out by the front gangs can be pulled in by the rear gangs, thus leaving the soil level.

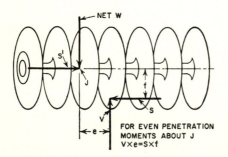

FOR EVEN PENETRATION
MOMENTS ABOUT J
$V \times e = S \times f$

Fig. 10-9   Moments of forces affecting the even penetration of a regular disk-harrow gang. For even penetration the net W (total weight minus any upward pull) must act off center. The number of disks in the gang has no effect on the distance e. (A. W. Clyde, University of Pennsylvania.)

[1] A. W. Clyde, Mechanics of Farm Machinery, *Farm Impl. News*, January-March 1944.

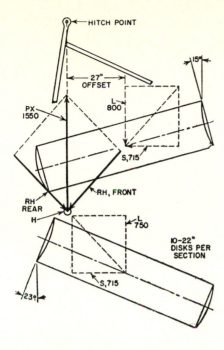

HITCH POINT

15°

27" OFFSET

L 800

PX 1550

S,715

RH, FRONT

RH REAR

H

L 750

10-22" DISKS PER SECTION

S,715

23°

Fig. 10-10 Moment of forces acting on the gangs of an offset disk harrow. (*A. W. Clyde, University of Pennsylvania*)

Soil Penetration of Disk Harrows   There are many factors within the harrow itself that will influence the depth to which it will penetrate the soil. They are enumerated as follows:

1   The angle of the disk gang
2   The weight of the harrow
3   The sharpness of the disks on the gangs
4   The size of the disks
5   The concavity of the disks
6   The angle of the hitch

All these factors are incorporated within the harrow itself. Other factors, however, that influence the depth of penetration have nothing to do with the harrow, such as the condition of the soil, the amount of moisture, whether plowed land or unplowed land, the amount of trash on the soil, and the amount of organic matter that may be in the soil.

Power Angling of Harrow Gangs   Some trailing, double-action, and offset disk harrows are equipped with remote-control, double-action hydraulic cylinders. The hydraulic cylinder is used both to angle the gangs for operation and to straighten the gangs for turns, crossing of grassways in fields, and transportation.

Draft of Disk Harrows   Richey[1] gives the draft and power requirements of a single-action disk harrow as 40 to 130 lb (18.1 to 51.0 kg) per ft of width,

---

[1] American Society of Agricultural Engineers Data: Crop Machines Use; I. Draft and Power Requirements of Crop Machines, *Agr. Engin. Yearbook*, p. 69, 1954.

and of a double-action tandem disk harrow as 80 to 160 lb (36.2 to 72.5 kg) per foot of width. The draft of a double-action harrow equipped with 22-in (55.9-cm) disks and spaced 9 in (22.9 cm) was given as ranging from 170 to 225 lb (77.0 to 101.9 kg) per foot of width or 90 percent of the weight of the harrow.

Tests conducted by Promersberger and Pratt[1] with a tandem disk harrow cutting a depth of 2 to 3 in (5.1 to 7.6 cm) in loam and clay fallow soils gave a draft of 57 to 150 lb (25.8 to 68.0 kg) per foot of width. The average hp-h per acre was 1.2 to 3.3. When the harrow was operated at a depth of 4 to 6 in (10.2 to 15.2 cm), the draft per foot of width ranged from 148 to 182 lb (67.1 to 82.6 kg). The average hp-h per acre was 3.3 to 4.0. An offset orchard harrow cutting a depth of 5 to 7 in (12.7 to 17.8 cm) gave a draft of 195 to 258 lb (88.3 to 116.9 kg) per foot of width. The average hp-h per acre was 4.3 to 5.7. The draft increased as the depth was increased.

Spike-Tooth Harrows

A typical spike-tooth harrow is shown in Fig. 10-11. It is commonly called a spike-tooth harrow because the teeth that stir the soil resemble long spikes. This harrow is also known as a *peg-tooth harrow*, a *drag harrow*, a *section harrow*, and a *smoothing harrow*. Its principal use is to smooth and level the soil directly after plowing. It will stir the soil to a depth of about 2 in (5.1 cm) if weighted. It can be used to cultivate corn and cotton and other row crops in early stages of growth.

The sections may range in width from 4 to 6 ft (1.2 to 1.8 m) and may have 25, 30, or 35 teeth. Several sections can be attached to a hitch bar and a wide swath harrowed. The sections may be either rigid or flexible.

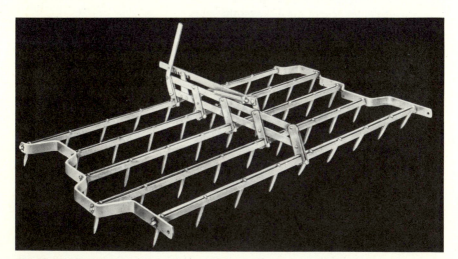

Fig. 10-11   Spike-tooth harrow section. (*International Harvester Company*.)

---

[1] W. J. Promersberger and G. L. Pratt, Power Requirements of Tillage Implements, *N. dak. Agr. Expt. Sta. Bul.* 415, 1958.

Sections that have guard rails across the ends of the bars are called *closed-end* harrows while those that do not have guard rails are called *open-end* harrows.

Spring-Tooth Harrows

Figure 10-12 shows a trailing or pull-type spring-tooth harrow. Its general appearance is similar to that of the spring-shank chisel plow, but the harrow tills the soil at a shallow depth. Trailing spring-tooth harrows are available in sizes ranging from 8 to 36 ft (2.4 to 10.9 m).

Tractor-mounted types of spring-tooth harrows are available in assembly units that can be lifted with hydraulic power.

Spring-tooth harrows are adapted for use in rough and stony ground. They are also used extensively to loosen previously plowed soil ahead of a grain drill seeding rice or small grains. The teeth will penetrate deeper than those on spike-tooth harrows, and they will give when obstructions are struck. The spring-tooth harrow is frequently advertised as a *quack-grass* and *Bermuda-grass eradicator*, since the teeth penetrate deeply and tear out the roots and bring them to the surface. Alfalfa sod is also cultivated with spring-tooth harrows. The teeth consist of wide, flat, curved, oil-tempered bars of spring steel, one end of which is fastened rigidly to a bar; the other end is pointed to give good penetration. The depth to which the teeth will penetrate the soil is controlled by adjusting the angle of the teeth by means of levers, as in the case of the spike-tooth harrow. Some spring-tooth harrows are provided with a power-angling hitch.

Harrow teeth for spring-tooth harrows are available with points of various widths and shapes and with detachable points for different types of work (Fig. 10-13). A rake-bar smoothing attachment is available, if desired, for smoothing the soil behind the spring-tooth harrow. The draft of a spring-tooth harrow may range from 75 to 150 lb (34.0 to 68.0 kg) per foot of width, depending upon the type of work and soil conditions.

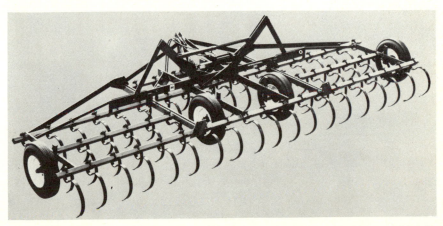

Fig. 10-12   Trailing or pull-type spring-tooth harrows. The outer sections can be folded for transportation. (*International Harvester Company*.)

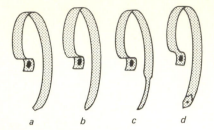

a   b   c   d

Fig. 10-13   Types of teeth on spring-tooth harrows: (*A*) regular; (*B*) quack grass; (*C*) alfalfa; (*D*) detachable point.

Special Harrows

A harrow that acts on the soil in a special manner is the weeder-mulcher.

Weeder-Mulcher   Weeders with coil-spring teeth are excellent tools for making a mulch, for breaking the soil crust over germinating seeds, and for controlling and destroying young weeds just after the field-crop plants have begun to grow. Fig. 10-14 shows a tractor-mounted weeder-mulcher. Fig. 9-16 shows mulcher attachments for a moldboard plow.

LAND ROLLERS AND PULVERIZERS

Land rollers or pulverizers are tools used for the further preparation of the seedbed. They can be divided into two classes according to the kind of work they do, the *surface packer* and the *subsurface packer.*

Surface Packers

There are several different kinds of commercial surface packers and pulverizers, named according to the shape of the roller surface: the V-shaped roller-

Fig. 10-14   Mounted weeder-mulcher with coil-spring teeth (*International Harvester Company.*)

pulverizer, the combination T-shaped and sprocket-wheel pulverizer, and the flexible sprocket-wheel pulverizer. The subsurface packers consist of a V-shaped packer and the crowfoot packer.

The surface roller is coming into more general use each year because it has a varied number of uses. The most important is as a clod crusher; at this it has no equal. Another very important use is to finish preparing the seedbed by thoroughly pulverizing and firming the loose soil so that there will not be any large air spaces or pockets. It presses the upper soil down against the subsoil, making a continuous seedbed in which moisture is conserved and given to the roots of the plants as it is needed. Also, when meadow, wheatland, and pasture land have heaved badly from freezing, the land roller is good to press the soil back down around the roots.

V-shaped Roller-Pulverizer The machine shown in Fig. 10-15 is a roller-pulverizer constructed of a number of wheel sections, so that when they are strung on a shaft the surfaces of the rollers form a kind of corrugation. It is from the shape of the surface it leaves that it gets the name *corrugated* roller. Each wheel or section is about 5 to 6 in (12.7 to 15.2 cm) thick and varies in diameter from 10 to 18 in (25.4 to 45.7 cm). The roller is hollow. It may consist of one or two pieces and is cast out of semisteel. When placed upon a shaft and rolled across the soil, it leaves small ridges. If only one set of rollers is used, these ridges will be rather large, being 5 to 6 in (12.7 to 15.2 cm) from one crown to the other. The common method, however, is to use a rear set of rollers so arranged that they will split the ridge made by the front pair, leaving a number of very fine ridges. It is claimed that this type of roller to a certain extent will prevent wind erosion. It also rolls, pulverizes, packs, levels, cultivates, and mulches the soil in one operation. Figure 10-16 shows a roller-pulverizer spring-tooth harrow combination. The front and rear roller gangs are spaced a few feet apart, and spring-harrow teeth are mounted between the sections to harrow the soil and bring clods to the surface so they can be crushed by the rear roller gang.

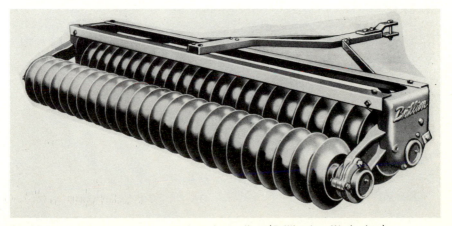

Fig. 10-15 Land roller with V-shaped cast-iron rollers. (*Brillion Iron Works, Inc.*)

Fig. 10-16   Combination land roller with crow-foot roller in front, followed by spring harrow teeth and V-shaped rollers in rear. (*Brillion Iron Works, Inc.*)

Combination T-shaped and Sprocket-Wheel Pulverizer   This type of roller-pulverizer has alternate T-shaped and sprocket-wheel sections assembled on a shaft (Fig. 10-17). T-shaped wheels crush and pulverize the soil, while the sprocketlike wheels give a mulching effect, leaving loose pulverized soil on the surface.

Treader Packer   The packer shown in Figure 10-18 is what might be called a heavy-duty rotary hoe or skew packer. The packer wheels have heavy curved teeth that will penetrate the soil when the machine is operated with the points run forward. It serves as a packer or treader when operated with points run so that the rounded or back part of the points strike the soil first. When the wheels are run backward, they serve as packers and, if used on wheat stubble, will tread part of the straw into the soil. When the wheels are run forward, they will break up and pulverize soil and uproot weeds. The two gangs of rotary hoelike wheels are operated at an angle like an offset disk harrow (Fig. 10-18).

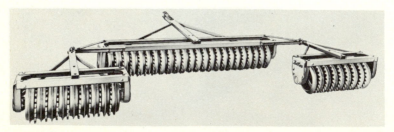

Fig. 10-17   Squadron hitch arrangement for land roller that has alternate T-shaped rollers and sprocket-wheel-type crushers. (*Brillion Iron Works, Inc.*)

Fig. 10-18   Treader-packer composed of two gangs of heavy rotary hoes set at an angle like an offset disk harrow. (*Deere & Co.*)

Subsurface Packer Rollers

It is often desirable to pack the subsurface of the soil. Special tools for doing this are called *subsurface packers*. They consist of a number of wheels with V-shaped rims strung on an axle with an overhead frame. The rims of the wheels setting close together, there is an interval of several inches between them. The rims of these wheels are also rather narrow. Their V shape allows them to go below the surface, pressing the soil together and leaving a good mulch on top. Another roller which can be classed as a subsurface packer is shown in Fig. 10-19. Because of the shape of the packer wheel rims it is called a *crowfoot packer*.

SUBSURFACE TILLAGE TOOLS AND FIELD CULTIVATORS

The need for control of wind and water erosion and for the conservation of moisture in the Great Plains region of the Middle West has brought about the development of new farming practices and new farm tools. The objective is to till the soil in such a manner that the crop residue will be left on the surface. This method of farming is known by several different names, such as *plowless farming, trash farming, stubble mulch, residue management, subsurface tillage*, and *minimum tillage*. In the opinion of the author, "subsurface tillage" appears to be the most appropriate name, as all the tools used in the operations stir and till the soil beneath the surface and under the trash.

The advantages listed for this method of farming are:

1   Increase in the capacity of the soil to absorb water
2   Reduction of runoff
3   Reduction of water and wind erosion
4   Reduction of rate of surface evaporation
5   Reduction of crop cultivation to kill weeds

Fig. 10-19 A subsurface crowfoot packer being used behind a three-bottom moldboard plow. (*Brillion Iron Works, Inc.*)

On the other hand, many farmers say that the stubble mulch is hard to handle because it makes it difficult to destroy undesirable vegetation, to develop a good seedbed, and to plant, and under some conditions it increases operation costs. Subsurface tillage requires considerable know-how and skill in operation of the tools used. Certain areas and soils are definitely not suited to this system of farming. For example, in the more moist areas, there is the danger of increased insect population.

Implements that operate under and do not materially disturb the trash on or near the surface are the most effective for subsurface tillage. The tools that meet these requirements are *sweeps* and *rod weeders.*

Subsurface Tillage Sweeps

To work under trash, sweeps must be set almost flat, mounted on strong, narrow standards that are staggered on the frame far enough apart to permit trash to flow around and between the standards. The complete tool is frequently called a *field cultivator* (Fig. 10-20). Tractor-drawn machines with subsurface tillage gangs and sweeps are provided with depth regulators and power lift and can be obtained in widths ranging from $5\frac{1}{2}$ to 15 ft (1.7 to 4.5 m), depending on the type of work to be done.

Where the tool is used for summer fallowing, the spring teeth have 6-in

Fig. 10-20 Wide sweeps are set flat to cut under surface vegetation and crop residue. The wings are operated hydraulically.

(15.0-cm) spacings, while the stiff standards have a 9-in spacing. If row crops are to be cultivated, the standards can be spaced to suit the row width. Trash must not be allowed to collect on the standards, to drag, or to cover small plants. To prevent this, large 16- to 22-in (40.6- to 55.9-cm) sweeps are used in connection with notched-edge rolling colters and concave-disk hillers. To prevent covering small plants, large rolling colters are used on each side of the rows to serve as fenders or shields.

Subsurface-Tillage Rod Weeders

Rod weeders are used extensively throughout the wheat-growing region of the Great Plains of the United States and Canada. They are used almost exclusively for controlling weed and voluntary vegetative growth on lands where summer fallowing is practiced.

The average rod weeder consists of a sturdy frame with four to five plowlike beams which have a shoe or slip nose on each point. Extending through a bearing in the shoes are round or square high-carbon-steel rods that revolve slowly. The rods are driven by a combination of sprocket chain and gear drive from one wheel. A lever or screw-type depth regulator is provided. The wheels on each end vary in diameter from 18 to 38 in (45.7 to 96.5 cm) depending on the width of the machine. The rod drive wheel is provided with lugs. The width of single units ranges from 8 to 12 ft (2.4 to 3.6 m), duplex units from 18 to 24 ft (5.4 to 7.3 m), and triplex units to 36 ft (10.9 m). In operation, the revolving rod runs a few inches beneath the surface, pulling up and destroying all vegetative growth. The rod revolves to prevent the roots of plants from hanging onto it. The front side of the rod moves upward to pull up and shed the roots of plants. Therefore, a reversing drive is provided.

Some farmers have substituted a sharp, heavy blade for the rod. This blade, fastened to middlebreaker beams, may be 8 to 10 ft (2.4 to 3.1 m) in length. It is also provided with an angling device for suction and penetration.

Rod-weeder attachments are available for chisel-type plows.

ROTARY HOE

When there is a large amount of crop residue on the surface in a fluffy condition, the rotary hoe, operated in reverse, is useful in packing the residue down into the surface soil. This treatment makes the use of field cultivators and rod weeders easier. See Chap. 12 for additional uses of the rotary hoe.

Table 10-1 shows the corn yields for eight methods of seedbed preparation for corn, where crop residues were used as a mulch.

Table 10-1   1949 Corn Yields for Eight Methods of Seedbed Preparation for Corn

Treatment prior to planting			Yield, bushels per acre	
Residue location	Depth of tillage, in	Implement used	Noble Co.* Miami silt loam	Throckmorton Farm & Carrington silt loam†
Left on surface	6 to 7	Lister bottom and spring-tooth cultivator in strip only	99	64
Left on surface	3	Sweeps	78	64
Left on surface	3 and 7	Sweeps	76	59
Mixed 0 to 3 in	3	Disk harrow	86	74
Mixed 0 to 6 in	6	Cover crop disk	79	79
Mixed 0 to 3 in	7	Special plow and disk	103	72
Under 4 in to 7 in	7	Ordinary plow	108	76
Mixed 0 to 3 in	7	Ordinary ‡ plow, special plow and disk	98	84

* Lowest significant difference, 5 percent, Noble County, 17.4 bu.
† Lowest significant difference, 5 percent, Throckmorton Farm, 14.5 bu.
‡ Ordinary plow in fall $1\frac{1}{2}$ in deep. Special plow in spring to mix the residue, 0 to 3 in with 7-in depth tillage.
Source: E. R. Baugh et al., *Agr. Engin.,* **31**(8):398-400, 1950.

MAINTENANCE AND CARE

1   Check equipment for proper hitching.
2   Check hydraulic units for proper operation.
3   On disk harrows, check gang angling system.
4   Store in off-season in shed on concrete slab, planks, or blocks.
5   Remove wheels with rubber tires and store in shed.
6   Tag wheels, listing implement and location on implement.
7   Apply rust-resistant grease or oil to disks and soil-wearing parts.
8   Make list of needed repairs.
9   Cover hydraulic hose ends with plastic tied securely.
10   If stored in yard, cover with plastic or canvas sheet, tied securely.

REFERENCES

Baugh, E. R., et al.: Some Results of Mulch Tillage for Corn, *Agr. Engin.*, **31**(8):398-400, 1950.

Bosworth, Douglas L., The Story Behind the John Deere Level-Action Disks, *Agr. Engin.*, **55**(4):14, 1974.

Bowers, Wendell, and H. P. Bateman: Research Studies of Minimum Tillage, *Amer. Soc. Agr. Engin. Trans.*, **3**(2):1-3, 1960.

Brarnes, K. K., and R. J. Rowe: Influence of Speed on Elements of Draft of a Tillage Tool, *Amer. Soc. Agr. Engin. Trans.*, **4**(1):55, 1961.

Cook, R. L., and F. W. Peikert: A Comparison of Tillage Implements, *Agr. Engin.*, **31**(5):211-214, 1950.

Harrold, L. L., and F. R. Dreibelbis: Machinery Problems in Mulch Farming, *Agr. Engin.*, **31**(8):393-394, 1950.

Hulburt, L. W.: Mulch Farming and Related Machinery Problems, *Agr. Engin.*, **31**(8):401-402, 1950.

Lillard, J. H., et al.: Application of the Double Cut Plow Principle to Mulch Tillage, *Agr. Engin.*, **31**(8):395-397, 1950.

Nutt, George B.: Machinery for Utilizing Crop Residues for Mulches, *Agr. Engin.*, **31**(3):391-392, 1950.

Poynor, R. B.: An Experimental Mulch Planter, *Agr. Engin.*, **31**(10):509-510, 1950.

Ryerson, G. E.: Machinery Requirements for Stubble-mulch Tillage, *Agr. Engin.*, **31**(10):506-508, 1950.

Tweedy, Robert H., Tillage, an Industry Viewpoing, *Agr. Engin.*, **55**(1):13, 1974.

PROBLEMS

1   Define and give the objectives of secondary tillage.
2   List the equipment used for secondary tillage.
3   Give the various uses of disk harrows, and make an outline listing the various types of disk harrows.
4   Explain the different structural and design features of trailing single-action, double-action, and offset disk harrows.
5   Compare the advantages of the wheel-lift trailing disk harrow and the tractor-mounted power-lifted disk harrow.
6   Explain how the various forces acting on a disk harrow gang can be balanced to obtain even penetration.
7   Explain the various uses of spike-tooth and spring-tooth harrows.
8   Explain the uses of the various types of land rollers and pulverizers.
9   Discuss the advantages of subsurface or stubble-mulch tillage, and describe the equipment used.

Chapter 11

Planting Equipment

The art of placing seed in the soil to obtain good germination and stands without having to replant is the goal of all who grow crops. There are a number of factors that influence the germination of seeds and the emergence of seedling plants. These are:

Quantity of seed planted
Viability of the seed
Treatment of the seed with chemicals to kill soil microorganisms
Uniformity of seed size
Planting depth
Type of soil
Moisture content of the soil
Type of seed-dropping mechanism
Uniformity of distribution of the seed
Type of furrow opener
Prevention of loose soil getting under the seed
Uniformity of coverage
Type of covering device
Degree of pressing and firming of the soil around the seed
Cleanliness and condition of the seedbed
Time of planting in relation to season
Temperature of the soil

Type of drainage
Condition of soil crusts
The good judgment, skill, and attention of the operator

History of Planter Development

Broadcasting seeds over the broken soil and covering them with some type of harrow was the common method of planting until about 1840. William T. Pennock of East Marlboro, Pennsylvania, was the first to start manufacturing grain drills,[1] although the first patent was granted to Eliakim in 1799.[2] The United States Census Report of 1880 estimated that about 53 percent of the wheat sown in 1879 was planted with grain drills.

The earliest type of row-crop planter was perhaps a wooden keg with holes around the center to permit seeds to drop out. A patent was granted to D. S. Rockwell in 1839 on a device for the planting of corn. About 1892, the Dooley brothers of Moline, Illinois, developed the edge-selection drop for corn planters. The check-row planter was patented by M. Robbins of Cincinnati, Ohio, in 1857.

The Dow Law cotton planter was developed about 1870.[3] The cell-drop and picker-wheel planting mechanisms for cotton planters were developed in the 1880s. The hill-drop attachment did not come into use until the 1920s. See Appendix Table 7 for dates of planter development.

Classification of Planting Equipment

Planting equipment is here considered to be any power-operated device used to place seeds, seed pieces, or plant parts in or on the soil for propagation and production of food, fiber, and feed crops. It is classified as follows:

Row-crop planters
    Trailing
        Drill
        Hill-drop
        Narrow-row
    Rear tractor-mounted
        Drill
        Hill-drop
        Transplanters or plant setters
Broadcast-crop planters
    Endgate seeders
    Narrow- and wide-track and weeder-mulcher
    Airplanes
Grain drills
Planting attachments for other equipment

---

[1] Leo Rogin, *The Introduction of Farm Machinery*, University of California Press, Berkeley, Calif., 1931.

[2] J. B. Davidson and L. W. Chase, *Farm Machinery and Farm Motors*, Orange Judd Publishing Co., Inc., New York, 1912.

[3] H. P. Smith and M. H. Byrom, *Tex. Agr. Expt. Sta. Bul.* 526, 1936.

ROW-CROP PLANTERS

Planters designed and constructed to plant seeds in rows far enough apart to permit cultivation of the crop are termed *row-crop planters*. Many row-crop planters are designed to plant seeds of only one certain crop, while others can be adapted to plant more than one crop by means of interchangeable hoppers, agitators, plates, and the speed-control mechanism of the seed-metering parts. Generally, row-crop planters can be divided into five classes, named according to the kind of crop the planter is specially designed to plant. The classes are corn; cotton; sorghum; vegetable, beet, and bean; and potato.

Equipment for placing growing plants or plant parts in the soil is called a *transplanter*.

Corn Planters

Corn planters are used in almost every state of the United States, the southern provinces of Canada, and all countries where corn is produced in sizable quantities. In the United States, they are used most extensively in the Corn Belt of the north Middle West region.

Trailing and tractor-mounted corn planters can be classified according to the manner in which the seeds are dropped: *drill* and *hill-drop*. *Lister planters* are drill planters designed to plant corn in listed furrows.

Trailing Drill Planters   Trailing drill corn planters are available in two-, four-, six-, and eight-row units. Where the contour of the land and soil conditions permit, there is a tendency toward the use of the larger multiple-row, flexible units. They reduce man- and tractor-power-hours required per acre and reduce costs (Figs. 11-1 and 11-2). Drill planters are used for corn because, when corn is harvested with the mechanical corn picker, drilled corn gives a more even flow of corn through the picker mechanism.

Fig. 11-1   Trailing or pull-type multiple-crop planter equipped with fertilizer, herbicide, and pesticide attachments. (*Deere & Co.*)

Fig. 11-2   Eight-row tractor-mounted planter. (*International Harvester Company*.)

Trailing Hill-Drop Planter   This type of corn planter has rotary valves in the boot to collect the seeds and drop them in hills at regular intervals along the row. The hills in one set of rows will not be in line across the rows. It is possible to easily attach fertilizer, herbicide, and pesticide units to trailing and rear-mounted planters, but not to front-mounted ones.

Central-mounted Planters   The central- or front-mounted planter consists of assembly units which can be quickly attached to or removed from the tractor. The front-mounted planter has become almost obsolete.

Rear-mounted Corn Planters   This type of corn planter has the same planting mechanism as the trailing-type corn planter. It is, however, directly connected to the tractor and is raised and lowered by the tractor hydraulic lift (Fig. 11-2). The planting mechanism is driven from the power takeoff.

Lister Corn Planters   Figure 11-3 shows a four-row direct-connected quick-attachable power-lift lister corn planter equipped with middlebreaker bottoms for planting in hard ground. The planter is regularly equipped with 14-in (35.6-cm) lister bottoms, disk coverers, and preemergence spray equipment. Shovel coverers can be used in place of the disk coverers. Loose-ground lister planters are equipped with runner double-disk furrow openers.

Component Parts and Accessories for Corn Planters   The accuracy of a planter depends upon the uniformity of kernels, shape of hopper bottom, speed of the plate, shape and size of the cells, and fullness of the hopper. A cone-shaped hopper bottom causes the seed to gravitate into the cells.

The yielding *cutoff* pawl (Fig. 11-4) acting under spring pressure pushes the extra kernels back as the cell passes under the plate cover, or it cuts them off from the cell and, at the same time, presses the kernel firmly into the cell. As the plate revolves to the point where the cell is over the seed tube, a yielding *knockout* pawl under spring pressure comes in contact with the kernel, knocking

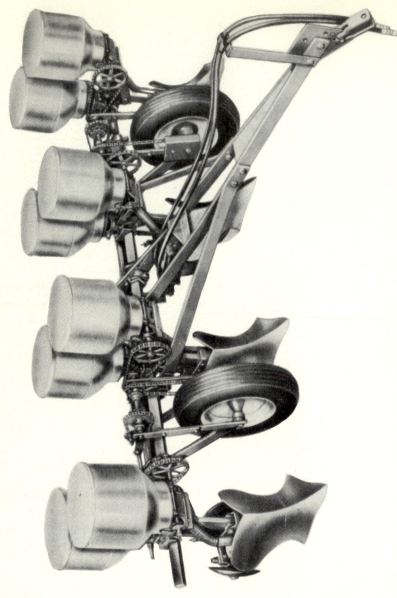

Fig. 11-3 Four-row trailing lister planter equipped with fertilizer attachment. (*International Harvester Company.*)

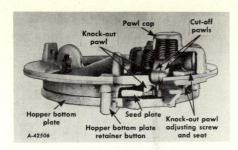

Fig. 11-4 Corn hopper bottom showing essential parts. (*International Harvester Company*.)

it through the cell into the seed tube, from where it is allowed to fall directly into the soil.

Four types of seed plates are used for planting corn, namely, the *edge-drop* and the *flat-drop*, which have the cells around the outer edge of the seed plate; the *flat-drop round-hole* type; and the *full-hill* plate. The *edge-drop* (Fig. 11-5) carries the kernel of corn on edge in the cell of the plate. The *flat-drop* (Fig. 11-5) carries the kernel flat in the cell of the plate.

The *full-hill plate* has cells around the outer edge large enough to admit several kernels at the same time. Sufficient kernels for one complete hill are dropped together (Fig. 11-5).

Kernels of corn do not vary greatly in thickness. They do, however, vary considerably in width and length. It is essential to select a plate having cells of sufficient thickness to prevent cracking the kernels as they pass under the cutoff cover plate. When the kernels are to lie flat in the cell, several plates are furnished, with cells adapted to small, medium, and large kernels. Both the edge-drop and the flat-drop plates do satisfactory work provided the size of the cell suits the size of the kernel. In each type, the corn should be graded to a uniform size. This is more important in the edge-drop than in the flat-drop.

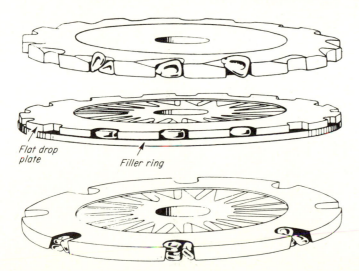

Fig. 11-5 Three types of seed plates for corn: (*top*) edge-drop; (*middle*) flat-drop; (*bottom*) full-drop.

Air-powered Metering System   This system consists primarily of a PTO-driven fan that delivers 10 oz of air pressure to the inside of a revolving drum having indented holes to hold seeds (Figs. 11-6 and 11-7). The company has given the system the trade name of CYCLO. In operation, seeds are fed into the bottom of a revolving seed drum. As the drum revolves, seeds are picked up by the holes or seed pockets in the drum (Fig. 11-7). The seeds are held in place by the 10 oz of air pressure. Just before the seeds reach the top of the revolving drum, the seeds pass a permanently mounted brush that knocks off any excess seeds. At the top of the revolution, the drum passes under air-cutoff wheels which momentarily close the holes and block the escaping air. The seeds are carried into the seed tube manifold. The air pressure then carries the seeds through the seed tubes into the furrow in the soil.

The drum usually has six rows of seed pockets or holes around the rim or circumference. There is an air-cutoff wheel for each row of holes so arranged that each row of holes can be blocked and two, four, or six rows can be planted. To change from corn or soybeans to milo, the drums are changed to provide holes of a size to suit the kind of seeds being planted. Figure 11-8 shows a six-row CYCLO planter.

The Cup-Pickup Seed-metering System   As shown in Fig. 11-9, seeds are fed from a hopper into a housing enclosing a revolving rim on the inside of which are pickup fingers or cups. These pickup fingers pick up seeds and carry them to an opening in the housing from where they are dropped into a seed tube.

Attachments for Corn Planters   Furrow openers are necessary to open furrowlike trenches in the soil for receiving the seeds as they are dropped by the mechanism of the planter. On planters, four types are used: the *cuved-runner*, the *stub-runner*, the *single-disk*, and the *double-disk*. The curved-runner type of opener is in most general use. The stub-runner is suited to rough and stony ground.

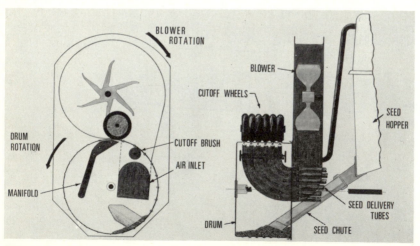

Fig. 11-6 Schematic view of the air-powered seed-metering system. (*International Harvester Company.*)

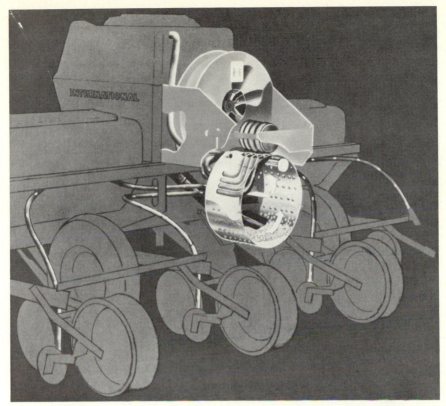

Fig. 11-7   Schematic view of the air-powered seed-metering system as mounted on a planter. (*International Harvester Company*.)

Fig. 11-8   Rear view of a six-row planter equipped with the air-powered metering system. (*International Harvester Company*.)

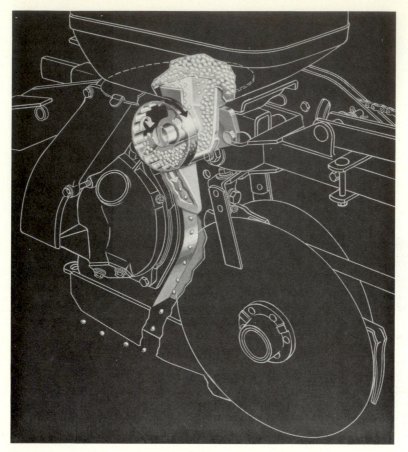

Fig. 11-9 Cutaway view showing revolving pickup cups that deliver seed to the seed tube. (*Deere & Co.*)

Various types of attachments are shown in Fig. 11-10. A furrowing and covering attachment is shown in Fig. 11-11. The first blades push away the rocks and clods, permitting the rear covering blades to scrape in a sufficient quantity of earth to cover the seeds.

Row markers are essential to keep the rows straight, parallel, and an equal distance apart (Fig. 11-1).

A fertilizer-distributing attachment can be mounted on any modern corn planter. The details of the types of feed and rate of distribution are discussed under "Fertilizing Equipment."

Plow-Plant and Minimum Tillage Attaching a soil packer and planter to the plow so that plowing and planting are done simultaneously is a growing practice by many farmers. This is what may be termed a minimum-tillage practice. Equipment is used to do as many jobs as possible in one operation.

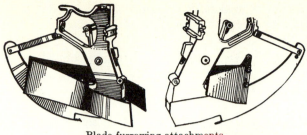

Blade-furrowing attachments

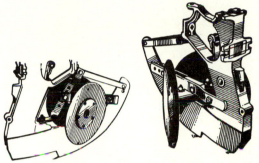

Nine-inch disks   Eleven- to thirteen-inch disks
Disk-furrowing attachments

Gage-shoe attachment

Fig. 11-10 Attachments for runner furrow opener on corn planter. (*International Harvester Company*.)

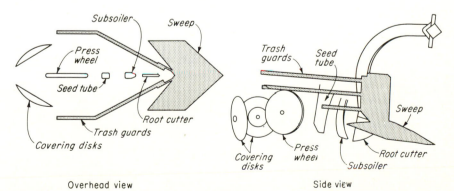

Overhead view   Side view

Fig. 11-11 Overhead and side views of a till-plant-type of lister planter developed in Nebraska.

Experiments have been conducted in Indiana[1] with equipment designed to plow, plant, fertilize, and apply preemergent chemicals for weed control in one operation.

Tests on 11 corn farms gave yields of 10.9 bushels more corn where the combined equipment was used in comparison with separate plowing, planting, fertilizing, and preemergent application.

Minimum Tillage   *Minimum tillage* is the minimum soil manipulation necessary for crop production or for meeting tillage requirements under the existing soil conditions.

In some areas farmers combine two or more operations into one, thus reducing the operations to a minimum. Where a rear-mounted planter is used on bedded land, a forward-mounted cultivator is used to cultivate the beds and kill weeds ahead of the planter. Fertilizer attachments, press wheels, and pre-emergent sprays are used with the planter. Figure 11-1 shows planter equipment that performs several operations at the same time.

Planting in the wheel tracks of the tractor is done largely to obtain a firm soil in which to place the seed to obtain moisture for germination. This is especially true where the farmer wants to plant in loose soil.

The combined use of equipment to do several jobs in one operation is limited only by the ingenuity of the farmer.

No-Tillage or Conservation Systems   The 1973 *ASAE Yearbook* defines *no-tillage planting* as a procedure whereby a planting is made directly into an unprepared seedbed. No-tillage is more often referred to as a *No-Til* system. It is also called *chemical fallow*. There are three general systems for no-tillage: (1) planting in unbroken land or a sod of bluegrass fescue or lespedeza; (2) planting in cover crops, such as small grains; (3) planting where the crop stubble or crop residue is on the soil surface (Fig. 11-12).

A fluted colter is generally used to open a narrow trench for the seeds. The narrow trench is closed by a suitable press wheel.

Weeds are largely controlled by the application of chemicals.

The advantages of no-tillage are listed as early planting of a following crop, reduced labor and machine costs, reduced danger of soil blowing, and soil and water conservation. Insect buildup is controlled by the application of chemicals.

Cotton Planters

Cotton planters are often termed *cotton and corn planters* as the seed-dropping or seed-metering mechanisms for cotton and corn are interchangeable for the same hoppers. Undelinted cottonseed is covered with a coating of short fiber. This lint fiber causes the seeds to cling together in the hopper, and they do not flow into the seed plate cells as do the hard, smooth seeds of corn. Therefore, the dropping mechanisms in the hopper must differ in design in order to handle

[1] C. M. Hansen, L. S. Robertson, and B. H. Grisby: Plow-Plant Equipment Designed for Corn Production, *Amer. Soc. Engin. Trans.*, 2:65-67, 1959.

Fig. 11-12 Planter designed to plant in stubble without previous seedbed preparation. (*Deere & Co.*)

the two types of seeds. Most cottonseed now used for planting is mechanically or chemically delinted.

Cotton is grown under climatic conditions that range from dry land conditions requiring irrigation to humid conditions receiving 70 to 80 in (1.77 to 2.01 mm) annual rainfall. The soil types range from light, sandy loams to heavy clay. There are many types of farming practices and methods of preparing the seedbed for the planting of cotton. These practices and methods vary greatly with the climatic conditions and the soil types. Consequently, cotton is usually planted on beds in the more humid regions and in the furrow in the subhumid areas. Farmers who have well-drained soils in the humid areas may plant on flat-prepared land. Some irrigated land is prepared flat for the planting of cotton.

These different planting practices require different types of planting equipment for planting on the bed, in the furrow, and on level, flat-prepared land. In an effort to furnish planters to meet these varied requirements, most manufacturers supply several models of planters. The principal differences in the models are in the manner of mounting the planter on the tractor, the equipment for making the seed furrow, and the method of covering the seed as required for the different farming practices.

Trailing Narrow-Row Cotton Planter   The narrow-row cotton culture requires planter units set 7 to 10 in (17.8 to 25.4 cm) apart (Fig. 11-13). Grain drills can be adjusted so that delinted cottonseed can be planted in narrow rows.

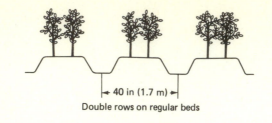

40 in (1.7 m)

Double rows on regular beds

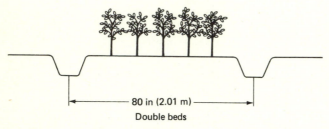

80 in (2.01 m)

Double beds

Fig. 11-13   Arrangements of narrow rows on beds. Plant rows are 7 to 10 in (17.8 to 25.4 cm) apart.

Rear-mounted Cotton Planters   This type of planter is available in two-, four-, and six-row sizes. Figure 11-14 shows a four-row rear-mounted cotton planter equipped to plant bedded land. Middlebreaker bottoms can be substituted for the sweeps to plant in the furrow. Rear-mounted planters are available with runner-knife seed furrow openers for planting land prepared flat. Hydraulic power lifts are used to raise and lower the equipment. Fertilizer herbicide and pesticide attachments can be used with this type of planter.

When the rear-mounted cotton planter is equipped with middlebreaker bottoms, hoppers, furrow openers, and covering devices so that the seeds can be planted in the furrow behind the bottom, it is termed a *lister planter*. When cotton is planted in the furrow, the land is usually relisted in the planting operation.

Cotton-dropping Mechanisms   Gin-run cottonseed is extensively used for planting. This is seed with lint adhering to it, just as it comes from the cotton gin. Delinted cottonseed is becoming popular in many sections. Two types of dropping mechanisms are used on cotton planters. They are the *cell drop* and the *picker-wheel drop*.

The Cell Drop   A typical cell drop is shown in Fig. 11-15. It consists of a plate with cells on the outer edge. As the plate turns, the agitators separate and stir the seeds, causing them to work down into the cells off a sloping collar and under feed springs which gently force more or less seeds into each cell. Then the yielding cutoff pushes back the surplus, and as the cells pass over an opening, a yielding spring-controlled knockout partially drops into the cell, forcing the seeds through the plate into the spout below. A small wheel with spurlike fingers projecting into each cell is also used as a knockout device.

Fig. 11-14 Front view of four-row cotton planter equipped with sweeps for planting on bedded land. It also has a fertilizer attachment. (*International Harvester Company.*)

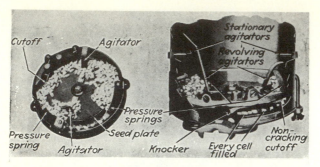

Fig. 11-15   Cell-drop cotton-dropping mechanism, showing the various parts.

With average-sized seeds, the quantity dropped by the cell drop ranges from 16 to 36 lb (7.2 to 16.3 kg) per acre. A pound of average-sized cottonseed will contain approximately 4500 seeds. Therefore if the planting rate is 30 pounds (13.6 kg) of seeds per acre, about 135,000 seeds are planted. Some authorities claim that under favorable conditions 70 percent of the seed planted will produce seedlings. If cotton is planted early, the expected percentage of seedlings may be reduced to around 50 percent. The quantity of seeds planted per acre is varied by changing the speed of the plate and by using plates which have different-sized cells.

The Picker-Wheel Drop   This type of drop is also called a *reverse-feed* drop because the agitator spider plate in the bottom of the hopper revolves in one direction while the small picker which is located under the outer edge of the hopper bottom revolves in the opposite direction (Fig. 11-16).

The quantity of seed is regulated by exposing more or less of the picker wheel to the seeds by means of a sliding-gate shutter. The quantity of average-sized cottonseed planted can be varied from about 14 to 90 lb (6.3 to 40.8 kg) per acre. The average planting rate ranges around 22 lb (10.0 kg) per acre. The standard weight of a bushel of cottonseed is 32 lb (14.5 kg).

Hill-Drop Mechanisms   The first hill-drop mechanisms used consisted of cells spaced at suitable intervals in the planter plate and large enough to hold

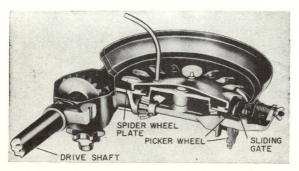

Fig. 11-16   Cross section of hopper bottom showing cotton plate and picker wheel in position. (*Deere & Co.*)

sufficient seeds for one hill, or of picker wheels having the notches in their surfaces so spaced as to drop the seeds in hills. These hill drops were located in the bottom of the planter hopper, and it was necessary for the seeds for each hill to fall from the hopper through the seed tube to the soil. In falling a distance of some 18 or 20 in (45.7 to 50.8 cm), the seeds became separated and scattered along the furrow to such an extent that it was difficult to distinguish one hill from another. Straight plastic seed tubes have less friction than steel, and there is less scattering of the seeds. Later, someone conceived the idea of placing a rotary valve in the lower part of the seed boot, low enough to the ground to prevent the seeds from scattering when they were dropped (Fig. 11-17).

Oates[1] found that cottonseed fell from the seed plate in a trajectory curve instead of straight down into the seed tube. The curvature of the trajectory changed when the speed of the plate was varied. A tapered seed tube $3\frac{1}{2}$ in (9.0 cm) at the top permitted bunches of seeds to fall a distance of 30 in (76.2 cm) without excessive scattering on a greased board. Bunches of free-falling cottonseeds from the hopper of a planter moving 4 to 5 mi/h (6.4 to 8.0 km/h) along the row are likely to scatter in the furrow because of kinetic energy.

Most cottonseed is hill-dropped by means of a rotary valve attached to the furrow opener. Rotary valves at the furrow permit hill-drop planting at high speeds.

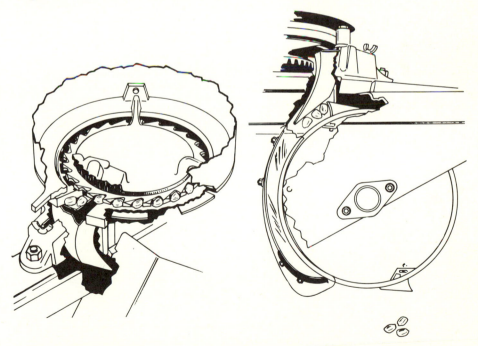

Fig. 11-17 Rotary-valve seed meter and drop: (*left*) seed grouped on accumulator plate; (*right*) shows seed dropping through seed tube just ahead of the rotor lug. (*International Harvester Company.*)

[1] W. J. Oates, *Okla. Agr. Expt. Sta.*, unpublished data.

Planting to a Stand   The term *planting to a stand* means that the rate of planting, or pounds of seed planted per acre, is selected to give the disired stand of plants with no thinning.

Research workers have collected considerable data as to the optimum plant population of various crops that will give the highest yield, low-cost weed control, and high machine-harvesting efficiency under different climatic and soil-fertility conditions. Under average conditions the plant population for cotton per acre should be between 40,000 and 50,000. For corn 15,000 plants per acre are considered to give optimum yields.

Planting seed should be tested for the germination percentage. It can not be expected that field conditions will give as high germination as was obtained in the germinator. Therefore, an allowance should be made in the number of seeds placed in the soil and the number increased to take care of the field mortality and poor field germinating conditions.

Attachments for Cotton Planters   The germination of cottonseed is greatly influenced by the type of *furrow opener* used. No loose soil should be permitted to fall into the furrow before the seeds come to rest in the furrow. The soil in the bottom of the furrow should be firm and moist. Tests conducted by the author[1] indicated that a narrow runner-type opener disturbed the soil to a lesser degree than shovel openers. Less loose soil is likely to fall into the furrow under the seeds. Generally, better stands were obtained with the runner opener, especially when a boatlike piece of metal was attached to the bottom of the runner in the notch just ahead of the point where the seeds drop between the side extensions of the runner (Fig. 11-18). This boatlike piece firms the soil in the bottom of the furrow. This type of furrow opener is called a modified runner opener.[2]

Wings attached to each side of a runner furrow opener (Fig. 11-19) aid in pushing extra soil, roots, and clods into the middle. This action permits and aids in the preparation of a drill for the application of pre- and postemergence herbicides.

Figure 11-20 shows a narrow, shovel-type furrow opener with rigidly attached soil shields and a narrow, soft-rubber-tired wheel to roll lightly on the seeds before they are covered. This is called a *seed press wheel*, as it rolls on the seeds before they are covered. The narrow furrow opener is used on lister planters to place the seeds in the furrow where the soil is moist. The ridges give some protection to the young seedlings against blowing sand.

Several kinds of *covering devices* are used on cotton planters. These include *open-center press wheels, shovels, disks*, and *scrapers*. The soil thrown over the seed by the covering devices should be pressed firm to hold the moisture in the soil around the seeds. Where the soil is sandy, steel open-center press wheels can be used as an attachment on the planter. A moist clay soil will usually stick to a

---

[1] H. P. Smith and M. H. Byrom, Effects of Planter Attachments and Seed Treatment on Stands of Cotton, *Tex. Agr. Expt. Sta. Bul.* 621, 1942.

[2] Rex. F. Colwick et al., Planting in the Mechanization of Cotton, *Southern Cooperative Series Bul.* 49, 1957.

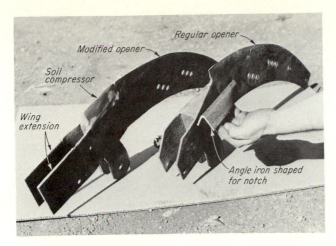

Fig. 11-18 Method of modifying a runner opener to firm the soil in the bottom of the furrow. (*Tex. Agr. Expt. Sta.*)

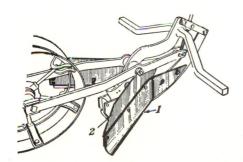

Fig. 11-19 Wings used on each side of a furrow opener to push sticks, clods, and extra soil to the middle and leave a level bed. A level bed is essential for the application of pre- and postemergence chemicals for weed control and for flame cultivation.

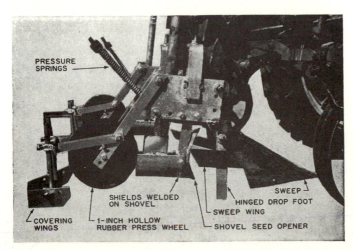

Fig. 11-20 A small rubber-tired wheel attached behind the furrow opener, to run on the seed before covering, compresses the soil around the seed to hold the moisture around it and thus aid germination in some areas. (*Tex. Agr. Expt. Sta.*)

Fig. 11-21 Two- and four-row seedbed rollers are used where the soil sticks to the steel press wheel of the regular planter. This type of roller is used after the soil dries. The method is termed *delayed rolling.* (*E. L. Caldwell & Sons.*)

steel wheel, and therefore, in this case, the press wheel is removed and the soil rolled in a separate operation with a two- or four-wheel roller (Fig. 11-21). This is termed *delayed rolling.* Figure 11-8 shows a planter equipped with a zero-pressure hollow-rubber tire on the press wheel. Where the soil is in good tilth for planting and not too moist, it does not stick excessively to the flexing rubber.[1,2]

Bed Shaping

The shaping of seedbeds has been associated with a wide variety of bed-forming equipment. It is the forming of seedbeds with precise dimensions in the horizontal and vertical planes. The implement used in forming the beds is basically a leveling and molding device which creates the desired profiles with smooth surfaces (Fig. 11-22). Forward components of the shaper accumulate soil from the high points along the sides and tops of the listed beds. Soil which is accumulated is then deposited in the low sections along the row, establishing beds with uniform profiles throughout the field.[3]

Fertilizer Attachments

These are available for most cotton and corn planters (Figs. 11-1 and 11-14). When granular fertilizer is applied as a part of the planting operation, the fertilizer should be placed in a band either on one side or on both sides of the seed row from 2 to $2\frac{1}{2}$ in (5.1 to 6.0 cm) to the side of and from $2\frac{1}{2}$ to 3 in (6.0 to 7.6 cm) below the seed level.[4,5,6]

[1] H. P. Smith and E. C. Brown, Mounting for Pre-emerge Press Wheel-rollers and Sprayer Nozzles, *Tex. Agr. Expt. Sta. Prog. Rpt.* 1520, 1952.

[2] ASAE standard—Agricultural Planter Press Wheel Tires, *ASAE Yearbook*, 1973.

[3] Lambert H. Wilkes and Price Hobgood, The Effects of Certain Precision Practices on the Efficiency of Cotton Production, *Tex. Agr. Expt. Sta. Bul.* 1074, 1968.

[4] H. P. Smith, M. H. Byrom, and H. F. Morris, Germination of Cottonseed as Affected by Soil Disturbance and Machine Placement of Fertilizer, *Tex. Agr. Expt. Sta. Bul.* 616, 1942.

[5] H. P. Smith, H. F. Morris, and M. H. Byrom, Machine Placement of Fertilizer for Cotton, *Tex. Agr. Expt. Sta. Bul.* 548, 1937.

[6] H. P. Smith and D. L. Jones, Mechanized Production of Cotton in Texas, *Tex. Agr. Expt. Sta. Bul.* 704, 1948.

Fig. 11-22   Cotton growing on a shaped bed. (*Tex. Agr. Expt. Sta.*)

Sorghum, Pea, and Peanut Planters

The corn and cotton planters can be used for the planting of sorghum, peas, and peanuts by removing the corn or cotton plates and putting in plates designed for the planting of the desired seeds

Hurlbut[1] found that more satisfactory seeding rates were obtained by using plates made especially for sorghum rather than by attempting to use a regular or revamped corn plate. He also found that the lower part of the plate seed hole should be taper-reamed to prevent sorghum seeds sticking in the hole and clogging it. A 15° bevel of the seed hole on the upper side helped to prevent the seeds from wedging between the sharp edge of the hole and the cutoff.

The number of seeds per pound varies with different varieties of sorghum, as shown in Table 11-1.

Beet and Bean Planters

The production of beets for sugar and of soybeans for oil and plastics has brought about new developments in planting equipment for these crops. Figure 11-23 shows a six-row beet and bean planter. Adjustments are provided to obtain a wide variety of row spacings—from 13 to 40 in (33.0 to 101.6 cm). The six-row machine can be adjusted to row spacings varying from 18 to 24 in (45.7 to 61.0 cm). When the machine is converted into a four-row planter, row spacings can be obtained ranging from 26 to 40 in (66.1 to 101.6 cm). The planter can be used for planting other field crops in narrow rows.

[1] L. W. Hurlbut, Adjusting Corn Planters and Listers for Sorghums, *Nebr. Agr. Expt. Sta. Cir.* 64, 1940.

**Table 11-1   Number of Seeds per Pound
for Six Varieties of Sorghum**

Variety	Number of seeds per pound
Sooner Milo	12,600
Feterita	15,900
Atlas Sorgo	20,900
Pink Kafir	22,400
Early Kalo	23,200
Early Sumac	33,400

Seed plates are obtainable for planting beans, soybeans, corn, beets, and other small-seed crops. The dropping mechanism for beets is designed to plant *segmented seed*, that is, to plant single beet seeds after the pods have been broken up. Some planters are designed to plant pelleted seeds. Special brush cutoffs (Fig. 11-24) and small, spurlike rollers with projections to punch the seeds through and out of the plate cell (Fig. 11-25) are available. Various-sized sprockets are provided to obtain up to 12 different plate speeds. Runner and double-disk types of furrow openers are interchangeable. The double-disk opener is provided with removable depth-gage bands or drums that fit on the sides of the disks. Narrow and pointed-shoe-type furrow openers for fertilizer are attached to the planter gangs so that the fertilizer can be placed to the side and below the seed level.

Fig. 11-23   Six-row rear-mounted beet and bean planter equipped with fertilizer distributors and rubber press wheels. (*International Harvester Company*.)

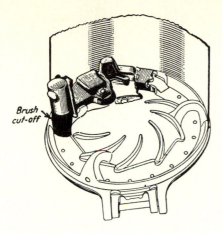

Fig. 11-24 Brush cutoff that can be substituted for the regular metal cutoff in seed hopper for planting beets.

Planter Hoppers

Manufacturers are making planter hoppers for seed, fertilizer, and herbicides out of fiber glass. In some cases the seed hoppers are being made of a translucent material so that the amount of seeds in the hopper can be determined at a glance.

The Application of Fungicides and Herbicides

Healthy stands of cotton seedlings are often severely damaged or almost destroyed by soil-borne organisms, such as soil fungi that attack the tender living root tissue. Such fungi can be controlled by spraying 5 lb (22.6 kg) of fungicide mixed and applied with 7.8 gal (29.5 l) of water per acre at a ground speed of about 4 mi (6.4 km)/h. The fungicide is applied as a part of the planting operation. The spray should cover a band $1\frac{1}{2}$ to 2 in (3.8 to 5.1 cm) wide from the bottom of the seed furrow to and including the surface of the covering soil.

Some herbicides are applied and incorporated or mixed into the soil before

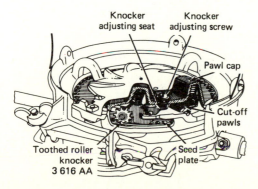

Fig. 11-25 Segmented beet seed plate assembled in hopper bottom. Part of the pawl cap is cut away to show assembly of toothed roller-knocker.

planting. This is referred to as a *preplant* application. Herbicides are also applied by spraying a chemical on the soil surface behind the press wheel as a preemergence operation.

Potato Planters

The slow, laborious hand-dropping method of planting potatoes has been largely supplanted by mechanical planters that open a furrow, drop and space the seed pieces at various distances, place fertilizer to the sides and below the level of the seed, and cover both seed and fertilizer to the desired depth (Fig. 11-26). There are two types of potato-dropping mechanisms: the picker wheel and the chain cup.

The Picker Wheel Drop   The picker wheel dropping mechanism consists of a picker wheel to which are attached from three to twelve picker arms (Fig. 11-27). The picker wheel revolving on the main axle causes the picker arms and picker head to pass through the picking chamber containing the seed. Each picker head is equipped with two sharp picking points which pick out a single seed piece, carry it over to the front, and, as the arm starts downward in its rotation, release the seed or force it off the points, dropping it into the seed spout, which guides it into the furrow made by the furrow opener. The seed piece is forced off the picker points by the opening of the picker arm when the base of the arm contacts a cam. The distance between seed pieces in the furrow is varied by changing the speed of rotation of the picker wheel. The quantity of seed flowing into the picking chamber is controlled automatically. A man rides on the two-row tractor-drawn planter to see that the hoppers are kept full and that the picking chamber does not choke. To plant at high speeds, the potato planter is equipped with two picker wheels, each having eight picker arms. These

Fig. 11-26 Two-row trailing tractor-drawn potato planter equipped with fertilizer attachment and hydraulic remote-control cylinder to raise and lower the furrow openers. (*International Harvester Company*.)

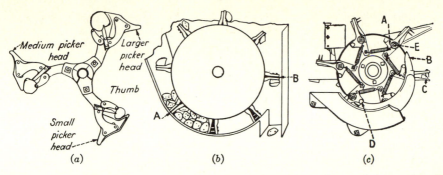

Fig. 11-27 Three makes of picker wheels for automatic potato planters. Only one size of picker head is used at one time: (a) picker wheel showing three sizes of picker heads; (b) picker wheel and arms, with housing cut away to show how seed pieces are picked from the picker chamber; (c) picker wheel showing how cams open the picker arm jaws to release the seed piece.

two picker wheels revolve only half as fast as a single picker wheel used at a normal speed. High-speed planting is done at twice the normal speed, but the picker arms do not revolve any faster than does the single wheel.

The Chain-Cup Drop As shown in Fig. 11-28, the chain-cup potato-dropping mechanism consists of an upward-traveling chain to which cups are attached. The cups are only large enough to pick up a single potato piece from the hopper. On the downward travel of the chain, the potato seed piece is held on the back side of the cup until it is dropped into the furrow. Two- and four-row tractor-mounted units are available.

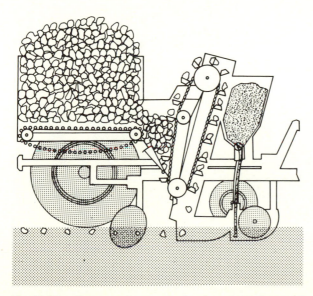

Fig. 11-28 Chain-cup drop mechanism for potato planter. (*Acme Machine Works.*)

Attachments for Potato Planters   The soil of some potato-growing areas is low in fertility, and fertilizer attachments are used in combination with the planter to save the cost of a separate operation to apply the fertilizer. The potato is easily injured if it comes in contact with fertilizer, and for this reason the fertilizer is placed in bands to each side of and below the level of the seed (Fig. 11-29). Double-disk fertilizer openers have a 12-in (30.0-cm) disk on one side to place the fertilizer about seed level and a larger disk on the other side to place the fertilizer deeper or lower in the soil. This method of applying fertilizer is called *hi-lo* (high-low) *application.* Stub-runner furrow openers are generally used on potato planters, but single- or double-disk openers can be obtained. Disks are used to cover and ridge the soil over the seed. Either the shoe or disk type of row marker can be obtained. A double spout is available for placing the seed in twin rows.

Tractor-drawn potato planters are made in one-, two-, and four-row sizes. The width between rows on the two-row planter can be adjusted for spacings ranging from 30 to 42 in (76.2 to 106.7 cm) at 2-in (5.1-cm) intervals.

Transplanting or Plant-setting Machines

When large quantities of plants such as cabbage, tobacco, tomatoes, and sweet potatoes are to be transplanted, time and labor can be saved by the use of a transplanting machine (Fig. 11-30). These machines have a device to open a small furrow, a tank to supply water., and disks or blades for closing the soil over the fertilizer and about the plants. With a transplanting machine, it is not necessary to wait for seasonable weather, because the machine automatically pours a small quantity of water around the roots of each plant as it is being set. Under favorable conditions, with a one-row machine, 3 to 8 acres (1.2 to 3.2 hectares) can be set to plants per day. Twice as much can be done with a two-row machine, or 6 to 16 acres (2.4 to 6.5 hectares), per day.

BROADCAST SEEDERS

Broadcasting is the oldest and simplest method of sowing seed. Broadcasting by machine is more accurate and rapid than by hand. Types of machine

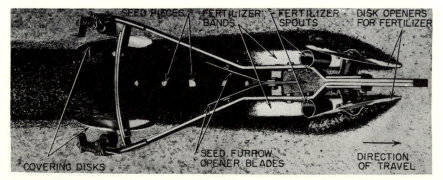

Fig. 11-29   Overhead view of potato-planter furrow-opener assembly showing disks for opening furrow for fertilizer, the opener blades for the seed piece, and the covering disks.

Fig. 11-30   Two-row tractor-mounted transplanter. (*Holland Transplanter Co.*)

broadcasters are the *knapsack, endgate, two-wheel, weeder-mulcher*, and the *airplane*. Broadcast planters drop the seed on the surface of the soil and do not have any covering attachments. If covering is desired, the seeds are usually covered by harrowing. The *knapsack* seeder consists of a good-sized canvas sack fastened to a seeding mechanism, the whole being slung over the shoulders. A crank turned by hand revolves a wheel having several different radial ribs for scattering the seeds. The ribs throw the seeds out to the front and sides in a steady stream. The quantity of seed is regulated by a sliding gate. The wider the gate is opened, the more seed per acre will be sown. This type of seeder is good for sowing clover seed and small grass seed on lawns and fields in the early spring.

Figure 11-31 shows a tractor-mounted power-takeoff-driven broadcast seeder. It consists of a hopper, a feeding device, and a distributing wheel. Attachments are available for spreading seed or fertilizer in various patterns (Fig. 11-32).

The *weeder-mulcher* broadcast seeder consists of a seedbox on a weeder-mulcher. The seeds are dropped on the ground and then covered by the long, spring-steel mulcher fingers.

Seeding attachments for land rollers can also be considered broadcast seeders.

The *airplane* can be considered as a type of broadcast seeder and is used to plant rice, clovers, and other crops whose seeds are sown broadcast. It is equipped with a special seeding tube that flares out at the rear and has curved sections which cause the seeds to spread over a wide swath.

Fig. 11-31 Tractor-mounted power-takeoff-driven broadcaster for seed, fertilizer, and dry chemicals. (*Avco Ezee Flow.*)

DRILL SEEDERS

Grain Drills

The grain drill is a machine designed and built to place the seeds of small grains and grasses in the ground in narrow rows spaced at 6 to 8 in (15.2 to 20.3 cm) apart a uniform depth. The principal parts are the main frame, transport and drive wheels, a box for the seed, a device to meter the seed out of the hopper in uniform quantities, furrow openers to open the furrows for the seed, and covering devices.

Tractor-operated grain drills can be divided into two types, namely, *trailing* and *mounted*. Most of the trailing types are supported by a wheel at each end of the drill (Fig. 11-33). This type is called an *end-wheel drill*. The wheels serve as ground traction drives to operate the moving parts of the drill. Large drills are divided so that each end wheel drives half of the feeds. Hydraulic remote-control cylinders raise and lower the drill on the end wheels. *Trailing press-wheel* drills are partly supported by the large press wheels at the rear and partly by the hitch bar in front (Fig. 11-34).

Mounted grain drills are mounted on the tractor by the three-point hitch. The drill feeds and other moving parts are driven by the power takeoff of the tractor.

Grain drills are also classified as *plain drills* and *fertilizer drills*. A plain drill has a hopper and feeds for the drilling of seeds only, while the fertilizer drill has a large seedbox which is divided lengthwise into two compartments, one for seed

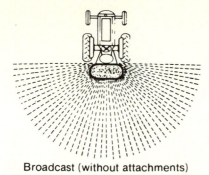

Broadcast (without attachments)

Band spreading (optional attachment)

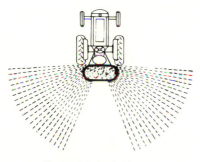

Spread to both sides
(optional attachment)

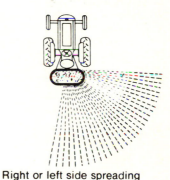

Right or left side spreading
(without attachments)

Limited spread (optional attachment)

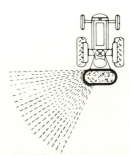

Right or left side limited spread
(optional attachment)

Fig. 11-32 Showing various patterns of spreading material by attachments on tractor-mounted broadcaster. (*Avco Ezee Flow*.)

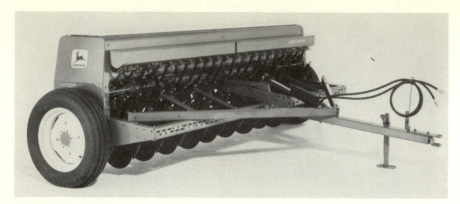

Fig. 11-33 Combination end-wheel-driven grain drill with fertilizer and grass seed attachment. (*Deere & Co.*)

and one for fertilizer. Some drills are provided with grass-seed attachments (Fig. 11-35). The fluted force feed and the double-run feed are used on both the plain and fertilizer drills.

Size of Drill The size of a grain drill is determined by the number of furrow openers and the distance they are spaced apart. The size is expressed as 12-6 or 18-7, which means there are 12 or 18 furrow openers spaced 6 or 7 in (15.2 or 17.8 cm) apart. Drills can be secured with the feeds and furrow openers spaced 6, 7, or 8 in (15.2, 17.8, or 20.3 cm) apart. A 27-7 drill will seed a strip $19\frac{1}{2}$ ft (5.9 m) apart. When grain drills are used to plant crops in rows spaced far enough apart to permit cultivation, some of the feeds must be covered (Fig. 11-36). Table 11-2 shows the feeds used for different-sized drills and two-row spacings.

Fig. 11-34 Trailing grain drill partly supported by large press wheels. (*Deere & Co.*)

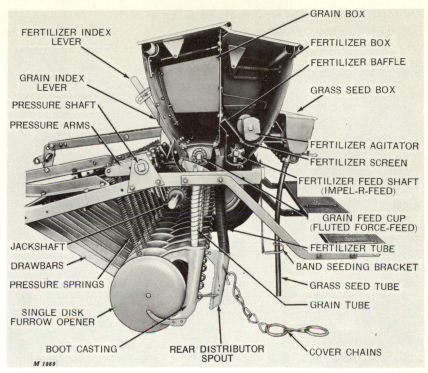

FERTILIZER INDEX LEVER

GRAIN INDEX LEVER

PRESSURE SHAFT

PRESSURE ARMS

JACKSHAFT

DRAWBARS

PRESSURE SPRINGS

SINGLE DISK FURROW OPENER

BOOT CASTING

REAR DISTRIBUTOR SPOUT

GRAIN BOX

FERTILIZER BOX

FERTILIZER BAFFLE

GRASS SEED BOX

FERTILIZER AGITATOR

FERTILIZER SCREEN

FERTILIZER FEED SHAFT (IMPEL-R-FEED)

GRAIN FEED CUP (FLUTED FORCE-FEED)

FERTILIZER TUBE

BAND SEEDING BRACKET

GRASS SEED TUBE

GRAIN TUBE

COVER CHAINS

M 1869

Fig. 11-35 Cross-sectional view of combination grain, fertilizer, and grass drill with parts named. Drag chain coverers are used. (*Deere & Co.*)

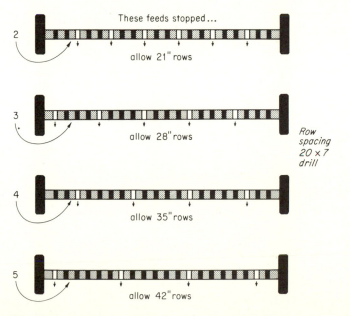

These feeds stopped...

2   allow 21" rows

3   allow 28" rows

4   allow 35" rows

5   allow 42" rows

Row spacing 20 x 7 drill

Fig. 11-36 Showing method of stopping feed cups to plant crops in rows with grain drills.

Table 11-2   Feeds* to Use for Wide Row Spacing
All other feeds must be covered with stops.

	7-in-spaced drills			
Drill	21-in row (52.3 cm)	28-in row (71.1 cm)	35-in row (88.9 cm)	42-in row (106.6 cm)
9 x 7	3-6-9	1-5-9	3-8	2-8
11 x 7	3-6-9	2-6-10	1-6-11	3-9
13 x 7	1-4-7-10-13	1-5-9-13	2-7-12	1-7-13
15 x 7	1-4-7-10-13	2-6-10-14	3-8-13	1-7-13
17 x 7	3-6-9-12-15	2-6-10-14	2-7-12-17	3-9-15

	8-in-spaced drills			
Drill	24-in row (61.0 cm)	32-in row (81.3 cm)	40-in row (101.6 cm)	48-in row (121.9 cm)
16 x 8	1-4-7-10-13-16	3-7-11-15	1-6-11-16	2-8-14

* The feeds are numbered 1 to 17.

Frame   The frame is usually made of angle steel, well braced and reinforced at the corners. It is necessary that the frame be strong enough to prevent sagging and to hold the parts in alignment, as all parts are connected to the frame. The axle is carried beneath, with the wheels on each end of it. The seedbox is carried above, while the furrow openers are suspended below. Roller bearings are usually used on each end of the axle.

Wheels   Most grain drills are equipped with rubber-tired wheels. These wheels are placed on the main axle of the drill. When smaller rubber-tired wheels are used on this equipment, they are placed on a stub jackshaft to elevate the drill to its regular height, so that the same drawbars and pressure rods and springs can be used with either type wheel. The implement tire size for grain drills is usually 6.70 x 15. The operator's manual should be studied to determine if a correction factor should be used in setting the seeding rate.

Seedbox   The seedbox is made of sheet metal or fiber glass and should have a large capacity. A tight-fitting lid should be provided to keep out rain. Inside the box near the bottom are one or two power-driven agitator rods that extend the full length of the box to agitate the seeds and prevent them from *bridging over* as the seeds are fed out. The grain feeds are set in the bottom of the box and are of two types: the fluted-wheel and the double-run feed.

Fluted-Wheel Feed   The fluted-wheel feed is considered the simpler of the two and is the more generally used. It has a greater number of seeding-rate settings and is easier to clean than the double-run feed. It consists mainly of a fluted-wheel feed roll, a feed cutoff, and an adjustable gate. Figure 11-37 shows

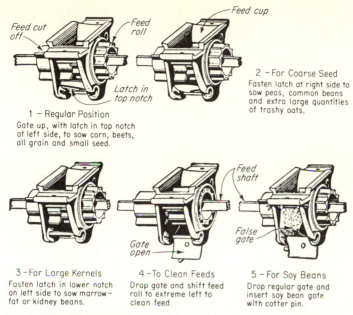

Fig. 11-37   Fluted-wheel grain feed showing various adjustments.

that the feed roll and the cutoff are mounted on a square shaft running through the feed cups. The feed roll turns with the shaft, forcing the grain out over the gate, where it falls into the seed tube. The gate is adjustable for different-sized seeds. Power is transmitted from the main axle to the feed shaft by gears or sprockets and chains.

The quantity of seed sown per acre is varied by exposing more or less of the feed roll to the seed inside cup and by adjusting the gate.

Internal Double-Run Feed   This feed, shown in Fig. 11-38, gets its name from its construction. It consists of a double-faced wheel having a small and a large side. The small side is used for planting small seeds, while the large side is used for planting larger seeds such as oats, wheat, peas, and beans. Figure 11-38 shows one side covered while the other is in use. The lid is hinged over the middle of the wheel so it can be reversed to cover either side.

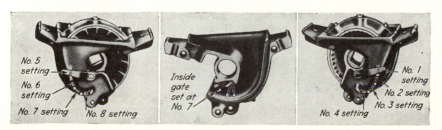

Fig. 11-38   Internal double-run type of grain-drill feed showing gate position for seeding rate.

The quantity of seed sown per acre is changed by varying the speed of the feed wheels. Figure 11-39 shows an arrangement for changing the speed. Special attachments to reduce the size of the outlets and adjustable gates also aid in regulating the quantity of seed sown per acre.

Land Measures   Grain drills are all equipped with a small device, similar to the one shown in Fig. 11-40, which is called a *land measure, acre meter*, or *surveyor*. This is an instrument so designed that it determines the number of acres sown. If the operator keeps a record of the number of bushels placed in the seedbox and the number of acres sown, a check can be made as to the accuracy of the drill in the amount of seed being sown per acre. This is *not* termed *calibration*, which is described below.

Calibration of Grain Drills   Many grain drills do not sow accurately, even though the indicator on the dial plate is set correctly. Some will sow more seed than the dial indicates, while others will sow less. Oftentimes the operator will attempt to check the drill in the field by measuring off a certain acreage, seeding it, and then determining the amount of seed sown. At best, this is a very poor method of checking a drill.

The method of calibrating a drill is as follows: First, find the width of the strip of drill will sow. Measure the distance between furrow openers and multiply it by the number on the drill. Next, find the length of the strip of that width necessary to make 1 acre (0.40 hectare). This is done by dividing 43,560—the number of square feet in 1 acre—by the width of the strip sown by the drill. The result will be the distance the drill must travel to sow 1 acre (0.40 hectare) of grain.

Fig. 11-39   The seeding rate per acre can be changed by using different sizes of drive sprockets (*Deere & Co.*)

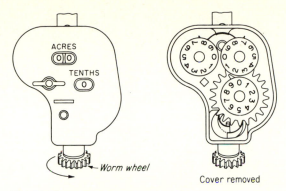

Fig. 11-40 Land measure or acre meter for grain drills.

Now find the number of times the wheels on the drill will turn in going this distance by dividing the distance to be traveled by the circumference of the wheel.

Fill the seedbox with grain.

Set the indicator on the scale to sow whatever quantity of seed is desired.

Jack up the drill and place a paper bag under each seed tube. Tie a rag around the tire of the measured wheel so each revolution can be counted.

Engage the clutch and turn the wheels, counting each revolution. Turn them at about the speed they would travel in the field. When the wheels have been turned the equivalent of $\frac{1}{4}$ or $\frac{1}{2}$ acre (0.10 or 0.20 hectare), collect and weigh the grain. The weight of grain sown by each feed should be recorded separately so that each feed cup can be checked. To figure on an acre basis, multiply the amount by 4 if $\frac{1}{4}$ acre (0.10 hectare) was selected and by 2 if $\frac{1}{2}$ acre (0.20 hectare) was sown.

If the indicator is set to sow 8 pecks, 8 pecks should also have been collected. If only 6 pecks of grain are collected, the drill is in error.

The percentage of error of the indicated quantity is calculated by dividing the difference between the quantity collected and the quantity the indicator was set on by the indicated quantity.

For example, the difference between the quantity collected and the quantity the indicator was set on in this case is 2 pecks. Dividing this by 8, the indicated quantity, gives an error of 25 percent.

Furrow Openers There are four types of furrow openers used on grain drills: the hoe, deep furrow, single, and double disk. A seed tube conducts the seed from the feed into the boot, from which it falls into the furrow. Furrow openers are attached to the frame of the drill by drag bars (Fig. 11-41).

The *hoe* furrow opener consists of a single- or double-pointed shovel fastened to the lower part of the boot. The grain drops into the furrow directly back of the shovel. A spring or pin trip is provided so that, when the hoe strikes an obstruction, no damage is done. This type of opener often gives trouble by clogging up when used in trashy ground.

Single-disk furrow openers consist of one disk slightly dished, securely fastened to the boot, and set to run at a slight angle (Fig. 11-41). The seeds are dropped from the boot on the convex side of the disk at a point below and to the rear of the center. A toe scraper is used on the convex side and a T scraper on the concave side to keep the disk clean. The single-disk opener gives good penetration, cuts trash well, and does not easily clog.

Single-disk semideep furrow openers have 14-in (35.6-cm) single disks to open deep furrows in subhumid areas. Half the openers are assembled with the concave side facing to the right and half facing to the left. They can be set either staggered or in a straight line. Penetration is aided by spring pressure. Since the disks revolve, they must be provided with bearings that are well designed, constructed, and lubricated. Some grain drills are equipped with a multi-lube system.

A *double-disk* opener is composed of two disks, having very little dish, set facing each other at a slight angle so as to form a bevel cutting edge where they penetrate the soil. In this position, the disks open a clean furrow and leave a small ridge in the center so that, when the seeds are deposited in the furrow, there is a tendency to make two distinct rows about 1 in apart. Saw-blade double-disk openers are designed to place the seeds in the ground with the downward movement of the disks ahead of the disk axle. This type of opener is suitable for high tractor speeds and for trashy land.

A *lister* or *deep-furrow* opener is shown in Fig. 11-41. This type of furrow opener is used to make deep trenches or furrows and ridge the soil so that snow and moisture will be caught and the soil prevented from blowing. The spacing of the openers is wider than that used for the regular grain drill and ranges from 12 to 16 in (30.5 to 40.6 cm) between openers.

Tractor Hitches

Where the ground is level and the acreage to be seeded is large, several grain drills can be arranged squadron fashion and hitched to one tractor. As many as five large drills can be hitched to the same tractor. A grain drill should be hitched to the tractor so that the frame and seedbox are level.

Attachments for Grain Drills

When a fertilizer attachment is used, the drill is usually known as a *fertilizer* drill, even though it is equipped with the regular grain-sowing feeds. A special

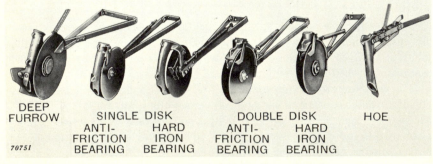

DEEP
FURROW   SINGLE DISK   DOUBLE DISK   HOE
            ANTI-   HARD   ANTI-   HARD
            FRICTION   IRON   FRICTION   IRON
70751   BEARING   BEARING   BEARING   BEARING

Fig. 11-41   Types of furrow openers for grain drills. (*Deere & Co.*)

fertilizer tube serves as a spout to conduct the fertilizer down to the soil and prevents the wind from blowing part of it away. Liquid-fertilizer attachments are available for grain drills.

A grass-seeding attachment can be secured for almost all grain drills. It is attached either in front of or to the rear of the main seedbox. The fluted-wheel type of feed is used in the feed cups. The seed tubes may either empty directly into the regular grain-seed tube or be clamped to the side so as to allow the grass seed to fall behind the furrow openers.

The most common type of covering device is the drag chain. Figure 11-35 shows how it is hooked to the boot and how it drags over the furrows to cover the seed without packing the soil.

In the subhumid regions, where the soil is dry and likely to blow, press wheels are used to cover the seed and press the soil around them. Figure 11-34 shows a drill equipped with large press wheels. The latter also drive the seeding mechanism. Small gang press wheels also can be obtained.

Multiple-Use Drills

Figures 11-42 and 11-43 show a special-type drill, sometimes called a *grassland drill*, that is designed to open furrows in unplowed land; in growing crops, such as alfalfa; and in pastures. The seeds are planted and fertilizer is applied as the

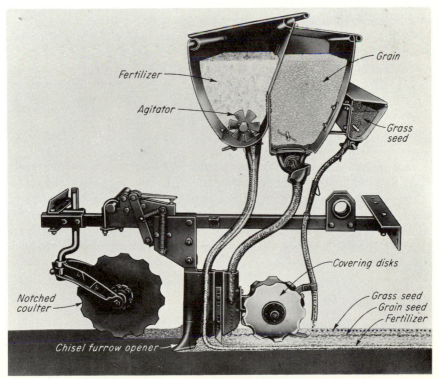

Fig. 11-42 Cross-sectional view of multiple-purpose or pasture drill showing how the grain, grass, and fertilizer are placed in the soil. (*Deere & Co.*)

Fig. 11-43   A heavy-duty pasture drill. It can be used in heavy sod. (*Midland Mfg. Co.*)

furrow is opened. Thus, three operations are performed simultaneously. Oats and other small grains and grass seed can be planted without first plowing, harrowing, and preparing the seedbed, as is necessary for the regular type of grain drill. If desired, fertilizer can be applied 3 or 4 in (7.6 to 10.2 cm) below the surface in pastures without materially disturbing the established sod.

The furrow openers are usually spaced 8 to 10 in (20.3 to 25.4 cm) apart. A sharp-edged, narrow, chisel-type furrow opener is recommended for pasture lands.

GRASS-SPRIG PLANTER

The grass sprigs are placed in a hopper provided with a feed device which feeds the sprigs into a rotating planting disc or wheel. The wheel passes through the hopper, picking up sprigs. As the wheel revolves, the sprigs are carried to the ground. The speed of the wheel is slightly less than the forward travel so that the friction of the soil pulls the sprigs from the "fingers" on the wheel, placing them in the soil. The soil is pressed together from the sides around the sprigs to hold moisture.

MAINTENANCE AND CARE

1   Check the seed-metering system to determine if the system is suited to the type of seed being used.

2   Calibrate the seed-metering system to determine if the desired amount of seed is being placed in the soil.

3   Check to determine if the seeds are being placed in the soil at the proper depth.

4   Check to determine if the seeds are being properly covered with soil.

5   Check to determine if the press wheels are functioning properly.

6   Check the performance of attachments.

7   Before storage at the end of the planting season, check over entire machine.

8   Make a list of any repair parts needed and order replacements.

9   Clean furrow openers and covering devices and apply grease or heavy oil to prevent rusting.

10   Thoroughly clean seed hoppers and coat moving parts with heavy oil.

11   Clean seed tubes.

12   Any fertilizer attachments should be thoroughly cleaned and coated with oil inside and outside. Liquefied fertilizer tanks should be flushed out.

13   All herbicide and insecticide attachments should be carefully cleaned and any waste disposed with care.

14   Roller chains should be removed, tagged, and stored in a can of light oil.

15   Detachable-link chains should be coated with oil.

16   Remove rubber press wheels and store in shed.

17   Store in a shed or cover with plastic or canvas sheets tied down tight.

REFERENCES

Andrews, William Baker (ed.): *Cotton Production, Marketing, and Utilization*, William Baker Andrews, State College, Miss., 1950, chap. 5.

Anonymous: No-Til Farming for Profit in the '70s, Allis-Chalmers booklet.

Bainer, Roy: Precision Planting Equipment, *Agr. Engin.*, 28(2):49-54, 1947.

Barmington, R. D.: The Relation of Seed, Cell Size, and Speed to Beet Planter Performance, *Agr. Engin.*, 29(12):530-532, 1948.

Collins, E. V., and C. S. Morrison: Mathematics of a Cumulative Drop Planter, *Agr. Engin.*, 29(1):28, 1948.

Colwick, Rex. F., et al.: Planting in the Mechanization of Cotton Production, *Southern Cooperative Series Bul.* 49, 1957.

Guelle, C. E.: Precision Planting of Beets and Corn, *Agr. Engin.*, 28(2):56-57, 1947.

Hansen, C. M., L. S. Robertson, and B. H. Grigsby: Plow-Plant Equipment Designed for Corn Production, *Amer. Soc. Agr. Engin. Trans.*, 2(1):65, 1959.

Harrold, Lloyd L., et al.: No-Tillage Corn: Characteristics of the System, *Agr. Engin.*, 51(3):120, 1970.

Hudspeth, Elmer B., et al.: Planting and Fertilizing, *U.S. Dept. Agr. Yearbook*, p. 147, 1960.

Jones, Jr., J. Nick, J. H. Lillard, and R. C. Hines, Jr.: Application of the Multiple-use Drill, *Agr. Engin.*, 32:417-419, 1951.

McBirney, S. W.: The Relation of Planter Development to Sugar-Beet Seedling Emergence, *Agr. Engin.*, 29(12):533-536, 1948.

Miller, H. F.: Planting Cotton to a Stand, *Agr. Engin.*, 30(10):487-488, 1949.

Miller, Vernon: Minimum Tillage Saves Fuel, *Progressive Farmer*, 89(3):35, 1974.

Narrow-Row Cotton Culture, Twelve papers in summary, Proceedings 1971-2 Beltwide Cotton Production-Mechanization Conferences, *The Cotton Ginners' Journal & Yearbook*, 1971-2.

Park, Joseph K.: Sweet Potato Planting Machinery, *Agr. Engin.*, **28**(9):415, 1947.

Poynor, R. R.: An Experimental Mulch Planter, *Agr. Engin.*, **31**(10):509-510, 1950.

Smith, H. P., and M. H. Byrom: Effects of Planter Attachments and Seed Treatment on Stands of Cotton, *Tex. Agr. Expt. Sta. Bul.* 621, 1942.

——, ——, and H. F. Morris: Germination of Cottonseed as Affected by Soil Disturbance and the Machine Placement of Fertilizer, *Tex. Agr. Expt. Sta. Bul.* 616, 1942.

—— and D. L. Jones: Mechanization of Cotton in Texas, *Tex. Agr. Expt. Sta. Bul.* 704, 1948.

Wittmuss, H. D., et al., Strip Till-Planting of Row Crops, *Trans. of the ASAE*, **14**(1):60, 1971.

QUESTIONS AND PROBLEMS

1 Enumerate the factors that influence the germination of seed.
2 Explain the following: (*a*) drilling, (*b*) hill-dropping, (*c*) broadcasting.
3 Explain the operation of the air-seed-metering system.
4 Discuss the differences in the dropping mechanisms for corn, sorghum, and cotton.
5 Give the advantages and disadvantages of the different types of seed-furrow openers and explain the functions of press wheels.
6 What is meant by segmented beet seed?
7 Explain the differences in potato-dropping mechanisms.
8 Explain how the size of a grain drill is determined.
9 Name and explain the different grain feeds used on grain drills.
10 Describe how a grain drill is calibrated.
11 A 40-acre (16.2-hectare) field of cotton is to be planted with a four-row central-mounted planter traveling at 5 mi/h (8.0 km/h). The row spacing is 40 in (101.6 cm). Allow 75 percent field efficiency for turns. The field is twice as long as wide. Calculate the time required to plant the field with the rows running in the longest and shortest directions.
12 Explain minimum tillage.
13 Explain the terms *plow-plant* and *wheel-track planting*.
14 Explain no-tillage.

Cultivation and Weed-Control Equipment

The poet can weave beautiful words about the wonders of grass and how the world would go to ruin if there were no grass. But the farmer's classic words about grass and weeds would be far from poetic. It is true that the livestock producer does need grass and plenty of it, but the farmer who grows crops would be much happier if it were possible to confine all the grass and weeds to the pasture. The farmers of the United States spend billions of dollars annually in controlling weeds that compete with crops for plant food and moisture from the soil.

It has been said that weeds are robbers: they rob the farmer of his profits by reducing yields, they rob by lowering the quality of the crop, they rob the farmer by harboring insects that damage his crop, they rob by reducing the land value, and they may become so thick that the crop has to be abandoned, so that they could rob the farmer of his home. Therefore, the farmer must fight to control weeds with every means he can command. The farmer's primary tools of war on weeds are (1) cultivation by stirring the soil, (2) the use of flame, (3) the use of chemicals, and (4) the laying of plastic strips over the row.

CULTIVATORS

Cultivation is an operation that requires some kind of tool that will stir the surface of the soil to a shallow depth in such a manner that young weeds will be destroyed and crop growth promoted. Cultivation to control weeds by stirring the soil may start on the prepared seedbed prior to planting. After planting, the soil can be cultivated before emergence of the plants for some crops. Cultivation usually begins soon after emergence of the young crop seedlings, as weeds generally emerge about the same time as the crop.

History of Cultivator Development

The first implement used for the control of weeds in crops was, perhaps, the hoe. In ancient times, most crops were planted broadcast, and the hoe was about the only tool that could be used to destroy weeds among the plants. Many Asiatic countries, even today, do not have hoes, and weeds are pulled from rice paddies by hand.

When crops were planted in rows, the plow could be used to destroy the weeds between the rows and throw soil over the young weed seedlings in the crop row, which smothered them to death. The single-shovel plow was changed to a double-shovel so that half the space between two crop rows could be stirred for weed control.

The next advancement in cultivation equipment was a V-shaped wooden frame to which wood or iron pegs were attached. Handles permitted the operator to control the tool manually. Early in the eighteenth century, Jethro Tull[1] invented the horse hoe. Davidson[2] states that "A patent was granted to George Esterly, April 22, 1856, on a straddle-row cultivator for two horses ...." Horse-drawn one-row walking and riding cultivators came into use in the late 1880s. The two-row riding horse-drawn cultivator came into use shortly after 1900. The first row-crop cultivators used with tractors were horse-drawn cultivators adapted for hitching behind the tractor.[3] The B. F. Avery Company built a tractor-mounted cultivator about 1918. The first integral-mounted cultivator attachment for tractors was developed about 1925 by the International Harvester Company. The gangs were lifted by manually operated levers.[4] Power lifts for cultivators were not developed until about 1932 or 1933.

Objectives of Cultivation

The primary objectives sought in the cultivation of a crop are:

1 Retain moisture by
   a Killing weeds
   b Loose mulching on surface
   c Retaining rainfall
2 Develop plant food
3 Aerate the soil to allow oxygen to penetrate soil
4 Promote activity of microorganisms

[1] Jethro Tull, *Horse Hoeing Husbandry*, 1822.
[2] J. B. Davidson and L. W. Chase, *Farm Machinery and Motors*, Orange Judd Publishing Co., Inc., New York, 1912.
[3] F. R. Jones, Large Scale Farming, *Tex. Agr. Expt. Sta. Bul.* 362, 1927.
[4] First picture shown in article by D. W. Watkins, Use of Machinery in Cotton Production, *Agr. Engin.*, 7(10):349, 1926.

Many types of cultivators are in use, ranging all the way from the small hand-pushed garden cultivator suitable for the family garden to large eight-row tractor-mounted cultivators capable of cultivating 100 to 130 acres (40.5 to 52.7 hectares) per day (Fig. 12-1). The type and size needed will depend upon the acreage, the kind of crop, soil type and conditions, rainfall, type of farming practiced and the kind of power available.

As tractors have largely supplanted horses and mules for farm power, only the tractor types of cultivators will be discussed. Many row-crop cultivators are mounted on tractors centrally but well forward. Tractors equipped with the three-point linkage hitch are usually provided with an assembly cultivator unit that operates behind the tractor. Cultivating equipment for track-type tractors is mounted either in front of or to the rear of the tractor.

Central-forward Tractor-mounted Cultivators

Central-forward-mounted cultivators are available in one-, two-, four-, six-, and eight-row sizes. Cultivator units can be mounted on both the four-wheel and the three-wheel or tricycle-type tractors. The one-row cultivator mounted on the one-plow-sized tractor is suitable for small farms and terraced fields with curving rows. The two-row cultivator is suitable for medium-sized farms and fields that have rows either straight or with gradual curves. The four- to twelve-row cultivator is suitable for large farms and level land where the rows are straight for long distances. All sizes can be used on either flat or bedded land but are not adapted for use in the listed furrow. The gangs are hydraulically raised and lowered and in some cases controlled hydraulically for depth of cultivation.

The older-style tractor-mounted planters and cultivators required several hours' labor to change from planting to cultivating equipment. Designers have devoted much thought to the design of equipment that can be changed quickly and easily. Therefore, most of the more recently designed cultivators are first

Fig. 12-1   Overhead view of front-mounted eight-row cultivator. (*International Harvester Company.*)

assembled into three units: one unit for each forward side of the tractor and one rear-section unit. The rear section carries three sweeps or spring-tooth sections which loosen the soil behind the tractor wheels to prevent soil compaction and leave a smooth middle.

Rear-mounted Cultivators

Rear-mounted cultivators are usually unit assemblies attached to tractors equipped with a three-point hitch as a single unit. Each gang is provided with a gauge wheel to control the depth of cultivation. The gangs are attached to a tool bar or frame bar (Fig. 12-2).

Gangs for Cultivators   The gang or rig consists of a beam to which is attached a shank or standard that has an adjustable foot set at an angle so a shovel or sweep can be bolted to it. There is a gang for each side of each row. A single-row cultivator has two gangs, a four-row cultivator has eight gangs, and an eight-row cultivator has sixteen gangs. There are numerous types of gang assemblies designed to suit different crops, soils, and farming practices. The number and type of soil-stirring members usually determine the gang style.

The *beam* may consist of a straight pipe, a shaped square bar, or a fabricated beam of flat bar steel. A pipe beam requires crossheads to support the shanks so the soil-stirring members can be adjusted laterally as the number of shanks and size of the sweeps vary. The square-bar beam is bent laterally so the shanks can be clamped directly to the bar. The fabricated-bar beam is shaped to permit clamping of the shanks at the desired positions.

Under severe conditions, the gangs are held apart a uniform distance by space bars, arches, or hobbles.

The gangs can be moved laterally on the frame bar to suit various row spacings. When the bolts are tight, the gangs are held solidly in place. No

Fig. 12-2   Rear-mounted six-row cultivator. (*International Harvester Company*.)

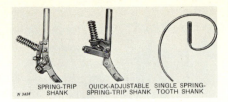

SPRING-TRIP   QUICK-ADJUSTABLE SINGLE SPRING-
SHANK   SPRING-TRIP SHANK TOOTH SHANK

Fig. 12-3   Types of cultivator shanks.
(*Deere & Co.*)

provision is made for shifting the gangs sidewise other than the regular steering of the tractor.

The rear section of a tractor-mounted cultivator consists of three beams, shanks, and sweeps. These sweeps are often termed *plow-out* sweeps, as they plow out the tracks made by the tractor wheels.

Three types of *shank trips* or *releases* are available for use in areas where rocks and roots are present in the surface soil. They are the *spring-trip*, *break-pin*, and *friction* (Fig. 12-3). Where there are no hidden obstacles in the soil, a solid, round-rod steel shank can be used. The rigid shank is used for *disk hillers* and *barring-off* disks.

Lifts for Cultivator Gangs   The gangs of the first tractor-mounted cultivators were lifted with hand levers. Later, gear-operated rocker arms were attached to the tractor and connected to the gangs by a linkage arrangement so the gangs could be power-lifted. All gangs on the cultivator were lifted at the same time.

Hydraulic power lifts are now used to lift all arrangements of cultivator gangs on all sizes and makes of field tractors. The use of delayed-lift valves and other hydraulic-control features makes it possible to lift the gangs on each side of the tractor separately and at different time intervals. This is helpful in cultivating point rows on terraced fields. The rear section is lifted and lowered slightly later than the front gangs to permit plowing out to the end of the row. Then the delayed lowering prevents the sweeps from entering the soil and tearing up headlands because the tractor moves forward a few feet before the rear gangs are lowered.

Shovels and Sweeps   There are numerous types and shapes of shovels and sweeps used for stirring the soil and killing weeds, as shown in Figs. 12-4 and 12-5. Shovels are available in widths up to about $3\frac{1}{2}$ in (8.9 cm) (Fig. 12.6), but sweeps can be obtained in widths ranging from 6 to 24 in (15.2 to 61.0 cm). The width or size is in even inches, such as 6, 8, 10, 12 (15.2, 20.3, 25.4, 30.5 cm), and up to 24 in (61.0 cm). The types of soil, crops, and weeds influence the shovel or sweep used. The sweeps designed for horse-drawn cultivators throw too much soil and cover small plants when used at high speeds on a tractor. New designs of high-speed sweeps have the crown[1] and wings set fairly flat to skim under the soil at a shallow depth without throwing excessive amounts of soil (Fig. 12-6). The ASAE[2] has set up standards and specifications for curved and straight-stemmed sweeps.

---

[1] The crown is the rounded area of the sweep between the wings and lower part of the sweep shank.

[2] Cultivator Sweep and Shovel Mountings, *Agr. Engin. Yearbook*, 1973.

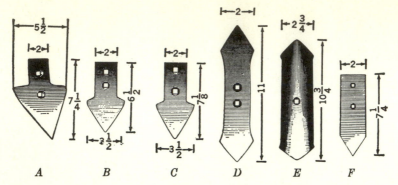

Fig. 12-4 Types of shovels for cultivators: (*A*) spear point for sleeve; (*B*) single point for spring tooth; (*C*) single point for sleeve; (*D*) double reversible point for spring tooth; (*E*) double reversible point for sleeve; (*F*) single point for spring tooth.

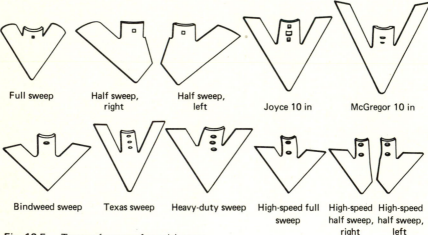

Full sweep   Half sweep, right   Half sweep, left   Joyce 10 in   McGregor 10 in

Bindweed sweep   Texas sweep   Heavy-duty sweep   High-speed full sweep   High-speed half sweep, right   High-speed half sweep, left

Fig. 12-5 Types of sweeps for cultivators.

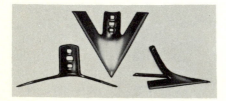

Fig. 12-6 Front, rear, and side views of a high-speed sweep.

Shovels and sweeps should be operated as shallow as possible to prevent pruning the roots from the crop plants and thereby injuring the plants. Sweeps should be set almost flat. When the point is resting on the floor or ground, the outer tip of the wing should be elevated only $\frac{1}{8}$ to $\frac{1}{4}$ in (0.3 to 0.6 cm) above the floor (Fig. 12-7).

Most tractors have sufficient power and gear ratios to operate cultivators at speeds of 5 mi/h (8.0 km/h) or more.

POINT TOO HIGH   POINT TOO LOW   CORRECT SETTING

Fig. 12-7   Wrong and correct settings for a cultivator sweep.

The shovels and sweeps should be set equal distances on each side of the row to keep the land well shaped, aiding in later cultivations and in the harvesting operation. They can be set uniformly before going to the field by using a line diagram on the floor (Fig. 12-8) or an implement-setting frame (Fig. 12-9). The lines are drawn for the row spacing desired, and the frame can be adjusted for any row spacing. The lines and frame can be used for setting the rear cultivator section and other rear- and front-mounted equipment.

Attachments for Central-forward-mounted Cultivators   In addition to the regular gang equipment that makes up the basic cultivator unit, there are a number of special-purpose attachments that add to the usefulness and improve the performance of a cultivator. A number of tractor-cultivator attachments are shown in Figs. 12-10 to 12-14.

The *rotary-hoe* attachment for tractor cultivators (Fig. 12-10) is a popular and useful attachment for breaking the soil crust over emerging plant seedlings and for destroying weeds in the early stages of plant growth. It is used on most of the row crops. The hoe wheels or spiders are mounted between the front

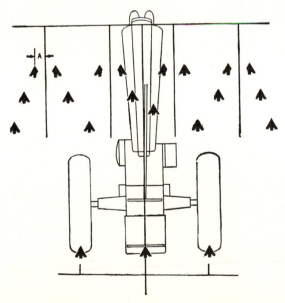

Fig. 12-8   Line diagram for setting the sweeps on a four-row tractor-mounted cultivator. Mark heavy lines on the floor at the desired spacing.

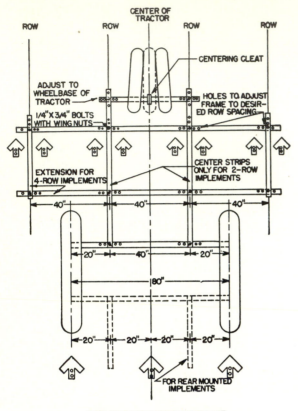

IMPLEMENT SETTING FRAME

Fig. 12-9 Frame for setting cultivator sweeps and other tractor-mounted equipment. The frame is built of light material. (*Tex. Agr. Expt. Sta.*)

Fig. 12-10 Rotary-hoe attachments for tractor cultivators.

shovels or sweeps. The *tines* of the wheel project into the soil for from 1 to 2 in (2.5 to 5.1 cm) directly over the plants. Flakes of soil crust are flipped up, and young annual weed seedlings are uprooted and killed. Young weeds are uprooted better if the rotary hoe is used after a soil crust has formed. Best results are obtained when the rotary hoe is operated at speeds of 5 mi/h (8.0 km/h) or faster. Rotary-hoe attachments can be used on bedded land, on flat land, and in the listed furrow.

Fertilizer Attachments   Two methods of applying fertilizer as a side dressing by the use of fertilizer attachments are shown in Fig. 12-11. In one method, the fertilizer is deposited behind the front sweep which is next to the row. The soil thrown by the rear sweeps on the gang covers the fertilizer. In the other method, a narrow chisel opener opens a narrow furrow to the desired depth and the fertilizer flows through the tube and boot into the furrow. An adjacent sweep throws soil over the furrow to cover the fertilizer. The types of fertilizer feed and hoppers used are discussed under "Fertilizing Equipment."

Disk-hilling and *barring-off* attachments are shown in Fig. 12-11. For hilling, the disks are set to throw the soil to the plant row. Where the crop is

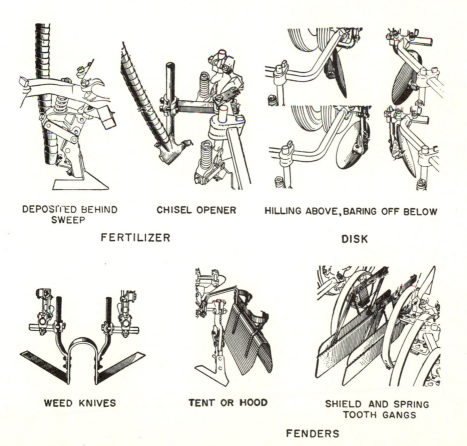

DEPOSITED BEHIND   CHISEL OPENER   HILLING ABOVE, BARING OFF BELOW
      SWEEP

          FERTILIZER   DISK

      WEED KNIVES   TENT OR HOOD   SHIELD AND SPRING
                                                        TOOTH GANGS

                              FENDERS

Fig. 12-11   Fertilizer, disk, knife, and fender attachments for tractor-mounted cultivators.

quite grassy, the disks are set to throw the soil away from the plant row. This leaves only a narrow strip to clean with the hoe.

The *knife* attachment shown in Fig. 12-11 is used as a barring-off tool, but it does not leave an open furrow as do the disks.

Fenders are used between the cultivator sweeps next to the plant row to prevent soil from covering small young plants. The *tent* or *hood* type allows soil to be thrown up on the fender so it will flow off the rear end and settle around the plants without covering them. The *blade* fenders hold the soil away from the plants. A *rotary-hoe wheel* or *spider* set to run on each side of the row, as shown in Fig. 12-10, also acts as a fender. It lets a small amount of soil shift through the tines and flow around the plants, but at the same time it prevents large masses and lumps of soil from being thrown over them.

Listed-corn, spread-bar, and *peanut-digger* attachments for tractor cultivators are shown in Fig. 12-12.

Disk-gang and *blade potato-hilling* attachments for cultivators are shown in Fig. 12-13. Disk gangs on cultivators are often used in extremely sandy soils.

Spray shields are used to hold spray nozzles close to the soil so chemicals can be sprayed on young grass in the plant row (Fig. 12-14).

Speed and Duty of Cultivators

The average speed of horse cultivation is usually given as 2.5 mi/h (4.0 km/h). When cultivators were first mounted on tractors, many operators used low gears or less than a full-throttle setting to hold the speed down to 2.5 to 3.0 mi/h (4.0 to 4.8 km/h), feeling that they could do a better job of weed control. There are still many farmers who operate their tractors at relatively slow speeds, even though the tractor has ample power for faster operation.

With rotary-hoe attachments and high-speed sweeps, it is more economical to operate the tractor at the higher speeds than at the lower speeds, as it reduces both labor and tractor hours

Special Cultivating Equipment

In addition to the regular cotton and corn types of cultivating equipment, there are several types of cultivators designed for special crops and conditions.

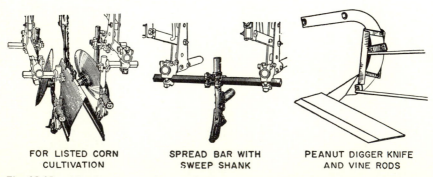

FOR LISTED CORN   SPREAD BAR WITH   PEANUT DIGGER KNIFE
CULTIVATION   SWEEP SHANK   AND VINE RODS

Fig. 12-12   Listed corn spread-bar, and peanut-digger attachments for tractor cultivators.

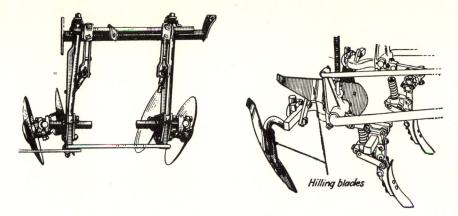

Fig. 12-13   Disk-gang and potato-hilling attachments for tractor cultivators.

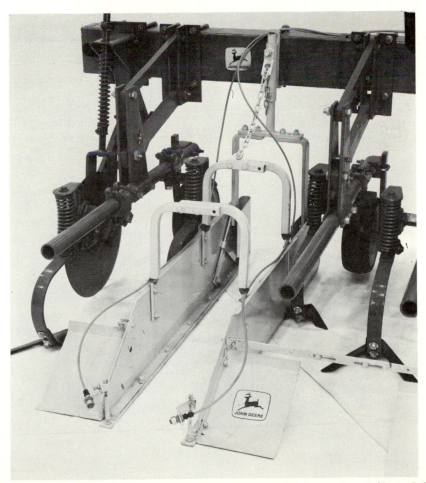

Fig. 12-14   Spray shield for cultivator to hold spray nozzles close to the soil. (*Deere & Co.*)

230   CHAPTER 12

Beet and Bean Cultivators   Beets and beans are usually planted in more closely spaced rows than cotton and corn and require shovels and sweeps for shallow cultivation. Gage wheels control the depth of penetration of the shovels. The weed-knife shanks are attached to tool bars and can be adjusted for row widths ranging from 12 to 28 in (30.5 to 71.1 cm). The tool bar can be mounted either centrally or on the rear of the tractor (Fig. 12-15). The sizes of beet and bean cultivators range from two to six rows, the four-row being the most popular size. Attachments are available for the application of fertilizer as a side dressing.

Lister Cultivators   Lister cultivators are particularly adapted to the cultivation of a listed crop in its early stages of development. Listed crops are those planted in the furrow or trench or below the general level of the ground. For the first cultivation, the disks are set to throw the soil away from the row of plants. For all later cultivations, the disks are set to throw the soil toward the plants.

Rod Weeders, Field Cultivators, Subsoil and Chisel Cultivators   These cultivators are generally used to control weed growth on fallow lands and are discussed under "Secondary Tillage Equipment."

Rotary-Hoe Cultivator   The rotary hoe is a cultivating implement used to cultivate and destroy weeds and grass around young plants. When rains cause a hard crust to form over the soil and hinder the emergence of young seedlings,

Fig. 12-15   Rear view of six-row three-point-hitch tractor-mounted beet-bean and vegetable cultivator. (*Deere & Co.*)

the rotary hoe is an excellent tool for pulverizing the crust. The tool can be used to advantage in young corn, cotton, soybeans, potatoes, and small grains.

The rotary hoe is made up of two gangs of hoe wheels. One gang is placed behind the other, and the wheels are spaced so that the wheels of the rear gang extend forward between the wheels of the front gangs. Some two- and three-row units have solid axles, while the larger units for tractor use are made up in sections so that each section can follow the contour of the soil (Fig. 12-16). The hoe wheels or spiders are usually made of malleable cast iron, but fabricated-steel wheels are also available.

Rotary hoes should be used at fairly high rates of speed. Good work can be done under some conditions at 10 mi/h (16.0 km/h). If the plants are large, there is a tendency for the tines to catch the foliage and pull up a few plants.

When a rotary hoe is run backward or with tines reversed, it makes a useful tool as a *treader* to tread down heavy stubble and other crop residues without clogging. It tears apart large clods, packs the soil from below, and at the same time treads the seed into the clean, firm seedbed through the protective residue which may be on the surface. It also becomes a good broadcast seeder when a seeder box is attached. Figure 12-17 shows a tractor-mounted rotary hoe.

Plant-thinning Machines When the seeds of cotton, beets, and other crops are drilled thick along the row to get good stands, it is sometimes necessary that the plants be thinned to obtain the best yields.

Fig. 12-16 A four-section trailing rotary hoe. (*International Harvester Company.*)

Fig. 12-17 Rear-mounted three-point hitch, 31-ft rotary hoe. Each pair of hoe wheels flexes independently. (*Deere & Co.*)

Small plant thinners have been used for several years. These thinners have either revolving knives or pendulum-swinging knives. They cut across the row, removing plants without regard to the stand.

The latest development in plant thinners is an electronic system which has a light beam that is directed across the row of plants and activates a knife to remove surplus plants by a selective system (Fig. 12-18).

Plants also are thinned and blocked by the use of sweeps set flat and run at right angles or across the row. This is called *cross plowing*.

FLAME WEED CONTROL

Flame has been used by railroads for many years to kill weeds. It has also long been used by ranchmen to burn the spines of cactus so that livestock could graze on the cactus during long droughts. The use of flame for the control of grass and weeds among row-crop plants is a comparatively new development. The equipment consists of a fuel tank, feed lines, control valves, and burners. The system is mounted on a tractor with skid supports for the burners (Fig. 12-19). Burners are provided for each side of each of two to four rows. A two-row system requires four burners, and a four-row system requires eight burners. The burners are mounted at about a 30 to 45° angle with the horizontal so that they will direct a hot, blue flame close to the ground on the grass among the plants. The flame should strike the ground about 2 in (5.1 cm) from the plant on the burner side. The width of the burner mouth should be 8 to 10 in (20.3 to 25.4 cm). Plants to be cultivated should be tougher and larger than the grass and weeds to be destroyed. The burner is provided with both vertical and horizontal adjustments.

Most flame burners are mounted on brackets supported by skids, but good

Fig. 12-18   Electronic plant thinner. (*Deere & Co.*)

Fig. 12-19   Flame burners mounted on brackets supported by skid shoes.

results have been obtained by mounting them on the regular cultivator gangs (Fig. 12-20). The burners should be mounted so that the flame of one burner does not meet the flame on the opposite side of the row (Fig. 12-21). This is termed *across-the-row flaming*. Recent research shows some advantages in the early stages of plant growth to setting the burners so the flame will be directed parallel to and alongside the row of plants.

The use of a water-shield attachment (Fig. 12-22) permits earlier flaming of cotton. The shield allows flaming of cotton 4 in (10.1 cm) in height without damage to cotton or soybean leaves.

Either butane or propane can be used for fuel for flame weeders. The fuel is carried in a high-pressure tank mounted on the tractor. The tank is fitted with control valves and fuel-line connections. The pressure on the fuel line leading to the burners may range from 30 to 40 lb. If the grass and weeds are dense and more than an inch in height, the tractor should be driven at the comparatively slow speeds of 2 to $2\frac{1}{2}$ mi/h (3.2 to 4.0 km/h). But if the weed seedlings are young and fairly thin, the tractor can be operated at around 3 mi/h (4.8 km/h).

CHEMICAL WEED CONTROL

The use of chemical herbicides for the control of weeds in crops is a relatively new development. In some broadcast crops, such as wheat and rice, the chemical used is selective. This means that the chemical will kill the weeds and not kill the growing crop plants. Some chemicals used on row crops must be applied in such a way that the chemical does not come in contact with the crop plants, or the plants will be injured or killed. Some herbicides are toxic to human beings and should be handled with all precautions to prevent injury to operators.

Commercial row-crop herbicides are classified as *pre-emergence* and *post-emergence*, according to the time of application and mode of action. The pre-emergence chemicals are applied before or after planting but before the crop plants emerge above the soil. Pre-emergence chemicals can be included as a part

Fig. 12-20   Flame burners mounted on the regular cultivator gang. (*Mississippi Agr. Expt. Sta.*)

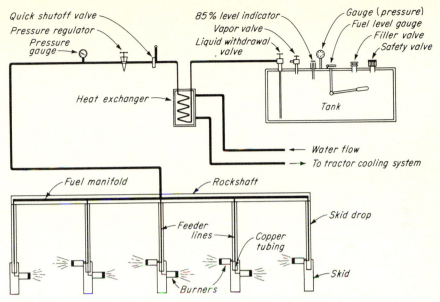

Fig. 12-21   Schematic diagram of a flame method of weed control.

of the planting operation. The post-emergence chemicals are applied after the crop plants have emerged. The plants must be allowed to obtain sufficient growth so the chemical spray can be directed on the young weed seedlings but below the foliage of the plants.

There are more than 30 companies producing herbicide chemicals under about 150 trade or brand names. Chemicals are available for field crops, vegetables, orchard fruits and trees, rangelands, and pastures.

Pre-emergence Applications

Several chemicals are used as pre-emergence herbicides. Formulations of dinitro-*o*-secondary-butylphenol have shown encouraging results. The manufacturer's

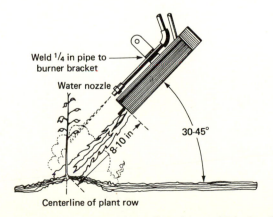

Fig. 12-22   Water shield for flame burner. (*Ark. Agr. Expt. Sta.*)

label gives instructions and cautions for the use of these materials under varying conditions.

Some herbicides are applied and incorporated or mixed into the soil before planting; these are referred to as *preplant*. To obtain the best results with pre-emergence herbicides, it is essential that plant residues from previous crops be thoroughly disposed of or removed. The seedbed should be well prepared so that the drill area will be left as smooth as possible behind the planter. The drill area or row should be slightly higher than the middle to prevent soil containing weed seed from being washed, blown, or pushed onto the treated area. A press wheel or roller 12 to 14 in (30.5 to 35.6 cm) wide should be mounted behind the planter furrow opener to smooth and firm the soil to be sprayed. A regular spray rig is mounted on the tractor in conjunction with the planter. A spray nozzle is mounted to the rear of the press wheel or roller. The mounting for the nozzle should be adjustable both vertically and laterally. A low-gallonage nozzle that gives a fan-shaped spray pattern is best for pre-emergence application of herbicides. The nozzle orifice should be constructed to give a wide fan-shaped spray of 80 to 95° with a pressure of 25 to 40 lb.

The best results with pre-emergence chemicals have been obtained where the spray has been applied on the moist soil directly behind the press wheel or roller. The frequency of rains should be such that rain will occur between the time of planting and emergence to seal the chemical in the surface soil. Where rain does not occur until after emergence of the crop, the raindrops will spatter the loose soil and chemical up onto the young seedlings, and damage is likely to occur. The chemical in the top $\frac{1}{8}$ in (0.3 cm) of the soil prevents annual weed seeds from germinating for a period of 2 to 3 weeks.

Post-emergence Applications

Post-emergence herbicides for cotton consist mostly of the nonfortified oils formulated especially for use in cotton. Other compounds, such as dinitro selectives, CIPC, and derivatives of 2,4-D (2,4-dichlorophenoxyacetic acid), can be used as post-emergence sprays for corn, grain, and certain grass crops.

For Cotton   The National Cotton Council of America in a 1961 *Progress Report* states that:[1]

> Certain herbicidal oils may be used as the principal method of post-emergence weed control, whether or not preemergence application has been made. They may also be used in conjunction with flame cultivation. Regardless of treatment, effective control requires application of herbicidal oil when weed seedlings first appear in the drill area. Timeliness of application is basic to the success of post-emergence control. Most weeds less than two inches high are controlled easily, but larger weeds are difficult to control.

[1] Cotton Pest Control Guides, Official Recommendations for 1961, National Cotton Council.

Herbicidal oils should be applied by means of directional spray equipment. The flat, fan-shaped spray patterns should be directed across the area to be treated in a horizontal manner at a height of one inch or less from the ground level. Two nozzle tips are used per row, one on either side. They should be set about 10 in (25.4 cm) apart and staggered to prevent interference in the spray pattern [Fig. 12-23].

It is usually possible to cultivate the middles and shoulders with appropriate ground-working tools at the same time the post-emergence herbicides are being applied. Herbicide chemicals are used at the last cultivation of the crop to control weeds and grasses from lay-by to harvest.

Isolated Johnson grass plants can be eradicated by spot spraying. The spraying can be done with a knapsack or hand guns attached to a long hose from a tractor- or trailer-mounted sprayer.

As the spray nozzles must be supported within 1 in (2.5 cm) of the soil surface and the foliage of the cotton plants must be protected, special applicator shields or shoes are required (Fig. 12-14). The shield supports the spray nozzle and protects the treated drill area from soil thrown by the sweeps which are used to clean the middle area between the rows. The shield applicator is attached to the cultivator gang by a parallel-linkage arrangement that permits the bottom edge of the shield to follow the ground surface.

Oil applications should be made at intervals of 5 to 7 days. The application should stop when the basic portion of the cotton stem begins to form a bark.

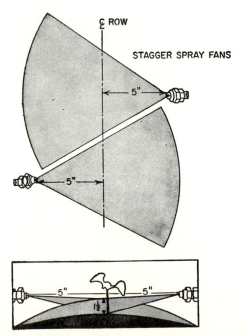

Fig. 12-23   Overhead and horizontal views of the correct method of setting nozzles for post-emergence application of chemicals for weed control.

Fig. 12-24   Method of applying weed and insect-control chemicals to corn. (*Tryco Mfg. Co., Inc.*)

The bark permits the penetration of the oil into the stem, which will cause it to swell and split, thus injuring the plant. A badly split plant stem may break.

The oil is applied at the rate of 5 gal (18.9 liters) per acre with rows spaced 40 in (101.6 cm) apart. Except for a few changes in arrangements to support the spray nozzles, the same spray equipment that is used for pre-emergence spraying or the application of insecticides can be used for spraying of post-emergence oils. Means for adjusting the height and the angle of the spray nozzles should be provided on the shields.

For Corn   It has been found that weeds in corn can be largely controlled by the use of pre-emergence and post-emergence sprays. As corn grows in height more rapidly than cotton, the spray pattern can be applied from multiple-nozzle heads attached to drops from booms that extend above and across several rows (Fig. 12-24).

A drop between the rows equipped with adjustable twin nozzles having a fan pattern of 65 to 80°, set at 30 to 35° from the vertical and 14 to 18 in (35.6 to 45.7 cm) above the ground surface, will usually give a uniform coverage for 40-in spaced rows (Fig. 12-25). As the corn grows taller, the boom must be raised and the drop lengthened to keep the nozzles near the ground.

For Grain and Pastures   The application of selective weed-killing chemicals to small-grain crops and pastures requires a long boom set at a height which

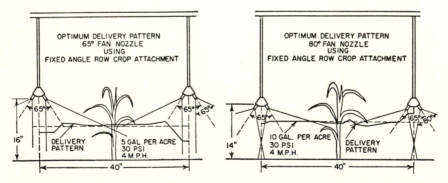

Fig. 12-25   Delivery patterns made by twin fixed-fan nozzles of 65- and 80-degree spread. The 80-degree spread gives the optimum delivery pattern for corn 14 in (35.6 cm) in height. (*Agr. Engin., 29(9):383, 1948.*)

Fig. 12-26   Spraying small grain to control weeds. (*Tryco Mfg. Co., Inc.*)

permits the nozzle pattern to give full coverage with fan-spray-pattern nozzles (Fig. 12-26). The angle of the spray fan may range from 60 to 90°, depending upon the conditions and height required for the boom. The nozzles can be attached directly to the boom and spaced to give full coverage according to the angle of the spray fan and the height of the boom. Figures 12-27 and 12-28 illustrate the influence of height of the boom and degrees of spray-fan width.

The equipment for applying herbicides should be accurately calibrated. The herbicides will usually cause the spray equipment to gum up, and it should be thoroughly cleaned and rinsed before and after each use.

The airplane is used extensively for the application of chemical dust and spray herbicides in the control of weeds in small grains and pastures.

Guides and Recommendations

Each state has developed guides and recommendations for the use of herbicidal chemicals. These can be obtained by writing to the State Agricultural Experiment Station or by contacting the local county agricultural agent. The National Cotton Council publishes a consolidated guide for all the cotton-producing states.

Electricity for Weed Control   Recent experiments with ultrahigh frequency (UHF) or microwaves of electricity show promise of their being able to

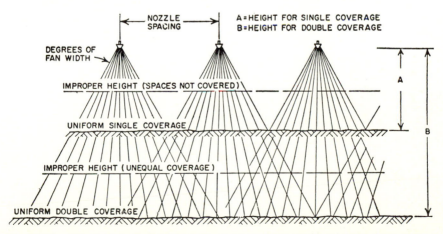

Fig. 12-27   Effect of nozzle height on the coverage obtained with the spray. (*Agr. Engin.,* *29(9):386, 1948.*)

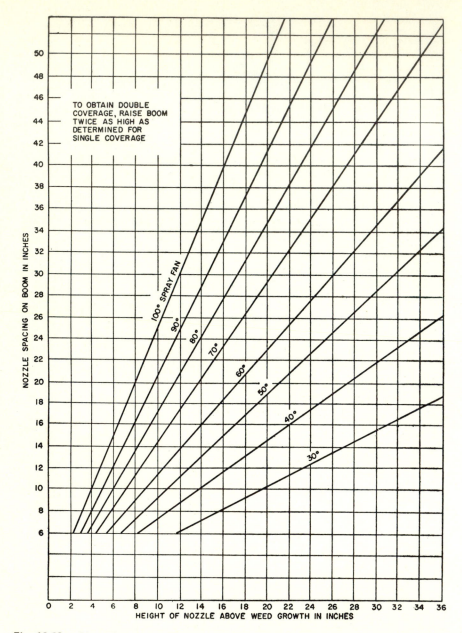

Fig. 12-28 Chart showing the height of nozzles above weeds required to give complete coverage for 6- to 48-in nozzle spacing on boom and for various nozzle spray-fan widths. (*Agr. Engin., 29(9):386, 1948.*)

kill weed seed in the soil. The microwaves are among the short waves in the electrical spectrum. The machine used in the field is called a "zapper" machine. It generates its own electricity and then converts it to microwaves. The microwaves create, inside the weed seed, heat of such intensity that it apparently kills the weed seed.

Foam Used in Weed Control   Many uses for agricultural foam have been suggested for study on the farm. Among these suggested uses is a chemical-laden foam for weed control in crops. A special foam-generating machine mounted on a tractor can spray a herbicide-laden foam in bands on both sides of a row of plants and kills weeds much like a chemical spray.

Plastic for Weed Control   Strips of plastic laid over the row after planting will delay the germination of weed seeds until the crop seeds have germinated.

MAINTENANCE AND CARE

1   Before going to the field, all cultivator gangs and sweeps should be carefully spaced and set.

2   At end of season, check over equipment for any needed repairs and place order for parts required.

3   Clean all ground-engaging parts, such as shovels and sweeps, and coat with heavy oil.

4   If not stored in shed, remove rubber-tired wheels, tag for location on machine, and store in shed.

5   Hydraulic hose connections should be covered with plastic and tied securely.

6   Chemical spraying eqipment should be carefully cleaned, especially the nozzles.

7   Flame cultivating equipment should be thoroughly cleaned and checked.

8   Store in shed or under plastic or canvas sheets.

REFERENCES

Akesson, N. B., and W. A. Harvey: Equipment for the Application of Herbicides, *Agr Engin.*, **29**(9):384-389, 1948.

Anderson, Earl D.: Engineering Developments and Challenges in Chemical Weed Control, *Agr. Engin.*, **33**(8):482-484, 1952.

Anonymous: Growers Report Experience Using Electronic Thinners, *Agr. Engin.*, **52**(3):127, 1971.

Barger, E. L., et al.: Problems in the Design of Chemical Weed-Control Equipment for Row Crops, *Agr. Engin.*, **29**(9):381-385, 1948.

Barr, Harold T., and Clair A. Brown: Weed Control with 2,4-D, *Agr. Engin.*, **28**(8):341-342, 1947.

Carter, L. M., R. F. Colwick, and J. R. Tavernetti: Evaluation Flame-burner Design for Weed Control in Cotton, *Amer. Soc. Agr. Engin. Trans.*, **3**(2):125, 1960.

Colwick, Rex. F., et al.: Weed Control Equipment and Methods for Mechanized Cotton Production, *Southern Cooperative Series Bul.* 71, 1961.

Cotton Production on Blackland Prairies of Texas, *Tex. Agr. Expt. Sta. Bul.* 984, 1961.

Creasy, L. E., and L. E. Cowart: Cultivation for Weed Control in Cotton, *Agr. Engin.*, **30**(10):490, 1949.

Holstun, J. T., et al.: Weed Control Practices, Labor Requirements, and Costs in Cotton Production, *Weeds*, vol. 8, no. 2, April 1960.

Huitink, Gary: *Flame Cultivation*, Ark. Ext. Ser. and USDA Leaflet 153, 1972.

MacDonald, W. P.: Field Spraying Equipment for Weed Control, *Agr. Engin.*, **29**(9):390-393, 1948.

Matthews, Edwin J., and Hix Smith, Jr.: Water-Shielded High-Speed Flame Weeding of Cotton, ASAE paper SWR 71-102, 1971.

McWhorter, C. G., O. B. Wooten, and G. B. Crowe: An Economic Evaluation of Weed Control Practices in the Delta, *Miss. Agr. Expt. Sta. Cir.* 203, March 1956.

Scarborough, E. N.: Cultivation and Weed Control of Cotton, *Agr. Engin.*, **30**(9):491, 1949.

Smilie, J. L., and C. H. Thomas: Discussion of Evaluating Flame-burner Design for Weed Control in Cotton, *Amer. Soc. Agr. Engin. Trans.*, **3**(2):127, 1960.

Stephenson, K. O.: Mechanized Weed Control for Cotton Production in Arkansas, *Ark. Agr. Expt. Sta. Rpt. Series* 87, 1959.

Weed Control Handbook for Mississippi, *Miss. Agr. Expt. Sta. Bul.* 612, 1961.

Williamson, E. B., O. B. Wooten, and F. E. Fulgham: Flame Cultivation, *Miss. Agr. Expt. Sta. Bul.* 545, 1956.

Wolf, Dale E.: Progress in Chemical Weed Control, *Agr. Engin.*, **30**(2):78-79, 1949.

QUESTIONS AND PROBLEMS

1 Discuss the importance of weed control in the culture of crops.
2 List the objectives of cultivation.
3 Explain the principal design features of central-forward-mounted tractor cultivators.
4 Explain the need for and action of delayed lifts for tractor cultivators.
5 List the various types and parts of a cultivator gang, and explain why a cultivator rear section is needed.
6 Explain how cultivator sweeps are set (*a*) with a line diagram and (*b*) with a setting frame.
7 Discuss the use of the rotary hoe as a cultivator attachment, and list six other cultivator attachments and explain their use.
8 Explain the use of flame for weed control.
9 Discuss and explain the difference between preemergence and post-emergence use of chemicals for weed control: (*a*) for cotton, (*b*) for corn, and (*c*) for small grains.
10 How many acres can be cultivated in a 10-hour day by a tractor and four-row cultivator traveling at the rate of 5 mi/h (8.0 km/h) if the field efficiency is 80 percent and rows are spaced 40 in (101.6 cm) apart? Calculate the cost per acre, assuming a charge of 50 cents/h for human labor and 75 cents/h for tractor.
11 Explain how weeds can be controlled by electricity, foam, and plastic.

Chapter 13

Spraying and Dusting Equipment

The problem of controlling insect pests and plant diseases makes it necessary for a large percentage of farmers and orchardists to include in their farm equipment machines for applying either dust or liquid insecticides and fungicides. It is estimated that insect pests and plant diseases cause an annual loss of $6.5 billion. In addition to these losses, there is the cost of purchasing spraying equipment and material, maintaining the equipment, and applying the sprays and dusts.

The selection of the proper equipment to combat a certain insect pest or plant disease is a problem that needs careful consideration.

SPRAYING EQUIPMENT

History of Development

Sprayers were probably first developed and used to apply fungicides for controlling diseases of grapes in vineyards in the vicinity of Bordeaux, France. The hand sprayer to combat insects was developed between 1850 and 1860 by John Bean of California, D. B. Smith of New York, and the Brandt Brothers of Minnesota.[1]

[1] *U.S. Dept. Agr. Yearbook*, p. 262, 1952.

Gasoline-engine power sprayers were developed about 1900. Tractor-mounted sprayers were not developed until several years after the introduction of the row-crop tractor in 1925. Spray booms were first attached to airplanes in the early 1940s.

Kinds of Sprays

Spray materials will usually fall within three classifications, as (1) inorganic compounds, (2) organic compounds, and (3) the oils.

The *inorganic* compounds are of mineral origin, mainly compounds of antimony, arsenic, barium, boron, copper, fluorine, mercury, selenium, sulfur, thallium, and zinc.

The *organic* compounds are synthetic compounds. Some organic compounds have been used as sprays for many years.[1] Carbon disulfide and naphthalene have been used for many decades. Other organic compounds that have been used for a quarter of a century are ethylene dichloride, ethylene dibromide, methyl bromide, and various thiocyanates. Newer groups of organic chemical spray compounds are the dinitro derivatives of phenol and cresol and the chlorinated hydrocarbons. Of this latter group DDT is the best-known. Other widely used organic compounds are Toxaphene, Chlordane, Aldrin, Dieldrin, Parathion, and 2,4-D.

Petroleum oils are used alone or to supplement the action of insecticides, fungicides, and herbicides. Oils are often added to a spray as stickers, stabilizers, and conditioning agents.

Function of a Sprayer

Bronson and Anderson in the 1952 *U.S. Department of Agriculture Yearbook* define the function of a sprayer as follows:

> The main function of a sprayer is to break the liquid into droplets of effective size and distribute them uniformly over the surface or space to be protected. Another function is to regulate the amount of insecticide to avoid excessive application that might prove harmful or wasteful.
>
> A sprayer that delivers droplets large enough to wet the surface readily should be used for proper application of surface residual sprays. Extremely fine droplets tend to be diverted by air currents and be wasted.

Types of Sprayers

There is available a type of sprayer for every use, in the home, garden, orchard, and field. Garden sprayers are manually operated, but field and orchard sprayers are power-operated.

Power Sprayers The term *power sprayers* in this discussion applies to sprayers operated with either internal-combustion engines or electric motors. They may be operated by gasoline engines of suitable size, or the sprayer may be

---

[1] *U.S. Dept. Agr. Yearbook,* p. 209, 1952.

operated by tractor power. On the following pages, the National Sprayer and Duster Association classifies and describes the various types of power sprayers.[1]

Hydraulic Sprayers
    Multiple-purpose Sprayers
    Small General Use Sprayers
    High-pressure, High-volume Sprayers
    Low-pressure, Low-volume Sprayers
    Self-propelled, High-clearance Sprayers
Hydro-pneumatic Sprayers
Blower Sprayers
Aerosol Generators

HYDRAULIC SPRAYERS

Most power sprayers in use today are the hydraulic type in which the spray pressure is built up by the direct action of the pump on the liquid spray material. The pressure thus developed forces the liquid through the nozzles, which break the spray into the proper size droplets and disperse them in the spray pattern desired; also, sufficient energy is imparted to the spray droplets to carry them from the nozzle to the surface to be treated.

The essential parts of the typical hydraulic sprayer are: pump (with air chamber, if required), tank containing an agitator, framework for mounting the sprayer, combined pressure regulator and unloader or relief valve, pressure gage, strainers and screens, control valves, piping and fittings, distribution system, and power source.

Pumps   Most hydraulic sprayers are equipped with positive displacement pumps capable of developing the pressures in the range required for many spray jobs. The discharge capacity of these pumps is approximately proportional to the speed. A pressure relief or by-pass valve is required to protect these positive-acting pumps from damage when the discharge line is closed and for the convenience of the operator.

The *reciprocating pumps—plunger* and *piston* types—have been the standard of the spraying industry for many years because of their satisfactory performance in pumping almost any spray materials, including wettable powder forms of pesticides and water-base paints in a wide range of pressures. The *piston* pumps [Fig. 13-1] are commonly used in the output range of less than two to eight g.p.m. [7.6 to 50.3 liters/min] and up to 400 p.s.i. [27.6 kg/cm^2] or more, whereas the *plunger* type is used in the range of about seven to sixty g.p.m. [26 to 227 liters/min] and pressures up to about 1000 p.s.i. [69 kg/cm^2]. Multiple units are used to achieve the higher capacities. These pumps are usually mounted on the sprayer frame although some piston types have been adapted for tractor power-take-off mounting and powering. An air chamber is supplied with *reciprocating* spray pumps to level out the pulsations of the pump and provide a constant nozzle pressure.

[1] This copyrighted classification and description of sprayers is reproduced from the *Sprayer and Duster Manual,* 1955, of the National Sprayer and Duster Association by special permission.

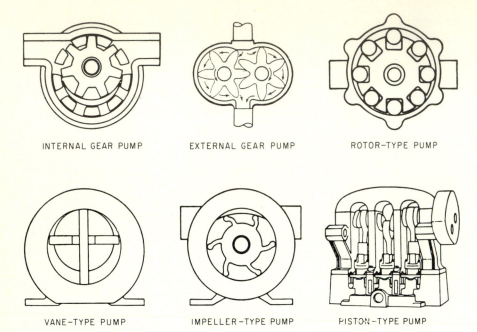

INTERNAL GEAR PUMP   EXTERNAL GEAR PUMP   ROTOR–TYPE PUMP

VANE–TYPE PUMP   IMPELLER–TYPE PUMP   PISTON–TYPE PUMP

Fig. 13-1   Types of sprayer pumps.

It should be emphasized that these pumps, which are capable of developing the higher pressures, are also used satisfactorily at the lower pressure range of 20 to 50 p.s.i. [1.41 to 3.45 kg/cm^2] and any pressure desired up to the indicated maximum.

Rotary pumps have been introduced into the spraying field in recent years for use in the lower pressure range. Compact and light in weight, these pumps are commonly mounted on and powered by the tractor power-take-off shaft. Oversize units are usually specified to provide sufficient by-pass liquid for hydraulic agitation in the tank. *Gear* pumps [Fig. 13-1] are classified as positive displacement and self-priming and are available in sizes up to about 20 g.p.m. [75.7 liters/min]. Certain design features limit their use to sediment-free types of spray materials. Their service life is greatly reduced if used to pump wettable powers or any other abrasive type of spray material or if attempt is made to operate in the upper range of spraying pressures. *Vane* [Fig. 13-1] or *roller impeller pumps* have about the same operating characteristics and range of use in spraying as the *gear-type pumps*. The flexible *impeller-type pump* has performance characteristics between the positive-displacement and the centrifugal pumps with a maximum operating pressure of about 50 p.s.i. [3.45 kg/cm^2].

Diaphragm-type pumps, suitable for spraying at pressures up to about 100 p.s.i. [6.9 kg/cm^2], have also been adapted for direct mounting on the power take-off of the tractor. They are relatively unaffected by abrasive-type spray materials.

Centrifugal pumps are being used extensively for sprayers. This type of pump can handle most any type of spray material. They have a long life and have a wide range of gallons per minute.

Tanks   Metal tanks are supplied on many models of power sprayers because they are easier to clean of spray residues when changing spray materials. Wood tanks, usually made of cypress, are also available, however. [Tanks are also made of coated steel, stainless steel, aluminum, fiber glass, and poly.] The size of tanks varies with the different models from about 5 gallons [18.9 liters] to 500 gallons [1892.0 liters] or more capacity to suit the wide range of spraying needs. A large covered opening, fitted with a removable strainer, is provided in the top for easy filling, inspection and cleaning. A drain plug in the tank bottom permits thorough drainage when cleaning.

Agitators   Positive agitation of the spray material in the tank is essential to permit using the full range of spray materials, including wettable powders, emulsions, fungicides, cold water paints or any other sprayable material. A propeller or paddle-type mechanical agitator is usually provided in sprayers equipped with reciprocating-type pumps. Hydraulic agitation is used in sprayers provided with pumps mounted on the power-take-off shaft. Hydraulic agitation usually is not as thorough as mechanical agitation.

Air Chamber   With the reciprocating-type pump, an air chamber is provided on the discharge line of the pump to level out the pulsations of the pump, thereby providing a constant nozzle pressure.

Pressure Gage   A pressure gage [Fig. 13-2] properly calibrated within the pressure range of the pump, is provided on the discharge line to guide the operator in properly adjusting the pressure for each spray job.

Pressure Regulator   The pressure regulator [Fig. 13-2] serves several important functions. It is the means of adjusting the pressure as required for any spray job within the pressure range of the pump. With the positive displacement type of pump, it also serves as a safety device in automatically unloading the excess pressure, directing the unused discharge flow from the pump back to the tank. When provided with an unloader, it permits the

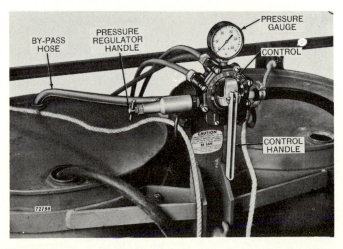

Fig. 13-2   Typical selective control valve. It has an off position, seven positions for selecting any one of several combinations of three-boom sections, and an outlet for a handgun. (*Deere & Co.*)

pump to operate at a greatly reduced load when the discharge line is closed.

Other Valves   One or more valves may be included in the piping system for use in connection with a tank filler [Fig. 13-3], also a ratchet-type, quick-acting, cut off valve is usually supplied to control the flow to the boom. Some boom-equipped sprayers have a special manifold-type control valve for quick flow control of any section or combination of sections of the boom [Fig. 13-3].

Strainers   A strainer is included in the suction line between the tank and the pump to remove foreign material which might affect the operation of the check valves, pump, and nozzles. These strainers, sometimes replaced with a sediment bowl, are easily removed for cleaning and are replaceable.

Distribution Systems   Several types of distribution systems are used with hydraulic-type sprayers [Fig. 13-2] —a hand-held spray gun, an automatic spray head and a field boom to take care of the various types of spraying jobs. The conventional hand gun contains one or more nozzles in a suitable mounting and a fast acting control valve which serves as a cut-off and means of adjusting the spray pattern.

Automatic *spray heads* have proved effective in recent years in materially reducing the labor required in fruit tree spraying. The devices consist essentially of a number of spray nozzles arranged on a vertical boom or series of booms attached to the sprayer. Most of these devices have an automatic oscillating mechanism to insure thorough spray penetration and coverage.

The *spray boom* or *field boom* consists essentially of a horizontal structural member on which the nozzles are properly spaced and mounted [Fig. 13-3]. This member, if tubular in shape, may be used as part of the piping system to supply the spray material to the nozzles, but usually it serves only as support and protection for the liquid supply line which may

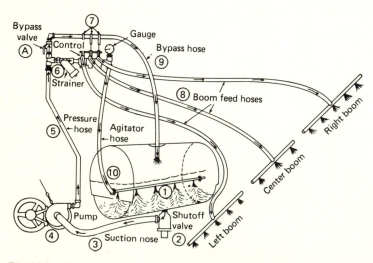

Fig. 13-3   Schematic view of typical power field sprayer operated by power takeoff of the tractor. The pump draws chemical out of tank and sends it through strainer and control valves on to nozzles. Bypass valve controls pressure by restricting flow through bypass hose. (*Century Engineering Corp.*)

be of brass or synthetic rubber mounted on or within the structural member.

The boom is adjustable vertically usually from about 18 to 72 in [45.7 cm to 1.8 m] for spraying plants of various heights and is divided into three or more sections connected by flexible joints to permit passage through farm gates and spraying uneven terrain, such as a highway shoulder and as a protection against damage from field obstructions. The most common length of boom for general field usage is about 21 feet [6.3 m] with an effective spray coverage of about 23 feet [7.0 m] or six rows spaced approximately 36 to 40 in [91.4 to 101.6 cm] apart. The outer sections of booms 30 feet [9.1 m] or longer are usually supported by auxiliary outrigger wheels. Counter-balanced hydraulically operated one side booms are also available for covering widths up to 40 feet [12.2 m] or more at the side of the sprayer.

The *boomless power jet sprayer* consists of an assembly of one to five nozzles supported on a single bracket [Fig. 13-4]. The bracket may be mounted on the drawbar, front, or side of the tractor. The sprayer is adapted for the broadcast spraying of grain fields, pastures, ditches, fence rows, and orchards [Fig. 13-4]. The spray can be directed in any desired direction and will cover a swath from 20 to 50 feet [6.0 to 15.2 m] wide, depending on the height of the nozzles and the pressure used. The pump pressure may be as high as 300 pounds [135.9 kg].

Nozzles   The nozzle is the all-important mechanism which breaks the spray liquid into the desired size of droplets for application to the surface to be sprayed [Fig. 13-5]. Since no single nozzle can meet all of the various spray requirements, they are now commonly manufactured with inexpensive replaceable nozzle tips or discs which can be selected to give the desired spray characteristics and volume for the specific job. The nozzles vary with respect to rate of discharge—gallons per minute or per acre, the angle of spray and the type of spray pattern, that is, hollow cone, solid cone or flat-fan [Fig. 13-5]. Built into some nozzles is a removable strainer [Fig. 13-6] with slightly smaller openings than the nozzle orifice to prevent clogging. Nozzle drops, pendants, or drop extensions are used for row crop work to replace the spray material more accurately [Figs. 13-7 and 13-8].

Fig. 13.4   Boomless sprayer spraying pasture land. (*Century Engineering Corp.*)

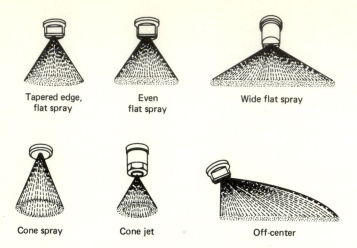

Tapered edge, Even Wide flat spray
flat spray flat spray

Cone spray Cone jet Off-center

Fig. 13-5   Various nozzle-spray patterns. (*Spraying Systems Co.*)

Selection of Nozzles   Nozzle manufacturers' data sheets give the discharge of various nozzles at different pressures. This information can be used to select the correct size for the spray job to be done. If not available, the manufacturer should be furnished the following information so that he can supply the proper nozzle:

1   Type of spray job—i.e. pasture, weed spraying, insecticide, etc.
2   Total amount of spray solution to be applied per acre for each spraying.
3   Row spacing and number of nozzles to be used per row, if application is to be on row crops.
4   Nozzle spacing if the entire area, as in pasture work, is to be sprayed.

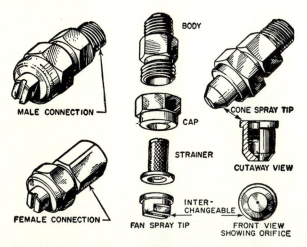

Fig. 13-6   Fan- and cone-pattern spray-nozzle assemblies for applying herbicides and insecticides. (*Spraying Systems Co.*)

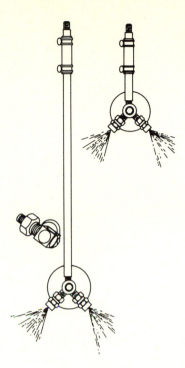

Fig. 13-7 Long- and short-nozzle extensions equipped with double-swivel nozzles. A single-swivel nozzle is shown at left.

5   Type of spray pattern desired, such as fan or cone type.

6   Approximate speed of travel. For tractor powered ground rigs this will usually be 3 to 5 mph [4.8 to 8.0 km/h].

7   Approximate pressure to be used in spraying. For example, in most weed work this will be 25-40 psi, while for defoliation the pressure range will be 40-60 psi.

If a spray rig, complete with nozzles, is to be purchased from an equipment manufacturer or dealer, the above information will aid him in outfitting your sprayer with the proper-size nozzles.

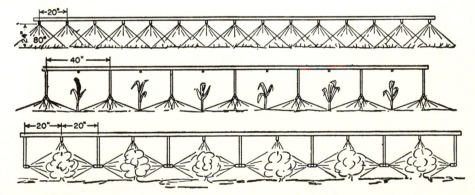

Fig. 13-8   Three methods of arranging nozzles on boom: (top) complete overtop coverage for weeds and between rows for weed control; (bottom) over and between rows for insect control. (Wyatt Mfg. Co.)

Effect of Nozzle Type and Arrangement   The author[1,2] and coworkers[3] began a study in 1952 to determine the effect of nozzle type and arrangement on the control of cotton insects.

The results covering a 5-year period show that a single hollow cone nozzle directly over the row gave better insect control and resulted in higher yields of seed cotton than two and three nozzles arranged about the row of plants (Fig. 13-9). The fan- and boomless-type spray nozzles did not give as good results as the hollow nozzle (Fig. 13-10). Figure 13-11 shows the relative size of spray droplets made by a boomjet nozzle across six rows spaced 40 in (101.6 cm) apart. The highest yields were produced on the rows nearest the nozzle and where a more complete coverage was obtained with the smaller droplets.

Types of Sprayer Mounting   Most sizes of hydraulic sprayers are available mounted either on skids or on wheels. The skid models, of course, are less expensive and preferred for many stationary spraying jobs. They can be transported in truck, wagon, trailer or jeep or the smaller models with 50-gallon [189.2 liter] or smaller tanks may be mounted on a platform on the rear of a farm tractor. The wheel mounted models usually have two wheels, although some are mounted on four. Some special use sprayers are designed for direct mounting on the tractor frame.

Power Source   Power is supplied to the skid models by an engine or electric motor which is an integral part of the unit. Wheel mounted models may be powered by a self-contained engine or by the power-take-off shaft of the towing tractor or truck. The size of the engine furnished varies with the pump capacity and pressure range. Most pumps of tractor-mounted

	CHECK PLOT NO SPRAY	ARRANGEMENT OF NOZZLE TYPES					
		CONE TYPE NOZZLES				FAN TYPE NOZZLES	
		One nozzle per row	Two nozzles per row	Three nozzles per row	Nozzles spaced 20" apart on boom	Nozzles spaced 20" apart on boom	One nozzle per row
Year	Yield per acre — pounds seed cotton						
1955	825	1,275	1,160	1,145	1,315	1,280	1,365
1956	315	402	357	500	434	472	419
Av 52–56	759	1,244 (5yr)	1,226 (5yr)	1,219 (5yr)	956 (3yr)	907 (3yr)	921 (3yr)
		Cone pattern				Fan pattern	

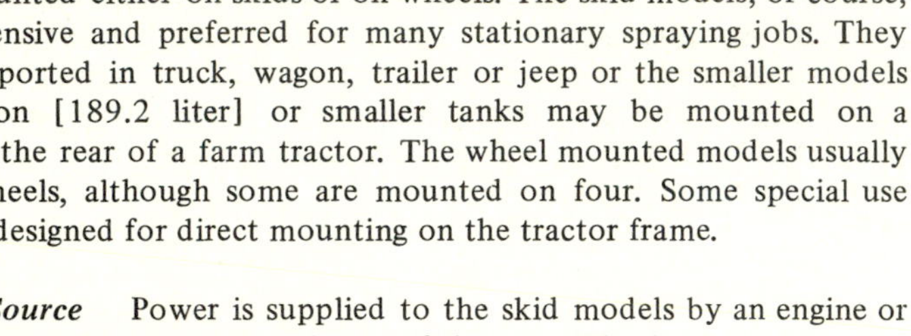

Fig. 13-9   Effect of nozzle type and arrangement on the control of cotton insects as measured by yield. (*Tex. Agr. Expt. Sta.*)

[1] H. P. Smith and R. L. Hanna, Effects of Type and Arrangement of Spray Nozzle on the Control of Bollworm and Boll Weevil, *Tex. Agr. Expt. Sta. Prog. Rpt.* 1752, 1955.
[2] H. P. Smith, C. M. Hohn and R. L. Hanna, Effects of Spray Nozzle Types and Arrangements on Cotton Insect Control, *Tex. Agr. Expt. Sta. Prog. Rpt.* 1906, 1956.
[3] L. H. Wilkes, P. L. Adkisson, and R. J. Cockran, Effect of Spray Nozzle Types on Cotton Insect Control, *Tex. Agr. Expt. Sta. Prog. Rpt.* 2078, 1959.

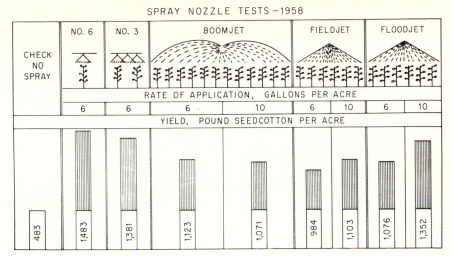

Fig. 13-10 Effect of nozzle types and rate of application on the control of cotton insects as measured by yield. (*Tex. Agr. Expt. Sta.*)

sprayers are powered by direct connection to the power-take-off shaft or by the belt pulley. Others are engine powered.

Tank Filler This is a useful device which utilizes the sprayer pump through an injector in filling the tank rapidly from any convenient source of clean water, such as a stock watering tank, cistern, pond or stream. A typical unit consists of an injector, a 15′ or 20′ length [4.5 or 6.0 m] of

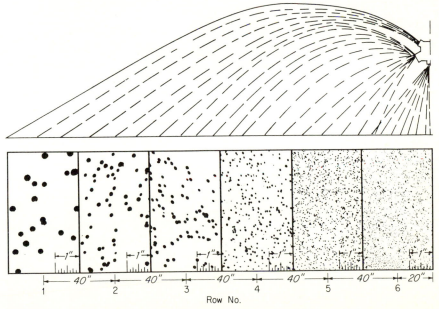

Fig. 13-11 Showing the relative size of spray droplets made by a wide-jet spray nozzle across six rows spaced 40 in (101.6 cm) apart. (*Tex. Agr. Expt. Sta.*)

suction hose equipped with a strainer and foot valve and connection for attachment to the piping system of the sprayer. For best operation the source of water should be within 10 to 12 ft [3.0 to 3.6 m] vertical distance from the sprayer pump.

Multiple-purpose Farm Sprayers   This type of sprayer has been developed to provide in one machine the versatility required to meet the many spraying needs on diversified farms. It has the necessary pressure range to be used for weed spraying or for spraying fruit trees or livestock. Because of the type of pump and agitation provided, the user has a free choice of spray materials since wettable powders, cold water paints and other spray materials which are abrasive or difficult to keep in suspension can be used satisfactorily in this type of sprayer.

Small General Use Sprayers   Small general use sprayers are available for those spraying jobs which are just too large or too tedious for coverage with hand equipment. They are popular for use on golf greens, estates, acreages, large gardens or greenhouses. Included in this class are the power wheelbarrow sprayer, the estate or small wheel mounted sprayer and the small skid mounted units.
Power wheelbarrow sprayers have reciprocating-type pumps which deliver $1\frac{1}{2}$ to 3 g.p.m. [5.7 to 11.3 liters/min] and develop pressures up to 250 lb/in^2 [17.3 kg/cm^2]. Power is furnished by an air cooled engine. For greenhouses or other interior stationary jobs, the sprayer can be operated by a $\frac{1}{2}$ h.p. heavy duty electric motor. The tank capacities of the different makes vary from $12\frac{1}{2}$ to 18 gallons [47.2 to 68.0 liters]. Standard equipment includes mechanical or jet agitator, pressure gage, pressure regulator and relief valve and an adjustable hand gun and hose. Some makes can be provided with a spray boom.

High-pressure, High-volume Sprayers   These sprayers used by the commercial growers of fruit and truck crops are the product of many years of successful development by the manufacturers in cooperation with the growers. The machines combine the high pressure and volume delivery desired to get complete spray coverage of high growing fruit and shade trees in full foliage, as well as the dense growth of vine and other truck crops.
The essential parts of these sprayers are basically of the same design as those described for the multiple-purpose farm sprayers with a few important exceptions as follows:
Plunger-type, multiple-cylinder reciprocating pumps are used almost exclusively with these larger sprayers, delivering from about 8 to 60 g.p.m. [30.2 to 227.0 liters/min] at maximum pressures in the range of 400 to 1000 p.s.i. [27.6 to 69 kg/cm^2]. Tank sizes are also correspondingly larger. Sprayers having pumps discharging up to 50 or 60 g.p.m. [189.0 or 227.1 liters/min] can be furnished with tanks up to 600 gallons [2271.0 liters] capacity.
Field booms designed for use with these sprayers may be of an adjustable type or of special design for use primarily on one specific crop. Nozzles commonly used are of the large volume type producing a cone shaped pattern. For many crops the nozzles are suspended on pendants or

nozzle drops to permit complete underleaf coverage of foliage. The hydraulically operated one side boom is often used with these sprayers for such crops as potatoes and tomatoes.

For fruit tree work the spray may be applied by a hand gun or an automatic spray head. The modern guns have been improved in design for quick shutoff, easy handling and for adjustment of the spray pattern. Safety towers mounted on the top or rear of the sprayers are available for more accurate spray placement and the protection and convenience of the operators. Automatic spray heads can be supplied with new equipment or furnished to convert older sprayers to one-man operation.

Low-pressure, Low-volume Sprayers   These sprayers have been designed primarily to meet the specific requirements for low-volume field spraying. Because of their relatively low cost, they have proved popular for use in controlling weeds and insects in field crops and for other spray jobs for which pressures under 100 pounds are satisfactory. Their use also is usually limited to the sediment-free type of spray materials due to the type of pump and agitation provided.

Two different models of these sprayers are available—*tractor mounted* and *wheel mounted*. The *tractor-mounted* unit consists of a kit of components parts which are mounted by the user on the tractor [Fig. 13-12]. Included in the kit are a pump, strainer, control valves, pressure gage, pressure regulator and relief valve, spray boom and brackets for mounting the boom and one or more metal drums. A pump is usually supplied for direct mounting on the tractor power-take-off shaft. Hydraulic agitation is provided by the by-pass from the pump.

The *wheel-mounted* or *trailer-type* sprayers [Fig. 13-13] in this class have the same performance characteristics and field use as the tractor-mounted units discussed above. They are more quickly readied for use, however, since the tank and boom are permanently mounted on the trailer. The tractor-powered units require mounting the pump on the tractor and connecting it to the power source for each use. Some models, however, have a self-contained engine. A conventional sprayer tank of from 50 to 250 gallons [189.0 to 945.0 liters] capacity is supplied with some units. Others are provided with one or more 50 gallon [189.0-liter] metal drums mounted on the trailer frame. These trailer tank units also can be used to convert a tractor-mounted, tractor-powered sprayer to the more convenient trailer-type machine.

Self-propelled High-clearance Sprayers   This sprayer has been developed as a special purpose machine to spray those field and row crops which are too high for conventional power sprayers and tractors. With the sprayer

Fig. 13-12   Boom-type sprayer mounted on rear of tractor. (*Spraying Systems Co.*)

Fig. 13-13   Trailer-type field sprayer showing tank and boom. (*Deere & Co.*)

removed, the carrier unit or chassis is useful for detasseling corn and for transporting vegetables and crates in the field.

The carrier to which the sprayer is attached may consist of a specially designed chassis which is usually powered by an air cooled engine [Fig. 13-14]. The units powered by an air cooled engine usually have three wheels and the power is commonly applied to the front wheel. These machines usually are equipped with a low-pressure, low-volume sprayer, and a second engine is often provided to power the sprayer. The spray boom is usually mounted on the rear of the machine. The height of the boom can be varied to permit spraying either high or low growing crops.

HYDRO-PNEUMATIC SPRAYERS

Sprayers of this type have about the same range of use as the low-pressure, low-volume sprayers previously described. The spray liquid is carried in a pressure tank and the spraying pressure is developed by means of an

Fig. 13-14   Self-propelled high-clearance sprayer equipped with fenders, which are essential when spraying cotton, corn, and other tall crops in mid-late season. The boom can be raised and lowered to suit field conditions. (*Deere & Co.*)

engine-powered air compressor. The spray material therefore does not pass through a pump or contact any other moving parts.

These sprayers are available mounted on skids or wheels. Because of the cost and weight of pressure tanks, the tank size is usually limited to less than 300 gallons [1135.5 liters] and maximum working pressure to 75 to 100 pounds [33.9 to 45.3 kg]. The compressors are usually rated in cubic feet per minute volume at atmospheric pressure. One cubic foot per minute at the desired pressure, therefore, will equal about $7\frac{1}{2}$ gallons [28.3 liters] per minute. Agitation is provided either by means of a mechanical agitator or by an air tube discharging air below the surface of the liquid in the tank. These units are almost always powered by a self-contained engine.

BLOWER SPRAYERS

These relatively new sprayers [Figs. 13-15 and 13-16], also known as concentrate or mist sprayers, have been developed to apply pesticides in concentrated form. Substantial saving in labor costs is thereby effected since the quantity of water required as a diluent may be reduced from 20 to 80 percent or more as compared with conventional dilute methods of spraying. Further savings are also reported in the quantity of chemicals used, since runoff from foliage may be reduced when the equipment is properly operated.

These sprayers are used for treating large acreages of fruit trees, large shade trees, vegetables and certain other crops. Large fruit trees may require thinning and pruning to permit proper penetration of the small airborne

Fig. 13-15 Blower-type sprayer in operation. (*John Bean Division, Food Machinery and Chemical Corp.*)

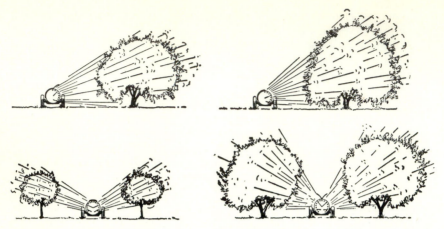

Fig. 13-16 Four spray patterns that can be delivered by sprayers. (*John Bean Division, Food Machinery and Chemical Corp.*)

droplets to the inner and top-most branches. The swath or strip covered will vary with wind direction and velocity.

These machines are basically similar to power dusters in that a blast of air is employed to carry the chemical from the machine to the foliage to be treated except, of course, that the chemical is in liquid, rather than dry, form. The typical concentrate sprayer of the blower type utilizes a low-pressure, low-volume pump which forces the spray material under low pressure to the fan where it is discharged into the airstream in small spray droplets by a group of nozzles or a shear plate. The air stream assists in breaking up the liquid into small particles, acts as a diluent to prevent the drops from coalescing and serves as the vehicle to carry these fine droplets to the surface to be treated.

AEROSOL GENERATORS—FOG MACHINES

These machines disperse the spray material in the form of extremely fine droplets (1-50 microns[1] in diameter) which remain airborne for a considerable period of time. The insect kill with this equipment is dependent upon the airborne insecticide contacting the insects since there is usually little or no residual action. This equipment is employed for the temporary control of adult mosquitoes, flies and other similar insects in buildings and in restricted areas such as ballparks, picnic grounds, resort areas or even large communities.

CONTROLLING THE SPRAY APPLICATION

The gallons per acre applied on a field will depend upon (1) the forward speed of the sprayer and (2) the number of nozzles and their rate of discharge.

---

[1] A micron is a thousandth of a millimeter; a section of a diameter between one-hundredth and one-millionth of a millimeter.

The importance of forward speed is readily apparent. If the forward speed is suddenly changed from 5 to 10 m.p.h. [8.0 to 16.1 km/h] or doubled, then the nozzles will have only half the time to deliver their spray in traveling a given distance, and the gallons applied in that distance will, therefore, be cut in half. Constant speed is especially important in applying sprays at high concentrations, that is, low gallonage rates. At a given throttle setting, tractor speeds will be fairly constant on level terrain. On the other hand there may be considerable variation in speed going up and down hills. For extensive spraying operations a tractor speedometer is recommended as insurance against the high cost and possible damage of overdosing and the ineffectiveness of underdosing. Changing from one constant rate of travel to another is a means of changing the rate of spray application on a field within the range of practical operational speeds. Thus patches of heavy weed growth can be given a heavier rate of application than the remainder of the field merely by reducing the forward speed over such areas.

[Tables 13-1 and 13-2 give data on the rate of travel and the time required to spray an acre.]

Effective spraying cannot be done when the wind velocity is above 10 m.p.h. [16.1 km/h]. A simple wind gage should be used to determine the wind velocity.

Assuming a fixed number of nozzles on a sprayer, the rate of discharge can be changed by increasing or decreasing the sprayer pressure within certain limits or by replacing the nozzles, nozzle tips or discs with similar units of a higher or lower capacity.

There are certain limitations to increasing the rate of delivery of nozzles by increasing the pressure. Increasing the pressure does not proportionately increase the discharge rate. For most nozzles the pressure must be increased about four times to double the delivery output. Increasing the pressure also tends to decrease the droplet size of the spray delivered and thus increase spray drift. The spray pattern may also be

Table 13-1   Spraying Time per Acre with Tractor-operated Sprayer*

		Width of boom						
Speed		10 ft (3.0 m)	15 ft (4.6 m)	20 ft (6.1 m)	25 ft (7.6 m)	30 ft (9.1 m)	40 ft (12.2 m)	50 ft (15.2 m)
mi/h	km/h			Minutes required to spray 1 acre				
2	3.2	24.8	16.5	12.4	10.0	8.3	6.2	5.0
3	4.8	16.5	11.0	8.3	7.0	5.5	4.1	3.3
4	6.4	12.4	8.2	6.2	5.0	4.1	3.1	2.5
5	8.0	9.9	7.0	5.0	4.0	3.3	2.5	2.0
6	9.6	8.3	5.5	4.2	3.3	2.8	2.1	1.5
7	11.2	7.1	4.7	3.5	2.8	2.4	1.8	1.4
8	12.8	6.2	4.1	3.1	2.5	2.1	1.6	1.2
9	14.4	5.5	3.7	2.8	2.2	1.8	1.4	1.1
10	16.0	5.0	3.3	2.5	2.0	1.7	1.2	1.0

* No allowance made for turning or for filling or servicing sprayer.

Table 13-2   Travel Speed

		Time required to travel*								
		100 ft		**500 ft**		**660 ft** **(40 rd)**		**1320 ft** **(80 rd)** **($\frac{1}{4}$ mi)**		**2640 ft** **(160 rd)** **($\frac{1}{2}$ mi)**
mi/h	ft/min	min	s	min	s	min	s	min	s	min	s
1	88	1	8	5	39	7	30	15	0	30	0
2	176	0	34	2	50	3	45	7	30	15	0
3	264	0	23	1	55	2	30	5	0	10	0
4	352	0	17	1	25	1	53	3	45	7	30
5	440	0	14	1	8	1	30	3	0	6	0
10	880	0	7	0	34	0	45	1	30	3	0

* To closest second.

affected. Generally speaking the pressure should not be varied more than 25% above or below the optimum pressure for the nozzle. Instead, changing to a nozzle tip or disc of the desired rated capacity is the preferred method of changing to a substantially different rate of spray application.

In some spraying operations it is desirable to change the rate of application by changing the number of nozzles on the boom. Thus in the case of some row crops the number of nozzles covering each row is increased from one to three or more as the size of the plants increase and present a larger surface area of foliage to be covered.

The practical relationship of these various factors of forward speed, pump capacity, nozzle size and length of boom is demonstrated in the several computations which follow for determining length of boom, size of pump and size of nozzle required for specific jobs.

FIXED-WING AIRPLANE SPRAYER

The use of the airplane to apply and distribute spray material has become one of the most popular and economic methods of applying insecticides and fungicides to agricultural crops. Figure 13-17 shows a commercial-type agricultural airplane.

Extensive studies have been made on the influence of nozzle spacings on the spray patterns. The arrangement and location of the nozzles on the boom under the airplane were studied. It was found that the center of deposit for each nozzle did not directly follow the arrangement of the nozzles. As shown in Fig. 13-18 the center of deposit for each nozzle 16 ft (4.9 m) from the center of the aircraft (dotted line) was farther out than for a nozzle located 17 ft (5.2 m) from the center of the aircraft. It is noted that there was no cross wind in these tests.

Crosswinds influence the spray pattern and uniform distribution of spray materials applied by aircraft, as shown in Fig. 13-19. The top section of the illustration shows the weighing units used to collect the spray material across the

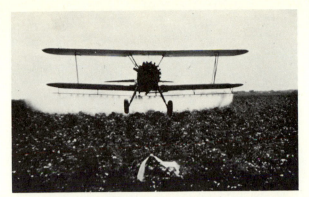

Fig. 13-17 Front view of airplane applying liquid spray. (*Texas A & M University*.)

spray pattern made by the aircraft. The center and bottom sections of the illustration show the volume of material collected by each of the weighing units. The height of the column represents the load on the scale as deposited in a single swath. It is noted that the cross wind of 2 mi/h (3.2 km/h) caused a considerable drift to the left.

Effect of Droplet Size on Coverage and Drift

Table 13-3 shows the effect of droplet size on the coverage and drift for a dispersal rate of 4 gallons (15.1 liters) per acre. The table shows that rain-sized

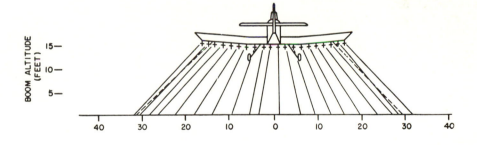

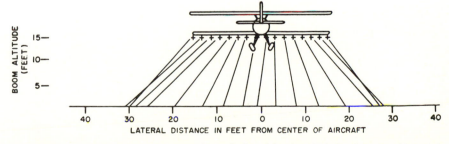

Fig. 13-18 Some nozzle locations on the agricultural (*top*) and Boeing Stearman (*bottom*) aircraft with their respective centers of deposit at ground level (zero crosswind). (*Texas Engineering Expt. Station, Aeronautical Laboratory*.)

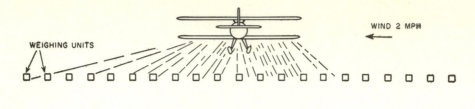

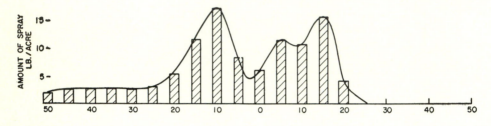

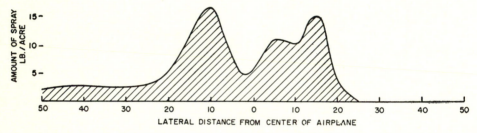

Fig. 13-19 Typical spray pattern for a single swath for a Boeing Stearman plane. (*Texas Engineering Expt. Station, Aeronautical Laboratory.*)

Table 13-3   Effect of Droplet Size on Coverage and Drift for Dispersal Rate of 4 Gallons per Acre*

Diameter of droplets, microns	No./in^2	Distance between droplets, in	Time to fall 10 ft	Distance carried in 10 mi/h uniform drift	Notes
1650	1	1.0	1.0 sec	14 ft	Heavy rain. Diam. of pin head. Largest recorded at A. & M.
1000	4.6	0.47	1.1 sec	16 ft	Moderate rain
500	37	0.16	1.6 sec	23 ft	Light rain
200	570	0.043	4.2 sec	62 ft	Drizzle
100	4600	0.015	11 sec	170 ft	Mist
50	36,800	0.005	40 sec	592 ft	Smallest size recorded at A. & M.
10	4.6 million	0.0005	17 min	2.8 mi	Aerosol
1	4.6 billion	0.000015	28 h	280 mi	Aerosol

* Assuming uniform droplet size and spacing, and no evaporation loss.
Source: Personal Aircraft Research Center, Texas Agricultural Experiment Station, College Station, Texas.

droplets in a 10-mi/h (16.0-km/h) wind will drift from 14 ft (4.3 m) for large-sized raindrops to 62 ft (18.9 m) for small-sized raindrops. Fine droplets from aerosol sprays may drift as far as 280 mi (451 km).

Sprayer Nozzles for Aircraft

As shown in Fig. 13-20, there are two types of nozzle available for spraying liquids with aircraft. Each type of nozzle has a check valve to stop the flow of the liquid when the pressure is cut off. It is claimed that, when the spray droplets are discharged from the side of the nozzle into the air stream, the shearing action of the air at high velocity as it passes the nozzle tip will further reduce the size of the droplet.

ROTARY-WING HELICOPTER SPRAYER

The rotary-wing helicopter is used for spraying orchards and areas where fixed-wing airplanes have difficulty in maneuvering. The helicopter can move slowly at whatever speed is necessary to thoroughly dispense the spray material most effectively. The disadvantage of the rotary-wing machine is that it is expensive, costing much more than the fixed-wing airplane.

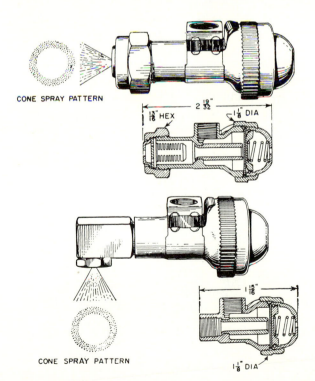

Fig. 13-20 Two types of sprayer nozzles for aircraft: (*top*) discharge at tip and in line with air stream; (*bottom*) discharge on side and at right angles to air stream. (*Spraying Systems Co.*)

DUSTING EQUIPMENT

Dust insecticides and fungicides were first applied to plants by placing the dust in a thinly woven bag. The bag was then shaken over the plants. Later, bags were attached to each end of a pole so that a man riding a horse or mule, at a trot, along the row of plants could hold the pole and suspend bags over the rows and shake the dust from the bags onto the plants. A hand duster equipped with a hopper, fan, and discharge tubes was invented by W. R. Monroe of Unionville, Ohio, in 1895. This duster (called Sirocco) was converted to a horse-drawn and traction-powered machine about 1897. A gasoline engine was adapted to power-operate this duster in 1911. About 1920, hoppers, fans, and discharge tubes were mounted on horse-drawn carts to apply chemical dusts to cotton. Long, flexible metal tubes were arranged on booms so the dust was discharged over six to eight rows of plants. Tractor-mounted field dusters were developed in the late 1920s and early 1930s.

Types of Dusters

Dusting equipment can be generally classified as field and orchard power dusters.

Tractor-powered Field Dusters   Figure 13-21 shows a dusting unit mounted upon a platform bolted to the rear of a tractor and operated by the power takeoff. Operating the tractor in high gear makes it possible to dust a larger acreage than with horse-drawn machinery of equal row capacity. Auxiliary engines are used on some tractor-mouted dusters to obtain more uniform power. The schematic drawing in Fig. 13-22 shows an agitator in the bottom of the hopper to keep the dust in a fluffy condition in order that it will be metered out uniformly through the feed system into the air stream. Tractor-drawn trailing dusters with power-takeoff drive and auxiliary-engine drive are available. Dusters of this type are easily attached to and detached from the tractor. They are suitable for small firms where one tractor is used to perform all cultural operations.

Fig. 13-21   Six-row tractor-mounted power duster in operation. (*Gustafson Mfg. Co., Inc.*)

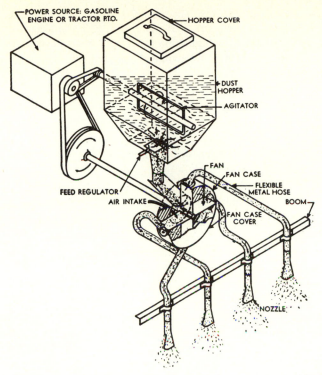

Fig. 13-22   Schematic drawing of typical power duster. (*National Sprayer and Duster Association.*)

Tractor wheel fenders are essential when either tractor-mounted or trailing sprayer and duster are used to apply chemicals to tall, wide-spreading cotton plants.

Orchard Power Dusters   Most orchard dusters are operated by an auxiliary engine. Others can be mounted on the rear of a tractor or on a trailer and driven from the power takeoff. Power units can also be mounted on the floor of a truck, thus saving the cost of a special sprayer chassis. Orchard dusters have only one large flexible metal hose, which can be turned to direct dust in any direction.

Electronic Duster   A company in Texas claims to have developed an electronic attachment for ground field dusters. The following is a statement from the company's trade literature:[1]

Transistor power supply runs off of 6 or 12 volt tractor battery. Draws 3 amps, steps voltage up to 12,000 volts. This voltage fed into a special nozzle creates an electric wind. Dust passing through picks up an electrical charge

[1] All Products Co.

which will magnetize the dust particles. The magnetized dust will go to the growing plant of opposite polarity, will coat plant with thin even coat of dust from top to bottom making it impossible for insects to crawl without making contact with the dust. Dust can be applied any time. Not necessary to wait for dew or calm weather.

Research is being conducted to determine if this method of applying insecticides will increase yields.

Airplane Dusters   Airplanes have been successfully used to apply dust to both field crops and orchards (Fig. 13-23). A hopper capable of holding 500 lb (226.5 kg) of calcium arsenate is built inside the fuselage in the space ordinarily occupied by the front seat. The opening in the top for filling is covered by a close-fitting lid, hinged in front. The dust in the hopper is stirred just above the outlet at the bottom by an agitator driven by a small propeller mounted on the wing (Fig. 13-24). The feed consists of an opening across the width of the fuselage. A slide covering the opening is operated by the pilot. The amount the feed valve is opened regulates the flow of dust and determines the poundage applied per acre. A venturi nozzle (Figs. 13-25 and 13-26) is mounted underneath the fuselage and the dust outlet. The rear end of the nozzle is tipped slightly downward. The blast of air created by the plane's propeller rushes through the venturi nozzle at a high velocity, catching the dust and discharging it in a whirling cyclindrical column that spreads and settles on the plants. It is claimed that the high velocity of air through the nozzle creates a partial vacuum in the feed opening, which aids the flow of dust.

An airplane can dust approximately 350 or more acres (141.7 hectares) per hour, which is many times the acreage that can be dusted with any other type of machine in the same length of time. Data kept on the time required for airplane operations show the average loading time to be 3 min 5 s, average flying time per load 14 min 30 s, and the average dusting time per load 4 min 45 s. About

Fig. 13-23   Airplane applying chemical dust. (*Texas A & M University*.)

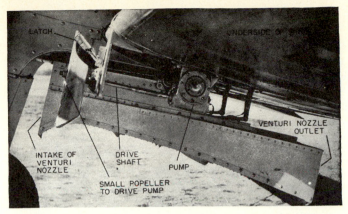

Fig. 13-24   Side view of venturi nozzle for aircraft showing propeller and wormgear drive for agitator in hopper. (*Delta Air Lines, Inc., Agricultural Division*.)

Fig. 13-25   Section of airplane fuselage showing location of venturi nozzle. (*Delta Air Lines, Inc., Agricultural Division*.)

Fig. 13-26   Close-up rear view of dusting-seeding-fertilizing venture with curved vanes to spread material being distributed (*Delta Air Lines, Inc., Agricultural Division*.)

one-third of the time is spend in actually dusting, the remainder being consumed in turning and in flying to and from the landing field.

Chemical dusts applied by airplane to control weeds in rice fields may drift for several miles and injure other crops, such as cotton. Wind velocity will affect the drift distance.

MAINTENANCE AND CARE OF SPRAYERS

1   Before operating sprayer, fill tank with water and test operation.
2   Check nozzles and screens to be sure they are clean and clear of sediment.
3   Check gallonage per acre.
4   At the end of each day's work, drain and flush tank.
5   Oil the boom and framework, especially before storage.
6   Do not allow chemicals to remain in tank for long periods of time or during storage.
7   Do not pour concentrated chemicals into empty tank. First fill tank about one-half full of water; then add chemical.
8   Remove tension from belts while in storage.
9   Grease all moving parts to prevent rust.
10   If possible, store sprayer under a shed. If shed is not available, cover with plastic or canvas sheets tied securely.
11   Dusting equipment should be cleaned thoroughly before storage.

REFERENCES

Adkisson, Perry L., L. H. Wilkes, and B. J. Cockran: Relative Efficiencies of Certain Spray Nozzles for Cotton Insect Control, *Jour. of Ec. Ent.,* **52**(5):985-991, 1959.

Akesson, Norman B., and W. A. Harvey: Chemical Weed-control Equipment, *Calif. Agr. Expt. Sta. Bul.* 389, December 1948.

Bode, L. E., et al.: The Effects of Flow Rate on the Distribution Pattern and Drop-Size Spectrum of a Spinning Atomizer, *Trans. of the ASAE,* **15**(L):86, 1972.

Brodell, Albert P., and Charles E. Burkhead: *U.S. Dept. Agr., BAE F.M.* 98, July 1951.

Creasy, Lawrence E., and Mansel M. Mayeux: The Sprayer and Its Use, *La. Agr. Col. Ext. Cir.* 1085, March 1951.

Edwards, G. J., et al.: Optical Determination of Spray Coverage, *Amer. Soc. Agr. Engin. Trans.,* **4**(2):206, 1961.

French, O. C.: Spraying Equipment for Pest Control, *Calif. Agr. Expt. Sta. Bul.* 666, May 1942.

Frost, K. R., and G. W. Ware: Pesticide Drift from Aerial and Ground Application, *Agr. Engin.* **51**(8):460, 1970.

Hedden, Orve K.: Spray Drop Sizes and Size Distribution in Pesticide Sprays, *Trans. of the ASAE,* **4**(2): 158, 1961.

Henry, James E.: Spreading Dry Materials by Airplane, *Agr. Engin.,* **42**(9):484, 1961.

Hull, D. O., and E. L. Barger: Using Farm Sprayers in Iowa, *Iowa Agr. Col. Ext. Ser. Pamphlet* 152, June 1950.

Luckmann, W. H., and H. B. Petty: Controlling Corn Borers in Field Corn with Insecticides, *Ill. Agr. Ext. Ser. Cir.* 768, 1957.

Morgan, M. J.: Report on Pump Selection for Agricultural Sprayers, mimeograph, Oklahoma State University, Agricultural Engineering Department.

Sprayer and Duster Manual, National Sprayer and Duster Association, 1953.

Wilkes, Lambert H.: Effect of Nozzle Types and Spray Application Methods on Cotton Insect Control, *Amer. Soc. Agr. Engin. Trans.*, 4(2):166, 1961.

QUESTIONS AND PROBLEMS

1 Explain the classification of spray materials.
2 Explain the functions of a sprayer.
3 Explain the differences in the various types of sprayers.
4 Describe the different types of pumps for sprayers.
5 Explain the functions of a pressure regulator.
6 A farmer wishes to apply 5 gal (18.9 liters) of spray per acre to a crop planted in rows spaced 40 in (1.02 m) apart, with a sprayer traveling 5 mi/h (8.0 km/h). What size nozzles should he use when using: (a) one nozzle per row, (b) two nozzles per row, and (c) three nozzles per row?
7 Discuss the effect of nozzle type and arrangement on the control of cotton insects.
8 Explain the various spray patterns that can be obtained with aircraft sprayers. What effect do cross winds have on the spray pattern?
9 Explain the various types of dusters.

Chapter 14

Fertilizing Equipment

Fertilizers are required where soils are deficient in plant food elements. When land is planted to crops over a long period of years, the plant food elements are reduced and yields of crops are lower. Sandy soils lose plant food elements rapidly because these are leached out by heavy rainfall or applications of irrigation water. Some clay soils in low-rainfall areas lose plant food elements much more slowly than the sandy soils. It is now recognized that higher yields can be expected from most soils in all areas if the right type of fertilizer is properly applied.

Fertilizer can be applied to the soil in several forms, such as barnyard manure, granular and pelleted fertilizers of various formulae, and fertilizers in liquid and gaseous form. Special equipment is required for the handling of these types of fertilizers, which are applied to the soil and crop in various ways at different stages of culture. For example, barnyard manure is usually broadcast over the land with a *manure spreader* before seedbed preparation. It is then worked into the soil, either by plowing or by disk harrow.

MANURE SPREADERS

The manure spreader is a machine for carrying barnyard manure to the field, shredding it, and spreading it uniformly over the land. This type of machine should be on every farm that produces several tons of manure per year. Manure spreaders can be classified as ground-driven and power-takeoff-driven.

Ground and Power-Takeoff-Driven Spreaders

The mechanism of a *ground-driven* manure spreader is operated by sprockets and chain from the wheels supporting the spreader. The power-takeoff-driven type is designed to operate from the tractor power takeoff having 540 r/min. The power-takeoff shaft is telescoping, so that it automatically adjusts for changes in length in turns. The shaft connects to the center of the spreader box. A chain transmits the power to a drive shaft that extends backward along the side of the box to a sealed gear drive connected to the drive mechanism.

The *tractor-drawn* power-driven spreaders are generally mounted on two rubber-tired wheels on an axle located slightly to the rear of the box so that part of the weight of the spreader will be carried by the tractor (Fig. 14-1). Some large tractor-drawn manure spreaders are equipped with dual axles and rubber-tired wheels.

Truck-mounted manure spreaders are available for handling large quantities of manure.

Side-unloading or flail manure spreaders deliver the manure to the side instead of to the rear. This type of spreader can handle most any type of manure condition from liquids to hard-packed or frozen manure.

The principal parts of the manure spreader are the frame, the box, the conveyor, the beaters, and the widespread.

The Frame

Since manure is very heavy and at least a ton or more is loaded on the spreader for each trip to the field, a substantial yet comparatively light frame is required.

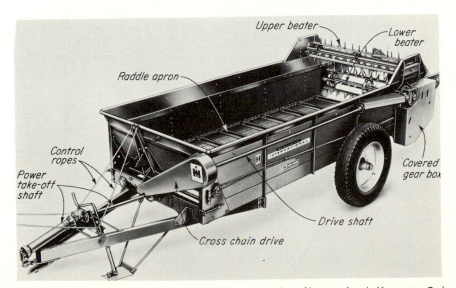

Fig. 14-1   Tractor-trailer power-takeoff-driven spreader. (*International Harvester Co.*)

The side rails on all spreaders should be made of a good grade of channel steel properly reinforced and braced.

The Box

The bottom and sides of the manure spreader box may be made of tongue-and-groove creosoted wood, of marine plywood, or of 12- to 14-gage sheet steel. The side flares consist of heavy sheet steel. One company used armored steel for the flares to resist damage by blows from the buckets of power loaders. The front of the box is closed by an inclined endgate, while the rear part of the box is open to the beaters. The rear end of the box is 1 to 2 in (2.5 to 5.1 cm) wider than the front end to reduce pressure on the sides. This also reduces friction as the load moves to the rear.

The *capacity* of a manure spreader box is rated in bushels (by volume), according to a standard formula adopted by the American Society of Agricultural Engineers (Fig. 14-2).[1] Manure spreaders are available in sizes ranging from 45 to 500 bushels.

Some manufacturers provide a high side so the box can be converted into a self-unloading or tight forage box.

The Conveyor

The manure in the box is moved to the rear by an endless double chain-and-slat conveyor or apron (Fig. 14-3). The angle iron bars used for the conveyor slats are riveted to the chains with the outside leg or high side facing to the rear of the box. The manure is deposited in the box on the conveyor; then as the conveyor moves, it carries the manure with it to the rear of the machine, where it comes in contact with the beaters.

The conveyor operates very slowly. The minimum travel per revolution of the main drive wheel is about 1 in (2.5 cm), while the maximum is about 3 in (7.6 cm). The rate of travel is controlled by a lever placed conveniently for the driver. From five to twenty loads can be spread per acre. The tension of the conveyor chain can be adjusted by a setscrew arrangement on each end of the front conveyor shaft.

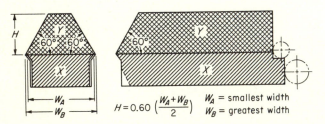

$$H = 0.60 \left(\frac{W_A + W_B}{2} \right)$$

W_A = smallest width
W_B = greatest width

Fig. 14-2   Method of measuring and formula for calculating the capacity of a manure-spreader box. (*ASAE Agricultural Engineering Yearbook, 1973.*)

[1] *Amer. Soc. Agr. Engin. Yearbook*, 1973.

Fig. 14-3   Rear view of manure-spreader box showing endless double chain and slat conveyor in place on box bottom. (*New Holland Machine Co*.)

Conveyor Drive

A ratchet-and-pawl arrangement has been the common device for driving the conveyor chain of a manure spreader (Fig. 14-4). As the feed cam raises the rocker arm, it causes the feed pawl to engage the teeth on the ratchet wheel and turn it. The number of teeth engaged by the feed pawl at a stroke is regulated by a stop pawl. This in turn regulates the speed of the conveyor and the volume of manure distributed. The adjustment of the feed pawl is controlled by a lever placed conveniently for the driver. The lever is connected to the feed pawl by a

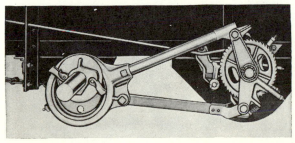

Fig. 14-4   Apron drive using eccentric device to operate the ratchets which slowly revolve the ratchet gear and thus move the apron over the box bottom at the rate desired.

long rod and can be shifted to any position without stopping the machine. Late-model manure spreaders have a worm-gear apron drive.

The power for driving the apron and beaters is derived from ground traction of the spreader wheels or from the power takeoff of the tractor.

The Lower Beater

The lower beater is placed just to the rear of the coveyor to beat, tear up, and spread the manure from the rear of the spreader (Figs. 14-5 and 14-6). It must be substantial because it must spread all kinds of manure in various states of physical condition. It should have good, strong bearings of the self-aligning or roller type. The beater has steel bars through which the teeth are fastened. Some teeth are riveted in, while others are held in place by nuts.

The beater revolves in the direction opposite to the main wheels. It is, therefore, necessary to have some arrangement to give it this reverse motion, which will be discussed under "Beater Drive." The beater should revolve at a comparatively high rate of speed; the ratio is usually about 6 or 7 to 1; that is, a beater revolves about seven times while the main wheel revolves once. Some manure spreaders are equipped with flail beaters to break up hard-packed or frozen manure.

Beater Drive

The chain is the common method of driving the beaters on ground-driven manure spreaders. A large drive sprocket is mounted rigidly on the main axle. In Fig. 14-7, the drive chain passes around sprockets on the end of the upper beater shaft and the main beater shaft and around a movable idler sprocket. The chain does not pass around the drive sprocket. As the movable idler sprocket is lowered, the bottom part of the drive chain is lowered onto the drive sprocket. This will cause the beaters and widespread device to turn in the direction opposite to that of the main-drive sprocket. The machine is thrown out of gear

Fig. 14-5   View of ground-powered manure spreader showing position of beaters and widespread.

Fig. 14-6   View of manure spreader with spiral widespread device and lower and upper beaters. (*International Harvester Company*.)

by raising the drive chain from the drive sprocket (Fig. 14-6). This is done by a control lever placed on the front of the box and connected to the idler sprocket by a rod.

The Upper Beater

Most manure spreaders have an upper beater placed above and a little to the front of the main beater (Fig. 14-5). This beater aids the lower beater in tearing up and pulverizing the large flakes that are encountered.

Fig. 14-7   Chain and sprocket beater drive.   (*International   Harvester Company*.)

Widespread Device

To prevent the manure being spread too thickly directly behind the center of the machine, a widespread device is used (Fig. 14-5). This also spreads the manure wider than the machine and makes it unnecessary to lap the loads. The device consists of spiral augerlike steel blades. One-half of the spiral is set to throw to the left, while the other half is set to throw to the right.

The manure is thrown backward by the beaters against the revolving spirals, which throw it backward and outward and spread it uniformly over a width of 7 to 8 ft (2.1 to 2.4 m). The widespread beater is driven by a chain from the main or auxiliary beater shafts.

Loading the Spreader

It is considered the better plan to start loading at the front end and finish at the rear end. The manure is torn up and broken to pieces more easily when the load is put on in this manner.

Mechanical Loaders

When manure is spread mechanically, more time is consumed in loading than in any other operation. This is also the hardest work. Mechanical loaders are available, which eliminate the necessity of loading with a pitchfork (see Chap. 23).

GRANULAR-FERTILIZER DISTRIBUTORS

Dry, granular chemical fertilizers have for many years been the most common type of fertilizer used by farmers. Such fertilizers are of many kinds and vary from highly concentrated chemicals, which must be used in small quantities, to rather low-grade mixtures used in large quantities. A fertilizer distributor is required which will distribute varying amounts of fertilizer in almost any physical and mechanical condition and place it in the soil so that it will not injure the seed. It is difficult to design a machine that will meet such a wide range of requirements. The proper application of fertilizer is the key to efficient fertilizer use.

Location of Fertilizer in Relation to the Seed

A committee on fertilizer application of the American Society of Agronomy recommended that all fertilizer attachments on planting and seeding machinery be designed to prevent contact between seed and fertilizer.

A joint committee on fertilizer application, representing the American Society of Agronomy, the American Society of Agricultural Engineers, The National Fertilizer Association, and the National Association of Farm Equipment Manufacturers, adopted the following statement on fertilizer application:

> Contact of fertilizer with the seed, except when fertilizer is used in very small amounts, tends to depress and delay germination and may even

prevent it. The extent of this delay or depression varies with the materials used in the fertilizer, with the moisture content of the soil, with the crop grown, and with the quantity of fertilizer applied. The recommended fertilizer placement is 2 to 3 inches [5.1 to 7.6 cm] to the side of the row and 3 to 4 inches [7.6 to 10.16 cm] below the soil surface. The location of the fertilizer will depend upon the amount and kind of fertilizer and the row spacing.

Accepting the above statements, we find that the fertilizer can be applied as hill applications with hill-dropping of the seed or can be drilled when the seed is drilled. For the small grains, it can be either drilled or broadcasted. To accomplish the desired results in each case, many different attachments are available.

Fertilizer Attachments

Operating costs can be reduced when attachments that permit dual operations are used in connection with various types of equipment. Fertilizer attachments are available for most tractor-mounted planters and cultivators. Attachments are also available for grain drills and some types of plows.

Attachments for Row Planters   Fertilizer attachments have been designed to work in conjunction with most tractor planters. Several illustrations in Chapter 11 show tractor planters equipped with fertilizer attachments.
Fertilizer attachments are available for bean, beet, and potato planters. Most potato planters are equipped with fertilizer attachments. A special-type fertilizer feed used to meter out fertilizer for application to potatoes and the placement of the fertilizer in relation to the seed is shown in Fig. 14-8.

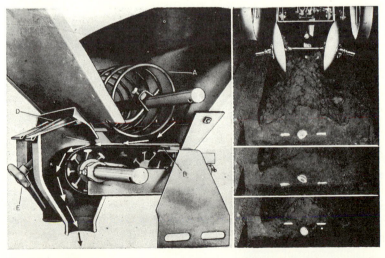

Fig. 14-8   The view at left shows a belt-type fertilizer feed. The view at the right shows how bands of fertilizer can be placed at the seed level, above the seed, or below the seed level. (*Deere & Co.*)

Grain-Drill Fertilizer Attachments   Fertilizer attachments for grain drills consist of a specially constructed hopper having a partition extending lengthwise through the middle. The planting unit is in the front half of the box, while the fertilizer unit is in the rear half.

Many fertilizer drills release both seed and fertilizer through the same tube. This is not good practice, because the seeds are in direct contact with the fertilizer and germination may be affected. A better method is to release the fertilizer through separate tubes which will place the fertilizer in the drills above the seed. A pasture drill equipped with a fertilizer attachment is shown in Fig. 11-43.

Fertilizer Attachments for Cultivators   Fertilizer attachments for tractor-mounted cultivators to apply granular fertilizer as a side dressing to growing crops are available.

Fertilizer Attachments for Chisel Plows   Fertilizer attachments are available for the chisel-type plow. Fertilizer can be applied as a preplanting operation and can also be applied to row crops as a side dressing and to pasture lands.

Broadcast Fertilizer and Lime Distributors

Figure 14-9 shows machines suitable for broadcasting either lime or fertilizer. Usually, a wire screen is used in the top of the hopper to remove large lumps and prevent clogging of the feeds. Some models drop the lime or fertilizer on a scattering board, which deflects and scatters the material so that it will be more thoroughly broadcasted. A tractor-mounted broadcast fertilizer spreader is shown in Fig. 14-10.

Fertilizer Feeds

The efficiency of any fertilizer-distributing machine depends upon the proper handling of the fertilizer by the feeding mechanism. A number of factors will influence the efficiency of the feed, some of which are:

1   Climatic conditions, based on temperature and rainfall
2   Amount of fertilizer to be applied
3   Kind of fertilizer:
    a   Chemical composition
    b   Physical state

Many attempts have been made to design a fertilizer feed that will handle any and all kinds of fertilizer, distributing any desired quantity. As a result, several different types are being used. The operator's manual supplied by the manufacturer should be studied so that the method of regulating the quantity of fertilizer applied per acre can be fairly accurately followed. The manuals give instructions on how to adjust the feed to change the rate of application and for the care and maintenance of the fertilizer equipment. A belt-type feed is shown in Fig. 14-8. Other types of feeds use metering notched discs, the star-wheel feed, and the bar reel feed.

Fig. 14-9   Trailing broadcast dry chemical distributor. (*Gandy Company, Inc.*)

EQUIPMENT FOR APPLYING GAS AND LIQUID FERTILIZERS

Liquid fertilizers are available in high- and low-pressure and nonpressure forms. Anhydrous ammonia is a high-pressure fertilizer, while a mixed fertilizer containing nitrogen, phosphate, and potash in solution is a nonpressure fertilizer. Most liquid fertilizers are applied below the surface of the soil, but some types can be sprayed on the foliage of the plants.

Fig. 14-10   Tractor-mounted broadcast fertilizer spreader. (*New Idea Farm Equipment Co.*)

Anhydrous Ammonia

The use of anhydrous ammonia as a source of nitrogen for cotton was started by some farmers of Mississippi in March 1947. By 1949, approximately 2 million acres (or around 80,000 hectares) in the Cotton Belt were fertilized with anhydrous ammonia.[1] It has become a popular source of nitrogen for many crops and for orchards, as it is the cheapest source of nitrogen available.[2]

Anhydrous ammonia is a colorless alkaline gas at normal temperatures and atmospheric pressure and contains 82 percent nitrogen. It, however, exists as a liquid at $-28°F$ $(-51.6°C)$ and boils at this temperature; as a liquid it has a density of 42.57 lb (19.3 kg)/ft^3. It is handled in commerce as a compressed liquid and stored under pressure. At higher temperatures the vapor pressure increases rapidly. For example, at $50°F$ $(10°C)$ it is 74.5 lb (33.7 kg)/in^2, at $100°F$ $(37.8°C)$ it is 197.2 lb (89.5 kg)/in^2, and at $125°F$ $(52.0°C)$ it is 293.1 lb (133.0 kg)/in^2. A gallon of anhydrous ammonia weighs 5 lb (2.27 kg) and contains 4.1 lb (1.9 kg) or 82 percent nitrogen.

Anhydrous ammonia is inflammable when mixed in proportions of 16 to 25 percent by volume with air. It is corrosive to copper, copper alloys, aluminum alloys, and galvanized surfaces. Ammonia vapors are stifling, and high concentrations may burn, blind, strangle, or kill. Therefore, it must be handled with caution. It must be shipped, stored, and handled on the farm in tanks in accordance with Interstate Commerce Commission regulations and accepted state safety codes. Bulk storage is usually handled by dealers.

Tanks for transporting anhydrous ammonia to the farm usually have a capacity of 500, 1000, or 3000 gal (1893.0, 3785.0, or 11,355.0 liters). They can be mounted on motor trucks, motor-truck trailers, or other types of trailers. These tanks must be handled in accordance with state rules and regulations.

Field Tanks for Anhydrous Ammonia   It is recommended that farmers who use propane gas for flame cultivators *do not* use these tanks for anhydrous ammonia, as the tanks usually have brass and bronze fittings. Ammonia may cause brass or bronze fittings to develop internal checks. A slight strain, blow, or twist to a brass fitting affected by ammonia might cause the fitting to break off, release the ammonia, and injure the operator. Tractor tanks should have steel fittings, valves, and gages to handle anhydrous ammonia safely. Tractor-mounted tanks (Fig. 14-11) should have a capacity of 80 to 110 gal (302.8 to 416.3 liters). Large trailer-mounted tanks may have a capacity up to 150 gal (567.8 liters). Figure 14-12 shows a typical small trailer tank and application parts. All tanks should be equipped with the following fittings:

1   Three-hundred-pound pressure gage.
2   Liquid outlet valve screwed directly into the shell of the tank and connected to a pipe extending down inside the tank to within $\frac{1}{2}$ to $\frac{3}{4}$ in (1.3 to

[1] William Baker Andrews (ed.), *Cotton Production, Marketing, and Utilization*, William Baker Andrews, State College, Miss., 1950.
[2] Harold T. Barr, Anhydrous Ammonia Equipment, *La. Agr. Expt. Sta. Bull.* 462, 1952.

Fig. 14-11   Tractor-mounted anhydrous ammonia applicator.

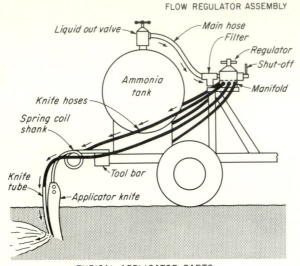

FLOW REGULATOR ASSEMBLY

TYPICAL APPLICATOR PARTS

Fig. 14-12   Typical small trailing anhydrous ammonia applicator.

1.9 cm) of the bottom. The bottom end of the pipe should have a fine-mesh strainer.

3   Safety or pop-off valve of $\frac{3}{4}$ to 1 in (1.9 to 2.5 cm) set for 250 lb (113.5 kg)/in^2 pressure.

4   A $\frac{3}{4}$- or 1-in (1.9- or 2.5-cm) automatic back-seating liquid-filling valve.

5   A liquid-level gage.

6   Vapor return valves, $\frac{1}{2}$ or $\frac{3}{4}$ in (1.3 or 1.9 cm), with a short pipe extending into the tank to prevent filling to over 85 percent water capacity. This valve is also used for gas escape in bleeding.

Metering Devices   A reliable type of meter recommended and guaranteed by a dealer should be installed on all anhydrous ammonia applicators. The meter regulator should be calibrated with ammonia and the same hose and applicator feet that will be used with the outfit. The metering device should be designed to withstand pressures of 150 to 250 lb (68.1 to 113.2 kg)/in^2 on the high side and 0 to 90 lb (0 to 40.9 kg)/in^2 on the low-pressure side. The meter regulates the flow of anhydrous ammonia to the desired rate per acre. The pressure in the tank will be reduced as the ammonia is used, but on the other hand midday sunshine will cause a rise in temperature and pressure.

A quick-cutoff valve should be installed between the tank shutoff valve and the meter regulator. This will permit closing the quick-cutoff valve without disturbing the meter setting, during turning at row ends.

Applicator Chisel Knives   The applicator is a ground-working, chisellike knife having a small $\frac{5}{8}$-in (1.6-cm) pipe welded to the rear side or a drilled hole to conduct the ammonia 5 to 6 in (12.7 to 15.2 cm) below the soil surface (Fig.

14-13). It is also called an *applicator foot, applicator shank,* or *applicator knife.*
There are many types with different shapes, curves, and accessories. The
applicators are attached to the regular cultivator frames or to special frames.
Applicators may be tractor-mounted or trailing.

Barr states that

> The applicator should be as narrow as possible, have sufficient strength to
> prevent breakage or be equipped with a single trip, be hard surfaced to
> prevent undue wear, provide for easy and secure attachment of anhydrous
> ammonia hose, and the final hole in the applicator should be constructed so
> as to deliver the ammonia without creating back pressure or without being
> plugged up easily by soil.

Some means, such as a wheel or disk blade attached to the beam slightly to
the rear of the applicator, should be provided to close the furrow made by the
applicator and prevent the escape of the ammonia from the soil. Some
applicators are designed so that liquid phosphate can be applied in the same
furrow but through separate tubes.

The applicators should be spaced to suit the crop-row spacing. Generally
there is one applicator per row. When ammonia is applied to small grains, the
applicators are spaced about 16 in (40.64 cm) apart.

Rubber hose should be used for the feed lines from the tank to the
applicator to prevent the possibility of ice forming and plugging the lines.
Expanding ammonia causes ice to form on iron pipes, and the feed line should
extend as near the soil as possible.

Fig. 14-13   Applying liquid fertilizer between rows of corn. (*New Idea Farm Equipment
Co.*)

Cautions   Anhydrous ammonia should be handled with all safety rules and cautions in mind, as it is irritating, it will burn, and concentrated quantities will strangle and kill.

Buy equipment from well-established concerns who make products that will comply with all safety codes and state regulations.

Leave a small amount of ammonia in trailer and tractor tanks during periods they are not in general use.

Protect all valves on tanks and do not leave the valves open to outside air. Do not fill the tank with water or with any material other than anhydrous ammonia.

Do not use the ammonia tank for propane, or the propane tank for ammonia.

Do not use brass or bronze fittings. Use only extra-heavy steel fittings. Do not weld on an ammonia tank.

Aqua Ammonia

When water is added to anhydrous ammonia to reduce the vapor pressure to atmosphere pressure it is termed *aqua ammonia*. This, then, is a low-pressure liquid fertilizer. It contains only about 20 to 25 percent nitrogen. It must be handled with the same caution as anhydrous ammonia. Gravity feeds can be used if the material is discharged on the surface of the soil to be covered with soil-working blades. Pumps are required if the material is applied below the soil surface.

Mixed Liquid Fertilizers

When nitrogen, phosphorus, and potash are applied as a complete fertilizer, the combination is near chemical neutrality and is termed a *nonpressure* liquid fertilizer. It can be metered out by gravity flow, pump, or air pressure. It can be sprayed either with ground equipment or by aircraft. Figure 14-13 shows liquid fertilizer being applied between rows of corn.

MAINTENANCE AND CARE

1   Before operating in the field, check over equipment, and set rate of application.

2   At end of season, thoroughly clean manure spreaders and granular and liquid applicators. Make a list of any repair parts needed and place order for replacements.

3   Apply heavy oil to manure-spreader conveyor chains, widespread blades, and beater spikes.

4   After cleaning granular applicators, apply heavy oil to all parts that may rust.

5   At the end of the season, the tanks and hose on liquid fertilizer equipment should be thoroughly flushed out.

6   All types of fertilizer equipment should be protected from the weather by storage in a shed or covered with plastic or canvas sheets.

REFERENCES

Andrews, William Baker (ed.): *The Response of Crops and Soils to Fertilizers and Manures*, William Baker Andrews, State College, Miss., 1947.

Anhydrous Ammonia Handbook for Agriculture, Agricultural Ammonia Institute, 1961.

Barr, Harold T.: Anhydrous Ammonia Equipment, *La. Agr. Expt. Sta. Bul.* 462, 1952.

Futral, J. G.: Precision Fertilizer Placement, *Agr. Engin.*, **42**:424-425, 1961.

QUESTIONS AND PROBLEMS

1   List the various forms of fertilizers.
2   Explain the essential features of a manure spreader and the function of the following: (*a*) conveying system, (*b*) upper and lower beaters, (*c*) widespread device.
3   Explain the drive mechanism for the conveyor on a manure spreader.
4   Explain the drive mechanism for the beaters and widespread device.
5   Discuss the placement of granular fertilizer in relation to the seed for row crops. What is the recommended location of the fertilizer in relation to the seed?
6   List and explain the various types of fertilizer attachments for planting equipment.
7   Define anhydrous ammonia, and explain its properties.
8   Discuss the equipment required for the application of anhydrous ammonia.
9   List several cautions in regard to the handling of anhydrous ammonia and the equipment.

Chapter 15

Hay-harvesting Equipment

Hay-harvesting equipment consists of machines necessary in the various steps of dry hay making, while forage harvesting equipment consists of machines required for placing green succulent material into silos.

The principal machines required in making hay are mowers, rakes, and balers. In some areas, loaders, stackers, and barn hay-handling equipment are used.

MOWERS

The mower is designed to cut meadow grasses and special crops for hay; however, it has many other uses. Horse-drawn and tractor types of mowers are available in sizes to suit almost any requirement.

History of Development

Miller[1] stated that the development of the mower for cutting hay was closely associated with the development of the reaper. The first machines were used to cut either grain or grass. William F. Ketchum was the first to put mowers on the

---

[1] M. F. Miller, Evolution of Harvesting Machinery, *U.S. Dept. Agr. Expt. Sta. Bul.* 103, 1902.

market as a machine distinct from the reaper. Ketchum's most important patent was dated July 10, 1847. The cutter bar of an endless chain of knives was soon abandoned, and Hussey's rigid bar substituted. Cyrenus Wheeler obtained a patent Dec. 5, 1854, on a machine that featured two drive wheels and a cutter bar joined to the main wheels. A patent was granted Cornelius Aultman, on July 17, 1856, containing basic principles of mowers, such as the ratchet-pawl drive. By 1860, the mower was considered a practical machine.

Horse-drawn mowers were first used with tractors about 1910. Tractor-mounted mowers were available about 1930.

Tractor Mowers

There are several types of tractor mowers, classed largely by the way the mower is attached to the tractor. The principal types are the trail-behind, the semimounted tractor mower and the integral rear-mounted, and the side- or central-mounted.

The Trail-behind Tractor Mower   This type of tractor mower is built as a unit assembly that can be easily attached to and detached from the tractor drawbar or three-point hitch (Fig. 15-1). The arrangement gives flexibility and allows the cutter bar to follow the contour of the ground. Power may be transmitted by gears, chain, or V belt. The regular adjustment features are provided. The cutter bar is lifted by a hydraulic or mechanical power lift.

Fig. 15-1   Trailing or pull-type mower. Power is transmitted by V belt from the power-takeoff shaft to the pitmanless knife drive. (*Allis-Chalmers Mfg. Co.*)

Integral Rear-mounted Mowers   This type of tractor mower is direct-connected to the tractor and is mounted on the tractor drawbar. All the weight of the mower is on the tractor. The power is transmitted directly or indirectly by V belt from the power takeoff to the pitman shaft; no gears or universal joints are required. An automatic safety release allows the cutter bar to swing back if an obstruction is hit. This safety release is connected to the tractor clutch foot pedal, and as the bar swings back, it automatically disengages the tractor clutch. Provision is made for the adjustment of the alignment, registration, and tilting of the cutter bar. The cutter bar is lifted hydraulically.

The Semimounted Mower

This type of mower is similar to the integral-mounted mower. It is mounted on the three-point hitch but has a wheel behind to support part of the weight of the mower (Fig. 15-2).

Side- or Central-mounted Tractor Mower   This method of mounting is available for tractors that are equipped with widespread front wheels (Fig. 15-3) and for tractors with tricycle-arranged wheels. Tractor operators like side-front-mounted mowers because they can watch the cutter bar more easily. The cutter bar is mounted on the right side of the tractor between the front and rear

Fig. 15-2   Semimounted mower. It is mounted on the three-point hitch but has a wheel behind to support part of the weight of the mower. (*Allis-Chalmers Mfg. Co.*)

Fig. 15-3 Side- or centrally mounted mower. (*Allis-Chalmers Mfg. Co.*)

wheels. This arrangement makes it easy to use power lifts. Power is usually taken from the power takeoff by V belt and transmitted to the knife through shafting, gears, and pitman. Automatic safety releases and snap or slip clutches are provided. Means of adjusting the alignment, registration, and tilting of the cutter bar are also provided.

Safety Release Clutch   A safety release clutch is essential on tractor mowers to permit the cutter bar to swing back should the cutter bar hit an obstruction while the tractor is moving forward.

Knife Drives   Prior to 1952 all mower knives were driven by a pitman which served as a connecting rod from a counterbalanced wheel to the knife, thus converting rotary motion to a reciprocation motion. The development of the pitmanless counterbalance drive eliminates the need for a pitman. Some counterbalance drives use a single counterbalance (Fig. 15-4), while others use twin counterbalance wheels (Fig. 15-5). This type of drive reduces vibration, permits a faster knife speed, and allows the knife to operate in positions 10 to 15° below horizontal to positions almost vertical

Cutter Bar and Its Parts   For the knife to do its work, it must have aid from a number of other parts which go to make up the cutting mechanism (Fig.

Fig. 15-4 Pitmanless knife drive with single counterbalance. (*International Harvester Company.*)

15-6). These consist of the cutter bar, inside shoe, outside shoe, guards, ledger plates, wearing plates, knife clips, and grass board and stick.

The Cutter Bar The cutter bar (Fig. 15-7) is made of high-grade steel. All other parts included in the cutting mechanism are connected directly or indirectly to it.

The Inside and the Outside Shoes A large, shoelike runner (Fig. 15-7) supports the inner end of the cutter bar when in operation. A removable sole is placed underneath the shoe and is adjustable to regulate the height of cut. The outside shoe (Fig. 15-7) supports the outer end of the cutter bar. It also has an

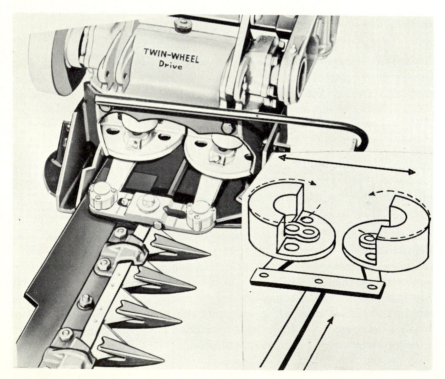

Fig. 15-5 Pitmanless mower knife drive using twin counterbalanced wheels. (*Allis-Chalmers Mfg. Co.*)

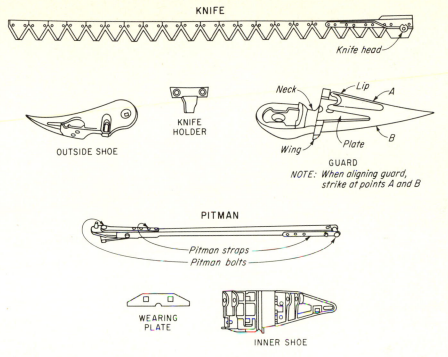

Fig. 15-6 Various parts of a cutter bar with knife and pitman.

adjustable sole to regulate the height of cut. The pointed front part of the outer shoe acts as a divider, separating the cut from the standing grass.

The Guards The guards serve to protect the cutting units (Fig. 15-8). They also provide a place for the ledger plates. They divide the material being cut so that the cutting units can do the best work. If one of the guards gets out of alignment, it should be hammered back into place. Special grain or pea- or bean-vine lifters are often used to facilitate the cutting of fallen material.

Ledger Plates The ledger plates are riveted to the guard (Fig. 15-8). They form one half of the cutting unit, the knife sections acting as the other half. The edges of the ledger plates are serrated to prevent stems of grass from slipping off the point of the shears. When a ledger plate becomes worn and dull, it should be replaced with a new one. A special anvil for removing and replacing ledger plates and knife sections is available for repair work from factories making mowers.

Wearing Plates Wearing plates (Fig. 15-6) are necessary to support the rear side of the knife. When they become worn, the rear side of the knife will drop down, causing the sections to kick up and make poor contact with the ledger plates. Heavy draft and poor cutting will result from such a condition. The wearing plate under the knife head should not be overlooked when a mower is being repaired.

Knife Clips Knife clips or holders (Figs. 15-6 and 15-8) are essential to hold the knife sections down close against the ledger plates. If they become worn and allow the knife to play up and down, making poor contact with the

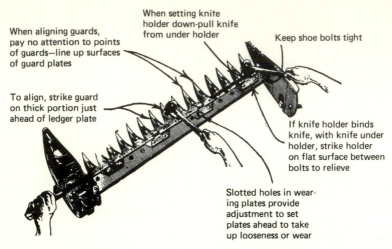

When aligning guards,
pay no attention to points
of guards—line up surfaces
of guard plates

When setting knite
holder down-pull knife
from under holder

Keep shoe bolts tight

To align, strike guard
on thick portion just
ahead of ledger plate

If knife holder binds
knife, with knife under
holder, strike holder
on flat surface between
bolts to relieve

Slotted holes in wear-
ing plates provide
adjustment to set
plates ahead to take
up looseness or wear

Fig. 15-7   Complete mower cutter bar with instructions for adjustments.

ledger plates, they should be hammered down. They are made of either malleable iron or steel.

Grass Board and Stick   These parts are attached to the outer shoe (Fig. 15-1). The board, with a yielding-spring connection, angles back and away from the uncut grass. Its purpose is to divide and rake away the cut from the uncut grass to give a clean place for the inside shoe on the next round. For long and tangled material, a *rotary* grass board can be secured that will leave a cleaner swath than the regular type.

Alignment of Cutter Bar   The outer end of the bar should be ahead of the inner end $1\frac{1}{4}$ to $1\frac{1}{2}$ in (3.2 to 3.8 cm) on 5-ft (1.5-m) cutter bars, $1\frac{1}{2}$ to $1\frac{3}{4}$ in (3.8 to 4.4 cm) for 6-ft (1.8-m) cutter bars, and $1\frac{3}{4}$ to 2 in (4.4 to 5.1 cm) for 7-ft (2.1-m) cutter bars. This setting is called *lead.* When the mower is in operation the friction between the cutter bar and the ground causes the bar to swing slightly backward, and this brings the knife in line with the pitman.

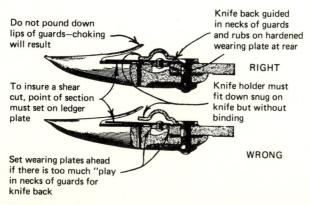

Do not pound down
lips of guards—choking
will result

Knife back guided
in necks of guards
and rubs on hardened
wearing plate at rear

RIGHT

To insure a shear
cut, point of section
must set on ledger
plate

Knife holder must
fit down snug on
knife but without
binding

Set wearing plates ahead
if there is too much "play
in necks of guards for
knife back

WRONG

Fig. 15-8   Right and wrong way for mower knife to fit and operate in the guards.

Figure 15-9 shows a method of measuring to determine the proper lead for the cutter bar.

Registration   "Registration" means that, when the wrist pin on the crank wheel pulls or pushes the pitman and knife to the extreme end of the in and out strokes of the knife, the center of the knife sections should be at the center of the guards for a pitman mower. Failure to register is a very common touble in mowers and should be looked for often.

Figure 15-10 shows that the knife sections do not quite reach the center of the guards at the end of the knife stroke for a pitmanless mower.

The results of failing to register are an uneven job of cutting, an uneven load on the entire mower, heavier draft, and, often, clogging of the knife. When an attempt is made to align the cutter bar by lengthening or shortening the drag

Fig. 15-9   Method of checking lead for mower cutter bar. Note that the outer end of the knife is about 1½ in ahead of the string, which is in line with the pitman.

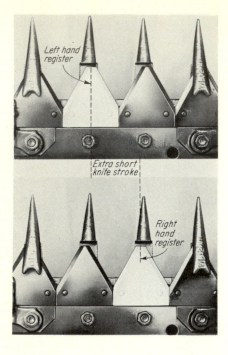

Fig. 15-10 Registration of the knife sections with the guards for a pitmanless mower. Note that the knife sections do not quite reach the center of the guards at the end of the knife stroke. (*International Harvester Company.*)

bar, it may, at the same time, disturb the registration of the knife sections with the guards. To adjust the registration on most mowers, it is necessary to move the whole cutter-bar assembly in or out to the point where the centers of the knife sections coincide with the centers of the guards at the ends of the stokes. On some mowers, the registration is adjusted by changing the length of the pitman by the addition or removal of shims between the pitman arm and the pitman head socket. As the knife sections move from the center of one guard to the center of another, the movement is forward and sidewise in a curve, as shown in Fig. 15-11.

Kepner[1] states that "the average cutting force during the cutting portion of the stroke may be at least as great as the maximum inertia force of the knife, and should be expected to cause vibration effects of considerable magnitude in a single-knife machine. For a given mower the average cutting force is directly proportional to the feed rate."

The knife on a pitmanless counterbalanced mower can be operated at a higher speed than on the mower with a pitman, as there is less vibration. The knife should be kept as sharp as possible because a dull knife may increase the draft 30 to 35 percent.

Cutter-Bar Attachments

The mowing machine is often required to operate under difficult conditions and to perform various kinds of special work. Attachments are available for the cutter bar to facilitate doing such jobs, and they are mentioned briefly.

---

[1] Robert A. Kepner, Analysis of the Cutting Action of a Mower, *Agr. Engin.*, **33**(11):693, 1952.

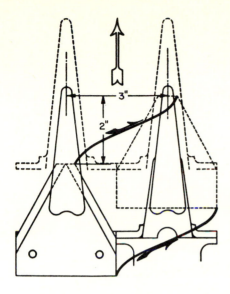

Fig. 15-11 Diagrammatic sketch showing movement of mower knife section in making single stroke. The knife is moved forward 2 in and sidewise 3 in with the tractor in low gear. The path of the knife section during the stroke is a curve. (*Allis-Chalmers Mfg. Co.*)

Cutter-Bar Power Lift   Most tractor-mounted mowers are provided with hydraulic remote-control cylinders to lift the cutter bar.

Grain, Pea, and Bean Lifters   These lifters consist of guards attached over the regular guards, which project quite a distance in front so that pea and bean vines can be lifted, allowing the cutter bar to slide underneath and cut off the stems below the heads.

Weed Attachment   This attachment consists of a wheel placed at the outer end of the cutter bar to carry it some 6 to 12 in (15.2 to 30.5 cm) off the ground so that the weeds can be cut without undue strain upon the mower parts.

Weed and Brush Bars   These bars are constructed with stub guards instead of the long, sharp-pointed guards used for cutting grass. Extra-heavy knives are also used with them.

Windrowing Attachment   The windrowing attachment consists of a number of bars attached to the cutter bar, curved upward at the rear end. The bars are about 3 ft (91.4 cm) in length at the outer end and gradually increase in length toward the inner end, where they are some 8 ft (2.4 m) long. The hay is allowed to slide to the side into a windrow. Some of these attachments can be folded to allow bunching. The windrow attachment is especially adapted for harvesting flax, clover, alfalfa, peas, and other crops.

Bermuda Grass Cutter Bar   Another special type of cutter bar being made has twice as many guards as the regular type. These guards are narrow, with ledger plates on each guard, and are so placed that the knife, in making an in or an out stroke, passes through two guards instead of one. It is claimed that this arrangement is effective in cutting Bermuda grass, which becomes very closely matted together.

Lespedeza Bar   The lespedeza cutter bar is equipped with special narrow guards, a seed screen, and a pan to receive the seed.

HAY CONDITIONERS

The conditioning of hay by means of crushing, crimping, or flailing is becoming increasingly popular with many haymakers. Boyd[1] gives the following advantages of conditioning hay:

1   Speeds field curing. Conditioning can reduce drying time by about 30 percent.
2   Reduces weather damage.
3   Field losses due to shattering are reduced as curing time and the amount of turning and tedding are reduced.
4   Conserves color and feed value through shorter exposure and less shattering.

There are three general types of hay conditioners: the smooth roll, the corrugated roll or crimper, and the flail-type forage harvester.

The smooth-roll type may use two rubber rolls of a combination of a rubber roll and a steel roll. The smooth rolls give a continuous crushing action to the hay, leaving no part not crushed. Most rubber rolls have spiral grooves to aid in picking up the hay and feeding it between the rolls.

The corrugated-roll- or crimper-type conditioner may be equipped with two malleable-iron rolls, with tapered flutes that mesh together similarly to gear teeth (Figs. 15-12 and 15-13), or with slatted-bar rolls. Some models use a fluted bar roll that presses against a smooth rubber roll. As the hay passes between the rolls, it is bent, crimped, cracked at intervals, and in some cases crushed.

Figure 15-14 shows a combination mower, conditioner, and windrower.

The flail-type conditioner is really a hay harvester, but when it is used as a hay conditioner, the shear bar is removed to reduce the cutting action. The hay is partially chopped by the swinging hammers or knives.

Figure 15-15 shows a self-propelled windrower that is equipped with hydrostatic drive.

Tests conducted by Person[2] showed that alfalfa hay dried faster when crushed with smooth rollers than when crimped with corrugated rollers (Fig. 15-16).

To obtain a quick rate of drying, hay should be conditioned simultaneously with the cutting or conditioned within 15 or 20 minutes after it is cut.

[1] M. M. Boyd, Hay Conditioning Methods Compared, *Agr. Engin.*, **40**(11):664-667, 1959.

[2] K. Person and J. W. Sorenson, Jr., Comparative Drying Rates of Selected Forage Crops, *Trans. of the ASAE,* **12**(3):352, 353, and 356, 1970.

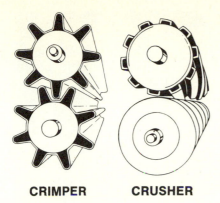
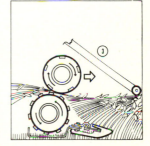

CRIMPER   **CRUSHER**

Fig. 15-12 Two types of hay conditioner rolls.

Fig. 15-13 A special bar pushes the hay crop forward so that when the crop is cut it is fed butt-first into the crushing rolls. (*International Harvester Company*.)

Fig. 15-14 Combination mower, conditioner, and windrower. (*International Harvester Company*.)

Fig. 15-15 A self-propelled windrower equipped with hydrostatic drive. (*International Harvester Company.*)

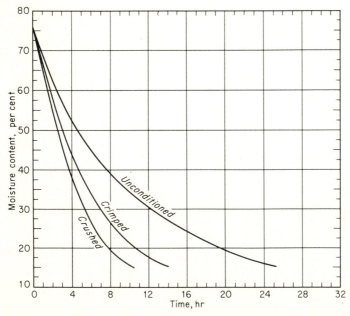

Fig. 15-16 Comparison of moisture content at various hours during the drying period for different methods of conditioning alfalfa. (*Tex. Agr. Expt. Sta.*)

Hay-Conditioner-Swather   This machine mows and conditions the hay and deposits it on the ground in a broadcast swath.

HAY RAKES

Hay rakes can be classified as *side-delivery* and *sweep*.

Side-Delivery Rakes

The use of hay loaders and pickup hay balers created a demand for a hay rake that would make a loose, fluffy, continuous windrow. Then, too, many haymakers rake the hay into windrows during cutting or directly after it is cut. Such windrows must necessarily be loose to allow the hay to cure. The side-delivery rake was developed to take care of this demand. It is also used to windrow peanuts.

Side-delivery hay rakes can be classified according to the types of reel construction, which are *cylindrical-reel, parallel-bar* or *side-stroke*, and *finger-wheel.*

The Cylindrical Reel   The cylindrical-reel type (Fig. 15-19) is equipped with a four- or six-bar cylindrical-type reel. The reel is suspended under a heavy angle-iron frame at an angle of about 45° to the direction of travel. As the rake is drawn forward, the reel is revolved so that the spring teeth on the rake bars move forward along the ground, thus raking the hay as the reel moves. The angle of the reel causes the hay to move along the front of the reel toward the trailing end and roll off in a loose roll. The direction of travel is the same as that of the mower, and the heads of the hay are rolled to the inside of the window, while the larger and juicier stems are left on or near the surface of the roll.

The reel bars are attached to three spiders. The rake teeth are curved forward near the end to aid in picking up the hay. Most teeth are made of spring steel and have a coiled section next to the reel bar. One make uses a rubber ball joint for flexibility.

While in contact with the hay, the teeth are held at an angle, with the points leading. This gives a pushing with a slight lifting action to the hay. The tooth bar is turned slightly to maintain the tooth angle. This action may be termed a *feathering* action (Fig. 15-17). The drive mechanism to hold and feather the teeth may use an eccentric spider, cam track, or planetary gears.

The hay is prevented from following or hanging on the teeth by stripper bars (Fig. 15-17). As the stripper bars partially surround the reel underneath, it is termed a *basket*. The reel and basket are suspended under a light- or heavy-duty frame.

Cylindrical-reel side-delivery hay rakes are available in four general types: the ground-driven trailing, the semimounted ground-driven, the tractor-mounted power-takeoff-driven, and the hydraulic drive.

The Parallel-Bar or Side-Stroke Hay Rake   This type of rake may have from four to six reel bars attached to two parallel plates or spiders at each end of

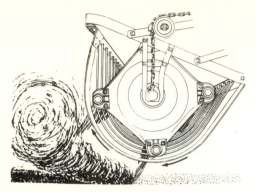

Fig. 15-17   Showing the action and angle of the cylindrical-reel hay rake teeth.

the reel (Fig. 15-18). The right front plate faces to the rear, while the left rear plate faces to the front. The plates are set at right angles to the direction of travel. The plates have short stub shafts fitted with sealed roller bearings for the reel bars. As the reel revolves, each bar rotates within the bearing mounts to keep the teeth in a vertical position at all times. When a bar approaches its lowest position, teeth come in contact with the hay, raking for a short distance, to be followed by the next bar. The angle of the teeth can be changed only by tilting the entire reel. The hay is moved to the side by parallel strokes of the reel bars with less agitation than when it is moved by the cylindrical reel. Figure 15-19 shows the action of the two types of rakes.

The width of swath raked by the parallel-bar rakes ranges from 7 to $8\frac{1}{2}$ ft (2.1 to 2.6 m).

Fig. 15-18   Semimounted hydraulically driven parallel-bar side-delivery hay rake. (*International Harvester Company.*)

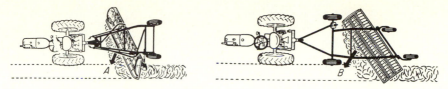

Fig. 15-19   The action of the parallel-bar rake (*left*) and the cylindrical-reel rake (*right*).

Figure 15-20 shows a side-delivery hay rake mounted on the front of the tractor so that two windrows can be moved into one and eliminate a separate raking.

The Finger-Wheel   The finger-wheel reel hay rake shown in Fig. 15-21 consists of five individually floating wheels set in a half-squadron arrangement. Fingers with a coil base are attached to the rim or periphery of the wheel. The fingers contact the ground lightly, and as the rake is drawn forward, the diagonally set wheels revolve and produce a drag-stroke action which moves the hay forward and to the side in a line about parallel to the axle of the wheel.

The wheels are attached to a crank arm and are partially supported by a tension spring so that they will follow the contour of the ground without excessive pressure on the ground. The wheels will "snake" over a terrace or irrigation ditch, raking the hay without damage to the rake.

Sweep Rakes

A typical tractor-mounted sweep rake is shown in Fig. 15-22. The raising and lowering of the teeth are operated by power from the tractor. The sweep rake is

Fig. 15-20   A hay rake mounted on the front of the tractor moves two windrows into one. (*International Harvester Company*.)

Fig. 15-21 Overhead front view of finger-wheel hay rake windrowing hay. (*New Idea Farm Equipment Co.*)

sometimes called a *buck* or *bull* rake. It is useful in collecting hay from the windrow and transporting it short distances to a stationary baler or to a stack. This type of rake is being displaced by more modern hay stackers.

HAY STACKERS

In some sections, hay is stacked in the field rather than stored in the barn. Many types of stackers are built, both commercially and locally.

Fig. 15-22 Tractor-mounted sweep rake that can also be used as a stacker.

Fig. 15-23 Firmly packed stacks of hay. The inset at top shows a large compression stack former. (*Heston Corp.*)

Where medium-sized stacks are put up, sweep-rake-type stackers attached to tractors can be used. A trip arrangement permits the operator to drop the load of hay on the stack.

A new system of stacking hay consists of a windrow pickup and a compression chamber (Fig. 15-23). To form the stack in the wagon compression chamber, the hay is picked up and blown into the chamber in stages. Each batch is compressed with a power packer until the chamber is full (Fig. 15-24). The density of the hay in the stack is about half the density of baled hay. Stacking wagon chambers are available to form stacks ranging from 1 to 6 tons (0.9 to 5.4 metric tons). Small stacks measure 7 ft (2.1 m) wide, 8 ft (2.4 m) long, and 8 ft (2.4 m) high. Large stacks are 8 ft (2.4 m) wide, 20 ft (6.1 m) long, and 11 ft (3.3 m) high. When the compression chamber is full, the stack is transported to the feeding area or storage site and power-unloaded. Special stack movers are available.

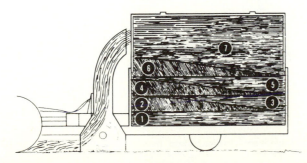

Fig. 15-24 Steps in forming compressed hay stack. (*Heston Corp.*)

Other systems of stacking hay from the windrow make either round or rectangular stacks.

Baled hay is also placed in stacks.

Feeding the hay directly from the stack reduces handling and costs.

HAY BALERS

When hay is being grown for commercial purposes and has to be shipped, it should always be baled so that as much as possible can be put into the average railway car or loaded onto a truck. Many hay growers prefer to bale their hay before storing it in the barn in order to conserve space.

Pickup Automatic Self-tying Balers Making Rectangular Bales

This type of baler is an automatic-pickup, self-feeding, and self-tying machine which requires only one person, who drives the tractor (Fig. 15.25). Most balers are operated by the power takeoff of the tractor. The size of the rectangular bale varies from 14 by 18 to 16 by 18 in (35.6 by 45.7 to 40.6 to 45.7 cm) and from 32 to 42 in (81.3 to 106.7 cm) in length. The weight of the bales produced will vary from 40 to 90 lb (18.2 to 40.9 kg). depending upon the kind of hay being baled and its moisture content. The average bale will weigh 60 to 75 lb (27.2 to 34.1 kg). Some makes of balers use a 15 gauge wire, supplied in rolls, while others use heavy, strong twine for tying the bales.

The pickup and feeder for the hay baler may be either on the right or the left side of the machine, according to the designer's preference. There are at least four methods of feeding the hay into the press chamber: auger and packer fingers, spring teeth and feeder arms, auger and feeder head, and carrier-roller feed.

Fig. 15-25 Tractor-drawn power-takeoff-driven pickup hay baler equipped with bale trailer. (*International Harvester Company.*)

Figure 15-26 shows a pickup cylinder that delivers the hay direct to a floating-feed cross auger, from which it passes to packer fingers, which feed it into the bale chamber.

Tying Mechanisms

The development of a pickup feeding mechanism, of tying mechanisms for heavy twine, and of twisting or tying mechanisms for wire, all of which operate automatically, makes the modern hay baler a truly automatic machine.

The twine knotters (for hay balers) are similar to the twine knotters used on grain binders. It is necessary, however, to have two needles and two knotters to tie the two twine bands simultaneously around the bale.

The wire-twisting mechanism is similar to a regular binder twine knotter. There are other types of wire-twisting mechanisms.

Manufacturers provide trouble-shooting charts for their tying mechanisms. These charts should be carefully studied to determine the cause of tying troubles and their remedies.

Pickup Automatic Baler Making Round Bales

This type of hay baler is termed a *Roto-Baler* (Fig. 15-27). The hay is raked into large, double windrows so there will be an ample volume of hay the full width of the pickup conveyor. The swath of hay is picked up by the pickup conveyor and fed into the bale-rolling mechanism shown in Fig. 15-27.

Another system of making round bales consists of an endless belt of raddles

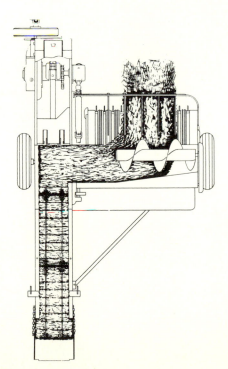

Fig. 15-26   Overhead view of auger to feed hay into bale chamber. (*International Harvester Company.*)

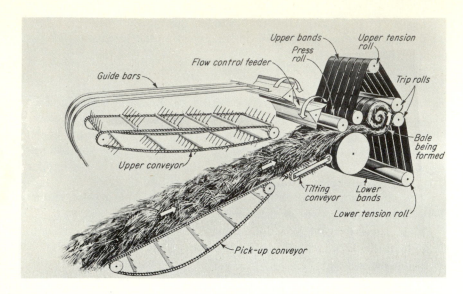

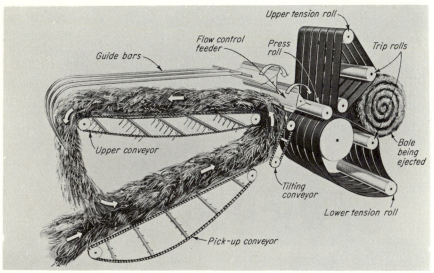

Fig. 15-27 Method of making round bales of hay: (*top*) feeding hay into baling chamber, where it is compressed between belts; (*bottom*) showing how the hay is deflected over the upper conveyor of the feeder while the completed bale is being wrapped with twine and ejected. This permits a nonstop operation. (*Allis-Chalmers Mfg. Co.*)

with coiled spring fingers (Fig. 15-28). As the machine moves forward, the raddle fingers pick up windrowed or swathed hay and roll it along the ground, forming the bale (Fig. 15-28). The tension on the raddle chain controls the density of the hay. The chain is driven by a hydraulic motor powered by a power-takeoff pump. When the bale is completed, the machine is stopped and the operator pulls a rope to open the rear gate. The machine moves forward and frees the bale. Bales of hay can be formed ranging in weight from 800 to more than 1200 lb (363 to 545 kg).

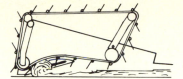

Raddle begins rolling windrow . . .

into circular bale on ground.

Bale's location keeps it compact . . .

until desired size is reached. Then . . .

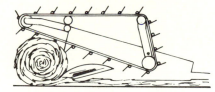

rear gate lifts and raddle freewheels.

Fig. 15-28 Steps in forming a round bale of hay on the ground by raddle chain. (*Hawk Bitt Co.*)

Mini-Rolls of Hay[1]   A system of packaging hay in rolls of 5 to 7 in (12.7 to 17.8 cm) in diameter and 3 to 8 in (7.6 to 20.3 cm) long has been developed. The rolls are formed at a density of 40 lb (18.2 kg) or more/ft^3. Rolls of less than 40 lb (18.2 kg)/ft^3 can be fed directly to livestock. Higher-density rolls must be broken up before being fed.

Bale Throwers

The mechanism shown in Fig. 15-29 is an automatic hydraulic-powered bale ejector that throws the tied bale of hay into a trailing wagon. Some bale throwers are powered by the power takeoff of the tractor, while others may be powered by a small gasoline engine mounted on the thrower.

[1] Staff report, Now, Hay in Mini-Rolls, *Progressive Farmer*, 90(4):45, 1975.

Fig. 15-29   Showing the action of an automatic bale thrower. (*International Harvester Company*.)

Fig. 15-30   An automatic pickup bale loader that can load rectangular or round bales (*Bush Hog*.)

Pickup Bale Loaders Several types of bale loaders are now available whereby the bales of hay are automatically picked up off the ground and elevated by an endless carrier chain to a height where a worker on a trailer wagon or truck can take the bales and stack them on the truck (Fig. 15-30).

Bale Accumulator Figure 15-31 shows how bales of hay are accumulated on a platform before they are dumped.

FARM HAY DRYERS

The artificial drying of hay with and without heated air is discussed in Chap. 22.

FIELD HAY WAFERING EQUIPMENT

In the late thirties and early forties commercial and cooperative plants were developed for the drying of alfalfa hay for the production of alfalfa meal to be used especially in poultry feeds. In the late forties there was considerable interest in the development of commercial plants for producing hay in pellets. The investment required for housing, equipment, and operation for each of the above systems was beyond the average hay producer.

The practice of lot-feeding livestock, particularly cattle, has increased and created a need for hay in a form that requires a minimum storage space and can be easily handled by mechanical equipment. This need has brought about the development of field equipment that will process loose hay into compact 1- by 2-in (2.5- by 5.1-cm) or 2- by 2-in (5.1- by 5.1-cm) cubes or 3-in^2 (19.4-cm^2) wafers. The wafers must have sufficient cohesive qualities to avoid crumbling and breakage in handling.

The hay is prepared for wafering by being placed in windrows with a flail-type hay harvester. It is cured to about normal moisture content. The field hay wafer machine has a flail-type pickup which delivers the hay to the hopper, from which it is fed into the compressing dies. Hydraulic pressure may vary from

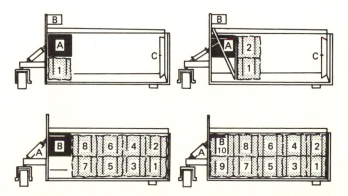

Fig. 15-31 Showing how 10 bales of hay are accumulated, then dumped. (*Schwartz Mfg. Co.*)

Fig. 15-32   A cutaway view of a hay cuber. (*Deere & Co.*)

100 to 1000 lb/in^2. The bulk density of the wafers is about 25 lb/ft^3. The capacity ranges from 5 to 6 tons/h. A 125-hp engine is required to operate the machine.

Figure 15-32 shows a cutaway view of a field hay-cubing machine. The first hay cuber was sold in 1961.

MAINTENANCE AND CARE

1   Practice regular inspection of all haying equipment for lubrication, adjustments, loose parts, and proper belt and chain tension.

2   Check for proper registration of knife sections with guards.

3   Check wear and adjustment of knife clips and wearing plates.

4   Test the roll tension on hay conditioners and adjust to suit field and crop condition.

5   On hay balers check tension for bale density.

6   Read and follow instructions in the operator's manual for all hay equipment.

7   At end of season clean and inspect for needed repairs and place an order for replacement parts.

8   Apply a coating of oil to any parts that may be damaged by rust.

9   Store in a shed or cover with plastic or canvas sheets. Tie sheets securely to machine.

REFERENCES

Anonymous: Pick a New System for Easier Hay Handling, *Farmer Stockman*, 86(4), 1973.

Bainer, Roy: New Concepts in Side-Delivery Rakes, *Agr. Engin.*, 32(5):266-268, 1951.

Bowers, Wendell, and Allen P. Rider: Hay Handling and Harvesting, *Agr. Engin.*, **55**(8):12, 1974.

Boyd, M. M.: Hay Conditioning Methods Compared, *Agr. Engin.*, **40**(11): 664-667, 1959.

Bruhn, H. D.: Performance of Forage-conditioning Equipment, *Agr. Engin.*, **40**(11):667-670, 1959.

Chancellor, W. J.: Energy Requirements for Cutting Forage, *Agr. Engin.*, **39**(10):633-636, 1958.

———: Formation of Hay Wafers with Impact Loads, *Agr. Engin.*, **43**(3):136-138, 1962.

Day, C. L.: Hay and Ensilage Harvesting Costs, *Mo. Agr. Expt. Sta. Bul.* 561, 1951.

Elfes, L. E.: Design and Development of a High-speed Mower, *Agr. Engin.*, **35**(3):147-153, 1954.

Floyd, Charles S.: Making Hay in the U.S.A. Part 1: Dollars, Acres and Methods, *Implement and Tractor*, **86**(8), 1971.

Giles, G. W., and C. A. Routh: The Finger-wheel Rake, *Agr. Engin.*, **32**(10):537-540, 1951.

Harbage, R. P., and R. V. Morr: Development and Design of a Ten-foot Mower, *Agr. Engin.*, **43**(4):208-211, 1962.

Johnson, Bill: Giant Bales and Stacks, *Progressive Farmer*, **88**(7), 1973.

Jones, T. N., and R. F. Dudley: Methods of Field Curing Hay, *Agr. Engin.*, **29**(4):159, 1948.

Kepner, Robert A.: Analysis of the Cutting Action of a Mower, *Agr. Engin.*, **33**(11):693-697, 1952.

Lorang, Glenn, and Gene Logsdon: Handling Hay Today, *Top Operator*, **5**(5):20–23, 1973.

Lundell, V. J., and D. O. Hull: Field Production of Hay Wafers, *Agr. Engin.*, **42**(8):412-415, 1961.

Pederson, Thomas T., and Wesley F. Buchele: Hay-in-a-day Harvesting, *Agr. Engin.*, **41**(3):172-175, 1960.

Ramser, J. H.: Harvesting and Storing Chopped Hay, *Ill. Ext. Ser. Mimeographed Sheet* 230, 1951.

QUESTIONS AND PROBLEMS

1 Explain the different ways in which a tractor mower can be mounted, and give the advantages of each.
2 Explain what is meant by register in a mower knife. Explain how register is determined and adjusted.
3 Explain what is meant by alignment of a mower cutter bar. How is it determined and adjusted?
4 What are the advantages of a pitmanless mower?
5 What are the advantages for conditioning hay?
6 Describe three types of hay conditioners.
7 Explain the differences in action of the cylindrical-reel, the parallel-bar, and the finger-wheel types of hay rakes.
8 Explain how round bales of hay are formed.
9 What are the advantages of round bales and hay stacks?
10 What are the advantages of a bale thrower?
11 What is meant by hay wafering, cubing?

Chapter 16

Forage-harvesting Equipment

Beef and dairy cattle and other types of livestock will remain in beter condition during the winter months if they receive a succulent feed. As there are no green pastures in the winter, such feed must be grown, harvested, processed, and preserved in a silo so that it will be available when needed. In the Northern states corn is the principal silage crop, while in the Southwest sorghum is used more than corn. The use of grasses for silage is growing in popularity. Some field-cured hay is chopped and stored as chopped hay in the barn.

The former method of putting up silage was to cut and bundle the crop with a row binder and then haul the bundles of stalks to the silo, where they were chopped up and blown in for storage. The silo may be either an upright cylinder 20 to 40 ft (6.1 to 12.2 m) in height or a long trench dug in the ground. The trench may be 8 to 12 ft (2.4 to 3.7 m) wide at the top and 30 to 60 ft (9.1 to 18.3 m) in length, the size depending upon the tonnage of the silage to be put up. Self-feeding trench silos are becoming popular. A few pit silos may be found.

The various methods and operations required in handling chopped hay and forage crops are shown in the following chart:

Row crops

Field operations	Barn or silo operations
Cut and bind, load manually, transport	Unload and feed manually, chop and blow
Cut and chop, transport	Unload, blow

Broadcast crops and grasses

Field operations	Barn or silo operations
Cut, cure, windrow, chop, transport	Unload, blow
Cut, windrow, chop, transport	Unload, blow
Cut and chop, transport	Unload, blow

ROW-CROP FORAGE HARVESTING EQUIPMENT

As shown in the above chart, there are two systems of harvesting and chopping green row crops to fill silos.

Row-Binder and Ensilage-Cutter System

The first and oldest system requires a row binder to cut and bind the crop into bundles. Horse-drawn row binders were supplanted largely by the tractor-drawn power-takeoff row binder. Usually the bundles of corn or sorghum were dropped on the ground, then manually loaded on wagons, trailers, or trucks and transported to the silo. Now this system has been supplanted by power-operated field forage harvesters.

Field-Harvester and Blower System

The second system of harvesting and chopping green row crops consists of a combination plant-cutting unit and a chopping unit. The complete machine is called a *field forage harvester* (Fig. 16-1). The field forage harvester performs the functions of both the row binder and the silage cutter, as it severs the standing stalks from the ground and chops them into silage in one continuous operation in the field. It does, however, require a blower at the silo.

There are at least ten advantages of the field forage harvester. They are as follows:

1   Eliminates the drudgery of lifting and loading 10 to 15 tons of heavy green bundles of corn per acre.
2   Provides ensilage at lower cost.
3   Provides more tons of feed per silo.
4   Permits filling the silo when the crop is at the right stage.
5   Makes ensilage with greater feeding value.
6   Provides more uniformity of feeding value from any part of the silo.
7   Provides a more uniform, solid pack without air pockets, thus preventing molds.
8   Causes no wilting of leaves or loss of previous moisture.
9   Leaves no mud or contaminating soil bacteria on butts.
10   Avoids soggy material which often occurs when the silo is filled too early with green, immature corn.

Fig. 16-1   Self-propelled field forage harvester with side-delivery spout. (*Heston Corp.*)

TYPES OF FIELD FORAGE HARVESTERS

Field forage harvesters can be divided into two general types according to the method or mechanism used to cut the growing crop. The most versatile type uses interchangeable units whereby windrowed hay, row crops, or broadcast crops can be cut and fed into a chopping unit. It is called a *field chopper*. The other type uses free-swinging chains, hammers, or knives attached to a 6- or 8-ft (1.8- or 2.4-m) horizontal rotor. This is commonly called a *flail harvester*.

Fig. 16-2   Field forage harvester with offset hitch. The chopped forage is being blown into a forage wagon. (*Allis-Chalmers Mfg. Co.*)

Fig. 16-3 Interchangeable attachments for field forage harvester. (*Kochring Farm Division*.)

Field Chopper-Harvesters

Field forage chopper-harvesters are available in tractor-drawn and self-propelled types. Some tractor-drawn types are driven by the power takeoff of the tractor, while others are driven by an auxiliary engine mounted on the chopper. The offset hitch permits the chopper to trail to the right of the tractor so the tractor does not run on or over crop rows, swath, or windrow (Fig. 16-2).

Several companies provide interchangeable harvesting units for the chopper-blower section (Fig. 16-3).

The row-crop attachment is suitable for harvesting row crops, such as corn and sorghum. It is equipped with two stationary side knives across which a single sickle section oscillates to sever the plants. The plants on most field harvesters are moved back by a gathering chain and power-fed into the chopping unit.

Figure 16-4 shows rubber gathering belts that grip the stalks and carry them to the feed rolls.

The cutter bar consists of a regular mowerlike cutter bar and a reel to throw the crop back onto an apron which conveys the material back to an auger, which in turn conveys the material to one side where it is fed into the chopping unit.

The pickup attachment has revolving fingers that lift wilted hay from the swath or windrow and move the material back to an auger which in turn conveys the material to the chopper throat.

Figure 16-5 shows two types of cutterheads used on field chopper-harvesters. Both types may be fitted with four or six knives. In the flywheel-type cutterhead, the knives for cutting and the impeller paddles for throwing and blowing are mounted on the wheel separately. The cylinder-type cutterhead usually has the knives designed both to cut and to blow (Fig. 16-5). Some cylinder cutterheads require a separate impeller blower. Each type of cutterhead has certain advantages and disadvantages. The cylinder type may have built-in sharpeners, but the knives must be removed from the flywheel for sharpening.

Fig. 16-4   Rubber gathering belts for field forage harvester. (*Deere & Co.*)

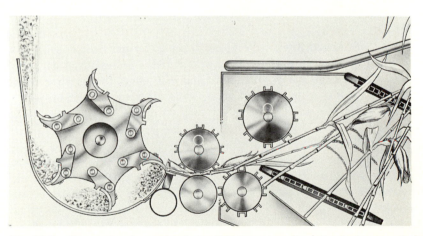

Fig. 16-5 Two types of cutterheads for field harvesters: (*top*) flywheel-type cutterhead; (*bottom*) cylinder-type cutterhead. (*International Harvester Company, New Holland Machine Co.*)

The length of cut or the length of the cut pieces can be varied from about ½ to 2 in (1.3 to 5.1 cm) when the stalks are fed straight into the cutting knives. The change in the length is made either by changing the number of knives on the cutterhead or by changing the speed of the feed mechanism. It is essential for efficient operation that the knives and shear plate be kept sharp and the recommended clearance between them maintained. These mechanical factors and the physical condition of the material being chopped will affect the total energy requirements and the horesepower requirements.[1]

The capacity of a field forage chopper-harvester is affected by:

1   The area of the throat opening
2   The speed and rate of feeding
3   The density of the material

The rate of feeding and density are in turn influenced by the yield of the crop.

The Field Harvester Blower Spout   A spout or delivery chute is necessary to conduct or convey the chopped material from the blower fan to the wagon or truck (Figs. 16-1 and 16-2). The blower outlet pipe for most choppers is round, except for the flail harvester, which has a trapezoidal duct over the flails. Extension sections are provided where additional height is required. The blower discharge pipe is usually about 8 in (20.3 cm) in diameter. The rectangular delivery spout or chute had a round base to fit the blower discharge. The spout can be set for either side or rear delivery. It is about 6 or 8 ft (1.8 or 2.4 m) in length and has a gradual curve. The bottom part of the spout is usually open.

The most important part of the spout is the swivel deflector attached to the top end. The lip can be controlled by the tractor operator by means of ropes so the material can be directed to any part of the wagon box.

Forage Wagons and Boxes   A typical forage trailer and self-unloading box is shown in Fig. 16-2. Most boxes are provided with a power-driven slat-chain carrier to move the silage to the unloading end. Some boxes have cross-conveyors at the end to unload the silage, while other types open the whole endgate.

Field Flail Forage Harvesters

As stated above, the flail-type field forage harvester uses free-swinging chains, hammers, or knives to sever the plants by a beating or cutting action. At the time the plants are being severed, the flails or knives travel in the same direction the machine is moving (Fig. 16-6). The flail chopper does not have chopping knives on the blower fan to chop the material into acceptable lengths for silage. Where just the flails are used for severing the plants, the harvested material can be blown into windrows for curing. The beating by the flails more or less conditions the hay.

[1] *Amer. Soc. Agr. Engin. Yearbook*, pp. 124-125, 1963.

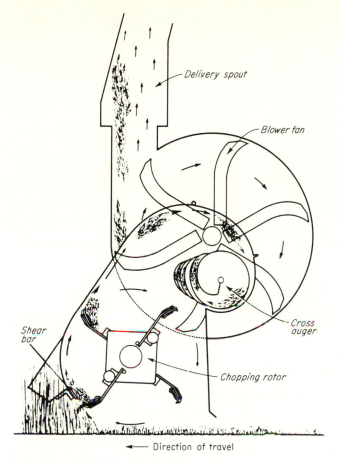

Fig. 16-6   The action of flail knives cutting forage against shearbar. The knives throw the cut material onto a cross auger which feeds the material into a fan. (*Gehl Bros. Mfg. Co.*)

The Self-propelled Forage Harvester

Figure 16-1 shows a self-propelled forage harvester equipped with row-crop attachment. Cutter bar and windrow pickup attachments can be interchanged for the row-crop attachment.

SILO FORAGE BLOWERS

Field-chopped forage is elevated into upright silos by stationary blower units at the silo. Some blowers are equipped with short auger conveyors for the side-unloading wagon boxes. Other blowers have long, belt-type conveyors for the wagons that open the endgate (Fig. 16-7). The blowers can be operated by the power takeoff or belt pulley of a tractor or by auxiliary engine. Blowers are equipped with a roller feed mechanism and an easily reached clutch lever for engaging and reversing the feed belt or auger.

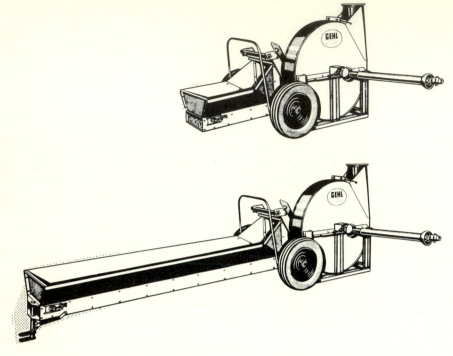

Fig. 16-7 Long- and short-hopper forage blowers. (*Gehl Bros. Mfg. Co.*)

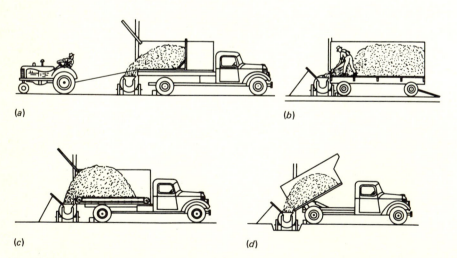

(*a*)   (*b*)

(*c*)   (*d*)

Fig. 16-8 Four methods of unloading chopped silage into blower: (*a*) pull off (tractor or winch attached by cable to false front); (*b*) pitch and push off (load is pitched or pushed through opening at back or side of trailer); (*c*) convey off (power-driven unloading conveyor built into truck or trailer); and (*d*) dump off (tilting body dumps load into hopper).

The combination suction blower can be used for filling both upright and trench silos. It is also useful for unloading seed cotton from trailers into piles. This type of suction blower can be integrally mounted on a tractor.

Figure 16-8 shows several methods of unloading truck and trailer loads of chopped forage into blower conveyors.

SILO UNLOADERS

A silo unloader installed in an upright silo will eliminate the hazards of climbing up into a silo and the labor involved in pitching the silage out manually (Fig. 16-9). An electric motor operates two knife-studded augers which feed the silage to a blower located at the center of the silo. The driving roller moves the augers around so that the silage is removed layer by layer. The whole unit is suspended by a cable from a special tripod at the top of the silo.

The machine shown in Fig. 16-10 is equipped with a power-operated digging and cutting reel to cut and loose packed silage in a trench silo. The loosened silage falls into a hopper and onto an elevating belt that conveys the silage up and deposits it in a truck or trailer box. The cutting reel is mounted on a hydraulically operated boom. Models are designed to mount on tractors or trucks and as self-propelled units.

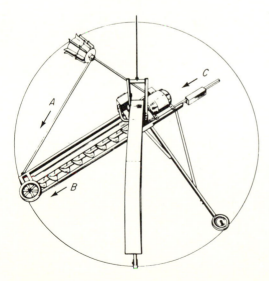

Fig. 16-9 An electrically operated silo unloader showing 9-in (22.9-cm) auger. (*Brillion Iron Works, Inc.*)

Fig. 16-10 Mechanical trench silo unloader that cuts loose the packed silage and elevates it into a truck or trailer box. (*Oswalt Industries, Inc.*)

MAINTENANCE AND CARE

1   Periodically inspect equipment for proper belt tension, hitch, and power-takeoff alignment.

2   Keep all cutting knives sharp.

3   Check chopper for proper length of cut.

4   Check shield on power takeoff.

5   Read and follow instructions in operator's manual.

6   At end of season clean and inspect for needed repairs and place order for replacements.

7   Store in a shed or cover with plastic or canvas sheets. Tie the sheets securely to the machine.

REFERENCES

Bookout, B. R.: Can You Afford a Field Chopper, *Mich. Agr. Expt. Sta. Quart.*, **33**(1):28-33, 1950.

———, and Karl Vary: Costs and Method of Harvesting Grass Silage, *Mich. Agr. Expt. Sta. Quart.*, **32**(4):582-588, 1950.

Chancellor, W. J.: Energy Requirements for Cutting Forage, *Agr. Engin.*, **39**(10):633-636, 1958.

Decker, Martin: Mechanical Silo Unloaders, *Kans. Agr. Expt. Sta. Bul.* 412, 1959.

Gass, J. R., R. A. Kepner, and L. G. Jones: Performance Characteristics of the Grain Combine in Barley, *Agr. Engin.*, **39**(11):697-702, 1958.

Guest, R. W.: Mechanical Equipment for Handling and Feeding Forage, *New York Agr. Engin. Ext. Bul.* 348, 1961.

Hoff, Paul R.: Combine Troubles, Causes, and Remedies, *New York Agr. Engin. Ext. Bul.* 335, 1959.

Liljedahl, J. B., et al.: Measurement of Shearing Energy, *Agr. Engin.*, **42**(6):298-301, 1961.

McLeod, H. E., and K. K. Barnes: Effect of Paddle Tip-Clearance on Forage Blower Performance, *Agr. Engin.*, **39**(8):456-457, 1958.

Richter, D. W.: Friction Coefficients of Some Agricultural Materials, *Agr. Engin.*, **35**(6):411-413, 1954.

Vary, Karl, and Byron Bookout: Farmers' Experiences with Grass Silage, *Mich. Agr. Expt. Sta. Quart.*, **32**(4):589-597, 1950.

Winkelblech, C. S., and Paul R. Hoff: Hay Conditioners, *New York Agr. Engin. Ext. Bul.* 339, 1960.

QUESTIONS AND PROBLEMS

1 Give several advantages of field forage harvesters.
2 Explain the differences and functions of row-crop and broadcast forage harvester.
3 List the factors that effect the capacity of a field forage harvester.
4 Explain the operation of unloaders for upright and trench silos.

Chapter 17

Grain-harvesting Equipment

In the preparation of many crops for the market, it is necessary that the seeds be separated from the stalk on which they grew. All the small-grain crops must have the seeds stripped from the straw, corn must be shelled from the cob, peanuts must be threshed or picked from the vines, and cottonseed must be separated from the lint. Different types of machines are necessary for the separation of the seed from the holding agent in the different crops. Generally, very large apparatus is necessary, incorporating a number of different operations in the same machine as the material passes through it.

Formerly, grain was cut with a binder and the bundles carried to a stationary threshing machine, where it was threshed and sacked. Under the present farming practices in the United States, the process is reversed; that is, the machine is carried to the crop in the field, where it harvests and threshes the crop, places the grain in a tank, and distributes the straw on the land.

HISTORY OF DEVELOPMENT

Development of the Binder

The biblical story of Ruth tells how grain was harvested with a hand reaper. The hand reaper was used in Europe and America until horse-drawn machinery was adopted. The long-handled scythe was developed toward the end of the colonial period. The cradle was introduced between 1776 and 1800. McCormick claimed to have demonstrated his first horse-drawn reaper in 1831 but did not obtain a patent until 1834. Obed Hussey obtained a patent on a reaper in 1833. McCormick built 50 machines in 1845 and about 800 in 1848.[1]

The self-raking reaper appeared about 1854. A platform for manual binding was introduced about 1850. The first mechanical wire-tying mechanism was introduced in 1873. Twine binders were introduced in 1880, but it was not until 1892 that Appleby obtained a patent on a twine knotter.

The horse-drawn grain binders were ground-driven. Auxiliary engines were mounted on some binders about 1920. The power-takeoff-driven binder was introduced in the late 1920s.

Development of the Thresher

Rogin quotes William Darling as follows: "In Bedford County, Pennsylvania, grain was still generally threshed with the flail in 1829." Much grain was trodden out by horses in the late 1830s. Other men were granted patents, but the patent granted to Hiram A. and John A. Pitts, Dec. 29, 1837, was the beginning of the thresher. It was horse-operated. The manufacture of the Case thresher was begun at Racine, Wisconsin, in 1844. By 1900, threshers were equipped with self-feeders, band cutter knives, weighers, and wind strawstackers.

Development of the Combine

A patent on what was termed a *combined harvester-thresher* was granted to Samuel Lane in 1828. The real beginning of the combine for harvesting, threshing, and cleaning was when A. Y. Moore, et al., of Kalamazoo, Michigan, obtained a patent in 1835. In 1854, 600 acres of wheat were combined in Alameda County, California, but the method was not truly initiated in California until about 1880. One of the earliest manufacturers of horse-drawn traction-driven combines was the Stockton Combined Harvester and Agricultural Works of California. Steam-tractor-drawn combines were introduced in the 1890s. Some of these machines were equipped with a 42-ft (12.8-m) header and harvested, it was claimed, from 90 to 125 acres (36.4 to 50.6 hectares) in a day.[2,3]

Gasoline-tractor-drawn combines were introduced on a large scale in the wheat areas of the Middle West as the result of labor shortages during the First World War, or about 1916. Combines were first introduced in northwest Texas in 1919, when seven machines were used.[2] The self-propelled combine was commercially introduced about 1938.[4]

[1] Leo Rogin, *The Introduction of Farm Machinery*, University of California Press, Berkeley, Calif., 1931.

[2] Ibid.

[3] H. P. Smith and Robert F. Spilman, Harvesting Grain with the Combined Harvester-thresher in Northwest Texas, *Tex. Agr. Expt. Sta. Bul.* 373, 1927.

[4] Chris Nyberg, Highlights in the Development of the Combine, *Agr. Engin.*, **38**(7):528-529, 1957.

GRAIN COMBINES

The combined harvester-thresher, or *combine*, heads the standing grain, threshes it, and cleans it as it moves over the field. Therefore, it takes the place of and eliminates from the harvest the grain binder, the header, the stationary thresher, and the tiresome tasks of shocking or stacking the grain and hauling the bundles.

The combine is adapted to harvesting all the small grains, soybeans, grain sorghums, and rice, as well as many other crops.

Types of Combines

There are two general types of combines, the *pull* or *tractor-drawn* and the *self-propelled.*

The Pull-Type Combine   The pull-type combine is drawn by a tractor. The smaller combines are driven from the power takeoff of the tractor, while the larger sizes have an auxiliary engine mounted on the combine to drive it. Pull-type combines range in size from a 4- to 8-ft (1.2- to 2.4-m) cut for the smaller sizes and from a 10- to 20-ft (3.0 to 6.1-m) cut for the larger sizes. Gatherers at each end of the cutter bar enable the average machine to cut a swath from 6 to 9 in (15.2 to 22.9 cm) wider than the actual length of the cutter bar.

The Self-propelled Combine   Self-propelled combines are powered with industrial-type engines of 60 to 150 hp. The self-propelled combine is operated by one man. It is easy to handle and transport from field to field and over the highway. A swath can be laid out without loss of grain. Sharp turns can be made to follow rice levees. It is provided with a gearshift or variable-speed drives, such as the hydrostatic drive, to give desired field and road speeds. There is also a reverse gear. Grain self-propelled combines can be obtained in sizes to cut swaths from 6 to 22 ft (1.8 to 6.7 m).

The operator of the self-propelled combine sits above and just behind the electrically or hydraulically controlled platform. The general operation is somewhat like that of a tractor. Many self-propelled combines are now equipped with air-conditioned cabs. The machine is steered by means of a large steering wheel connected to the wheels in the rear. If sharp or right-angle turns are to be made, wheel brakes assist in making the turn, similarly to the row-crop tractor. The engine can be started by pressing a button, which actuates the self-starter. Transmission and separator clutch levers are conveniently located to control machine travel and operation (Fig. 17-1). A slight movement of a lever on the steering post enables the operator to raise and lower the platform or header to meet changing conditions in the field. An electric sensor-header height control can be mounted under the platform to prevent the cutter bar from digging into the soil where there are high spots. It also holds the header at a set height. Field speeds range from $1\frac{1}{4}$ to 4 mi/h (2.0 to 6.4 km/h), while road speeds range from $2\frac{1}{2}$ to 13 mi/h (4.0 to 20.9 km/h). No changes need be made in the machine for short-distance transportation over highways other than raising the

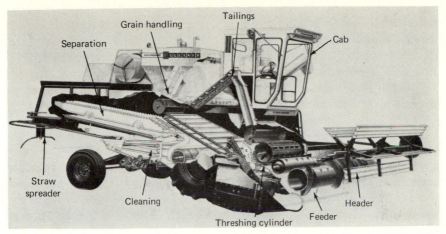

Fig. 17-1 Cross section of a self-propelled combine showing the functional areas for cutting, feeding, threshing, and cleaning and grain storage tank. (*Allis-Chalmers Mfg. Co.*)

platform and shifting into road gear. Some of the self-propelled combines are designed to operate on hillsides. The combine wheels are adjusted to suit the slope of the land by hydraulic rams. The cutterhead can be tilted to cut on slopes up to 55°. The threshing, separating, and cleaning units are kept level.

Functions Performed by a Combine

The basic operational functions of a combine can be divided as follows: (1) cutting the standing grain, (2) feeding the cut grain to the cylinder, (3) threshing the grain from the stalk or stem, (4) separating the grain from the straw, (5) cleaning the grain by removing chaff and other foreign matter, and (6) handling the grain from combine to tank and from tank to truck (Fig. 17-1).

The Cutting Mechanism The standing grain is handled by a cutter bar and a reel to sweep the grain back onto a canvas or auger table or platform (Fig. 17-1). The entire assembly is called a *header*. The knife usually extends the full width of the cutter bar, which ranges from 4 to 22 ft (1.2 to 6.7 m). It is usually operated by a rocker-arm pitman that gives a 6-in (15.2-cm) stroke across two guards. The knife sections are serrated. On some combines the cutting mechanism or header can be angled to cut on sloping land.

The reel is located above the cutter bar and ranges from 40 to 60 in (101.6 to 152.4 cm) in diameter. It may have four or six wood bats. Under some conditions canvas or rubber strips are attached to the bats to aid in sweeping the cut material back onto the table or platform. The entire reel can be adjusted up and down as well as back or forward of the cutter bar. Different crops and growth conditions require different positions of the reel. Gass[1] found that for most conditions in combining barley the reel peripheral speed should be 1.25 to

[1] J. R. Gass, R. A. Kepner, and L. G. Jones, Performance Characteristics of the Grain Combine in Barley, *Agr. Engin.*, 39(11):697-702, 1958.

1.5 times the forward speed of the combine. On some narrow-cut machines the header-feeder canvas is as wide as the cutter bar swath and conveys the cut material direct to the threshing cylinder. Most wide-cut combines have right- and left-hand augers that feed the cut material to the central elevator-feeder (Fig. 17-1). The augers vary in diameter from 16 to 20 in (40.6 to 50.8 cm).

Grain-lifting fingers can be attached to the cutter bar to lift fallen grain. A windrow pickup can be attached to pick up windrowed grain.

The entire cutting mechanism, or cutterhead, is usually controlled by hydraulic or electric lifts.

Threshing Mechanism The threshing mechanism, which separates the grain from the stalks, consists mainly of a revolving cylinder and the concaves (Fig. 17-1). A feeder-beater is usually located in front of the cylinder and at the upper end of the elevator-feeder to assist the elevator-feeder in feeding the grain to the threshing mechanism. Most combines are provided with the rasp-bar-type cylinder and concaves (Fig. 17-1). The grain is rubbed from the stems without materially cutting the straw. Tooth-type cylinder and concaves are available on some combines (Fig. 17-2). Adjustments are provided for varying the speed of the cylinder to suit the kind of crop being harvested. V belt variable-speed drives are used on most combines. The straw is thrown back onto the separating mechanism, while the grain falls through the concaves onto a grain pan or grain carrier and is conveyed to the cleaning mechanism.

On the small straight-through combines, the cylinder and concaves extend almost the full width of the cutter bar, while on the larger machines, the cylinder and concaves are only about 30 in (76.2 cm) in width and 18 to 24 in (45.7 to 61.0 cm) in diameter. All machines provide means of adjusting the concaves to the cylinder. The peripheral speed of combine cylinders ranges from 2000 to 7000 r/min, depending upon the type of crop and its condition.

Separating Mechanism The main separation of the grain from the straw is through the concaves. The loose grains, which are mixed with the straw as it

Fig. 17-2 Two types of threshing cylinders: (*left*) rasp-bar-type cylinder and concaves; (*right*) iron-tooth-type cylinder. (*Allis-Chalmers Mfg. Co.*)

leaves the cylinder, are separated by oscillating straw racks. These racks may consist either of one piece or of several sections which alternately move with a slight elliptical action to pitch the straw rearward with each movement (Fig. 17-1). If the straw racks are made in sections, a long, revolving crankshaft is used to operate them. The pitching action of the straw racks sifts the loose grain from the straw and lets it fall onto a grain pan underneath.

A straw beater is usually located just to the rear of the cylinder to beat out loose grains. One or two steel and canvas curtains hang down from the top housing to deflect and keep the straw in contact with the straw racks (Fig. 17-1).

Cleaning Mechanisms   The function of the cleaning unit is to remove chaff and other foreign matter from the grain. This is accomplished by passing the uncleaned grain over a series of oscillating sieves and screens through which a current of air is forced by a fan (Fig. 17-1). Different types of sieves and screens are available for different kinds of crops. Partially unthreshed heads of grain, termed *tailings*, drop into an auger which delivers the tailings to an elevator which in turn conveys the material upward to the rethreshed by the cylinder.

Attachments for Combines

A number of attachments can be obtained for combines. These include a straw spreader (Fig. 17-1), straw windrower, straw loader, straw chopper, windrow pickup, windrow spreader, bundle-topping vertical cutter bar, flax roller, bagger, grain bin, and cylinder-speed tachometer.

The grain tank is usually provided with a power-driven auger for transferring the grain from the tank to a grain cart or truck (Fig. 17-1).

Harvesting Corn with Combines

Most self-propelled combines can be equipped with a corn harvesting attachment (Fig. 17-3). The speed of the cylinder and the adjustments of the concaves and screens are changed to suit the conditions.

Harvesting Sorghum Grain and Soybeans

Sorghum grain and soybeans are harvested with the grain combine with minor adjustments in speed of the threshing cylinder and the sieves and screens.

Peanut Combines

Peanut combines differ in design from the grain combine as they must separate the peanut from a vine. This type of combine is described in detail in Chap. 20.

Windrowing Machines

In some areas, farmers find that cutting and windrowing the top portion of the plants with the grain attached permit earlier harvesting and protect the grain under the following conditions: (1) when the grain is unevenly ripened, (2) when fields are weedy, (3) when the straw is green but the crop is ripe, (4) when the grain is high in moisture, (5) when crop conditions are such that legume crops tend to shatter if left until ripe, and (6) when weather conditions delay direct

Fig. 17-3   Cross-sectional view of a rice combine. Note the large drive wheel equipped with mud lugs. (*Deere & Co.*)

combining. The windrowing machine consists of a power-takeoff-driven knife, platform canvas, and reel.

The heads of grain are clipped off and fall upon the traveling platform canvas, which delivers the grain over one end onto the stubble. Most windrowing machines deliver the grain over the end farthest away from the standing grain. Center-delivery machines are available.

The Rice Self-propelled Combine

The principal outside differences between a regular grain combine and a rice combine are in the size and arrangement of the wheels. The regular-grain self-propelled combine is equipped with medium-sized tires (Fig. 17-1). The main wheels of the rice self-propelled combine are equipped with large, deep mud lugs to enable the combine to climb over the narrow contour levees needed in flooding rice fields and to give traction in muddy, poorly drained fields (Fig. 17-3). Tires on the combines are more satisfactory for traveling from field to field and for transportation over the highway, as they are faster and do not shake the machine so badly. The rice self-propelled combine is equipped with an engine of greater horsepower than the grain combine because more power is required to travel over the soft rice fields.

Rice combines were used extensively in south Texas for the first time in 1943. The use of the combine has brought about revolutionary changes in handling this crop. Formerly all the rice was handled in sacks; now it is handled in bulk like wheat. Grain carts (Fig. 17-4) have been developed to get the rice from the combine out in the wet, muddy field to a truck located on a graded field road. Furthermore, when rice is harvested, there is too much moisture in it

Fig. 17-4 Special grain cart for transporting the rice from the combine in the field to trucks located on a field road. (*E. L. Caldwell & Sons.*)

for it to be stored, and consequently, it must be dried. There are now many large commercial rice dryers throughout the rice-growing regions of Texas, Louisiana, and Arkansas.

Design, Costs, Adjusting, and Duty of Combines

Basic Design Requirements of Combines Carroll[1] states that the basic design requirements for self-propelled units are as follows:

1 Accessibility
2 Simplicity, with easier and simpler adjustments
3 Complete ease of control and comfort for the operator
4 Capacity to harvest all crops under every condition encountered throughout the grain-growing countries of the world
5 Lighter weights and greater capacity in relation to width of cut
6 Working speeds from $\frac{1}{2}$ mi/h (0.8 km/h) to a maximum of $5\frac{1}{2}$ mi/h (8.8 km/h), with a road speed of 7 mi/h (11.3 km/h)
7 Sufficient engine power to take care of difficult ground conditions as well as to operate the combine mechanism
8 Proper weight distribution in relation to wheels
9 Attachments, drives, and straw-handling equipment
10 Necessary traction equipment for rice fields

[1] Tom Carroll, Basic Requirements in the Design and Development of the Self-propelled Combine, *Agr. Engin.*, **29**(3):101-105, 1948.

Cost of Combining   The various items of cost in harvesting with a combine are *operating* and *fixed* costs. Operating expenses consist of the costs of fuel and lubricants, use of tractor, labor, and repairs. Fixed charges are for depreciation, interest on investment, taxes, and insurance. The cost of housing may also be added. The profit which a self-propelled combine can return to the farmer is measured by the time and labor saved and the increased return resulting from the use of added power.

Adjusting the Combine   If losses are to be reduced to a minimum, the various units of the combine must be carefully adjusted. An explanation of these adjustments requires more space than can be taken in a text. Therefore, as each make of machine has different means provided for making the adjustments, the operator should carefully study the operator's manual and make the adjustments in order from front to rear of the machine, including the power mechanism. Adjustments will not cure all the troubles, because if the machine is driven too fast or the cutter bar set too low, the machine will be overloaded and excessive losses of grain will occur.

Acres Cut by Combines   Most people think of the capacity of a machine as the amount of work it can do in a day's time. The principal factors that influence the rate of cutting are the size of the machine, the rate of travel, and the yield of grain.

Reynoldson, Kifer, Martin, and Humphries[1] calculated, from 214 reports of combines equipped with auxiliary engines, that the rate of cutting would be increased 0.27 acre (0.11 hectare)/h by each foot added to the length of the cutter bar. The average cut per hour for each foot of width was approximately 0.23 acre (0.9 hectare). The rate of cutting depends upon the rate of travel and the size of the machine.

Self-propelled combines with a 14-ft (4.2-m) cutter bar can harvest small grain and sorghum at the rate of 40 to 50 acres (15.2 to 20.2 hectares) per day. When rice was harvested in soft fields and where contour ridges had to be crossed, 27 acres (10.9 hectares) were harvested per day with a 14-ft (14.2-m) self-propelled combine.

MAINTENANCE AND CARE

1   Before operating a grain harvester, inspect the machine to determine if all parts are functioning properly.
2   Check tension on all belts.
3   Check safety shields on power-takeoff shaft on pull-type combines.
4   At end of season, thoroughly clean equipment inside and outside.
5   Remove belts, tag for location, and store in a shed.
6   Apply oil to cutter-bar parts to prevent rust.
7   Clean engine, drain crankcase, flush, and add fresh oil.

[1] L. A. Reynoldson, R. S. Kifer, J. H. Martin, and J. H. Humphries, The Combined Harvester-Thresher in the Great Plains, *U.S. Dept. Agr. Tech. Bul.* 70, p. 14, 1928.

8 Inspect machine for any needed repairs and place order for replacement parts.

9 Store in a shed or cover machine with a plastic or canvas sheet. Tie the sheet securely to the machine.

REFERENCES

Barger, E. L.: Engineering-Management Aspects of Self-propelled Farm Machines, *Agr. Engin.*, **29**(3):106-108, 1948.

Carroll, Tom: Basic Requirements in the Design and Development of the Self-propelled Combine, *Agr. Engin.*, **29**(3):101-105, 1948.

Gass, J. R., R. A. Kepner, and L. G. Jones: Performance Characteristics of the Grain Combine in Barley, *Agr. Engin.*, **39**(11):697-702, 1958.

Johnson, H. William: Efficiency in Combining Wheat, *Agr. Engin.*, **40**(1): 16-20, 1959.

McCuen, G. W., and S. G. Huber: Harvesting with Combines, *Ohio Agr. Col. Ext. Bul.* 330, 1952.

Nathan, Kurt: An Economic Study of Combines, *Agr. Engin.*, **30**(6):274, 1949.

Nyberg, Chris: Highlights in the Development of the Combine, *Agr. Engin.*, **38**(7):528-529, 1957.

QUESTIONS AND PROBLEMS

1 Trace the development of the combine.
2 Explain what is meant by the term *combine.*
3 Describe the different types of combines.
4 Explain the various functions performed by a combine.
5 Explain why a windrowing machine is used in some areas.
6 What are some of the basic design requirements of combines?
7 How many acres can be harvested in a 10-hour day with a 14-ft (4.2-m) combine traveling at 2.5 mi/h (4.0 km/h) in harvesting 20-bushel wheat? Allow 85 percent for field efficiency
8 What are the basic design requirements for a self-propelled combine?

Chapter 18

Corn-harvesting Equipment

The corn picker is a single- or double-row machine equipped with snapping rolls to remove the ears from the standing stalks. As the corn picker is a great time and labor saver, it is being used extensively to replace the slow, hard hand method of harvesting. Only one man is required to operate any of the power-driven one-, two-, and four-row pickers shown in Fig. 18-1. Additional help may be required to haul the corn and place it in the bin. Most machines do not sever the stalks from the ground. The gatherer sides and chains guide the stalks into the throat between the downwardly revolving snapping rolls, which pinch and snap the ears from the stalk. The ears are deflected into an elevator system which conveys them to a wagon or trailer drawn either beside or behind the machine. The ears can also be snapped, husked, and shelled in a continuous operation.

History of Development

The corn picker was first invented by Quincy in 1850. WIlliam Watson of Chicago invented a corn picker shortly after Quincy's invention. The snapping-roller-type corn picker was developed by manufacturers about 1874, but it was not patented until about ten years later. The rollers were placed in an inclined position. Because of the development and the use of the corn binder, interest in corn pickers lagged until about 1920, when manufacturers introduced several

Fig. 18-1   A four-row self-propelled corn-picker-husker. (*New Idea Farm Equipment Co.*)

new machines. The early makes of corn pickers were ground-driven by means of a large bull wheel. Tractor pull-type, power-takeoff-driven corn pickers and the tractor-mounted picker were introduced about 1930. A self-propelled corn picker was made available about 1950.

Types of Corn Pickers

Machines are available for harvesting ears of corn in three different ways. The simplest machine snaps the ears from the stalks and does not remove the husks. This type of machine is called a *snapper*. Most machines used in the Corn Belt are also equipped with a husker attachment which, in addition to snapping the ears from the plant, also removes the husks. This type of machine is called a *picker-husker*. A more recent development is a machine that snaps the corn and shells it in the field. This type of machine is called a *picker-sheller*. Generally, however, corn pickers are classified according to the number of rows harvested and the way the machines are attached to the tractor. Figure 18-1 shows a four-row corn picker-husker. The pull and tractor-mounted types each have their advantages and disadvantages. The pull types are unit machines which can be easily attached and detached from the tractor. With them, however, the operator must look back and to the side to watch the machine. (Fig. 18-2). At least three unpicked rows of corn are broken down by the tractor and wagon in opening a lane through the field. Figure 18-2 shows a swinging hitch can move the corn harvester sidewise behind the tractor. With the tractor-mounted type, time is required to mount the machine, and the tractor cannot very well be used for other work while the machine is mounted. Some machines are more easily mounted than others. But with the tractor-mounted machine, the operator can steer the tractor and watch the machine without any neck twisting. With this type of picker, too, a lane can be opened through the field without breaking down extra rows. The wagon or trailer is drawn behind the machine.

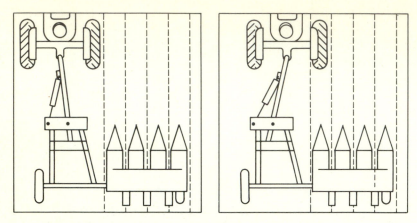

Fig. 18-2   Showing how a swinging hitch moves the pull-type corn harvester sidewise behind the tractor. (*New Idea Farm Equipment Co.*)

Figure 18-3 shows a combine equipped with an eight-row corn-head attachment. The ears are fed into the threshing cylinder, where they are shelled. The husks and cobs pass out over the straw racks, while the shelled corn flows through the cleaning unit and up to the grain tank. Under favorable conditions and high yields a large combine fitted with a four-row corn-head attachment can harvest 3000 to 4000 bushels of corn per day.

Self-propelled corn picker-shellers (Fig. 18-5) are available, but their use is somewhat limited, since the corn attachment for self-propelled combines has been developed. The corn-snapping attachments for combines can be purchased for much less than a corn picker-husker-sheller.

Fig. 18-3   Eight-row corn head on combine. (*Massey-Ferguson, Inc.*)

Power Transmission

The early makes of corn pickers were horse-drawn and ground-driven. The present pull types are power-takeoff-driven. The power is transmitted from the tractor by means of a long drive shaft to a gearbox on the picker, from which the various parts of the picker are driven. The tractor-mounted types require short drive shafts, countershafts, and chain and belt drives. Slip or jump safety clutches are provided for each of the functional parts.

The Gathering and Snapping Mechanism

There is no difference in the gathering and snapping mechanism of the one- and two-row pull types and tractor-mounted types, except that, for the two-row pickers, an extra unit is added. On a two-row picker, the middle divider point is always hinged, while the side gatherers on the one- and two-row pickers may or may not be hinged. The hinging of the points permits them to be set close to the ground and to follow the contour of the surface, thus slipping under and picking up stalks lying close to the ground. As the stalks are guided into the throat by the *gatherer* points, lugged gatherer chains assist in feeding the stalks in between the snapping rolls. Spiral lugs (Fig. 18-4) on the downwardly moving adjacent sides of the snapping rolls catch the stalks, pull them down, and discharge them

Fig. 18-4   Corn-picker unit with shields and gathering points removed to show gathering chains, snapping rolls, and conveyors. (*International   Harvester Company.*)

under the machine. The design and shape of the corrugated lugs on the snapping rolls differ with the various makes of machines (Fig. 18-5). As the stalks are pulled down between the snapping rolls, the ears are pinched off. Ordinarily, the outside roll is set a little higher than the inside roll, and this aids in deflecting the ears into the conveyor elevator to one side if it is a one-row picker and in between the units if a two-row picker.

The snapping rolls should be adjusted as close together as conditions of the crop and the design of the rolls will permit.

Two-row corn pickers are designed so the units can be adjusted to harvest row spacings from 38 to 42 in (96.5 to 106.7 cm) apart. One make has an arrangement whereby the snapping rolls can be adjusted closer or farther apart by means of levers within the convenient reach of the operator. The adjustment can be made while the machine is idle or while in operation.

Some makes of corn pickers have trash knives installed under the snapping rolls to cut off vines, grass, and stalks that tend to wrap around the rolls under

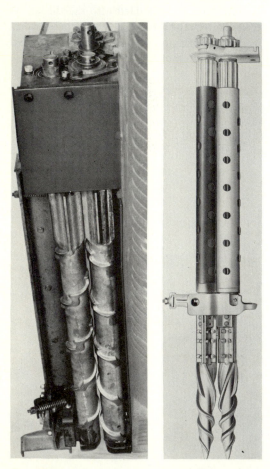

Fig. 18-5   Two types of snapping rolls for corn pickers: (*left*) scalloped rolls with vine knives; (*right*) spiral lug. (*International Harvester Company*)

damp conditions. Figure 18-5 shows rotary knives to cut vines from the snapping rolls. Other makes have trash rollers at the rear end of the snapping rolls to remove trash and broken stalks.

The Conveying and Elevating Mechanism

As the ears fall from the snapping or husking rolls, they are conveyed back and elevated so they can be dropped either through an air blast into the wagon elevator or into the husker or sheller units.

The Husking Mechanism

The husking rolls may be set in line with the snapping rolls, or they may be part of an interchangeable attachment set across the rear of the machine. The husking rolls operate in pairs, with each pair held together under a spring pressure which can be regulated. One of each pair of husking rolls usually is made of rubber or has a rubber covering. There should be just enough tension on the rolls to cause them, with the aid of corrugations and husking pins on the rolls, to grasp the husks and pull them through the rolls so that the ears are stripped clean with a minimum amount of shelling. Pressure on the retarding plates can be adjusted for large and small ears. The number of rolls in a husker ranges from two to four. Some corn pickers are equipped with a fan to blow the husks and trash off the husking rolls. Any corn shelled by the husking rolls is gleaned as it drops into the grain saver.

The Shelling Mechanism

Figure 18-6 shows a typical corn-sheller unit for a tractor-mounted corn picker-sheller. It consists of a peg-studded cylinder about 8 in (20.3 cm) in diameter rotating inside a cylindrical screen. The shelled corn drops through the

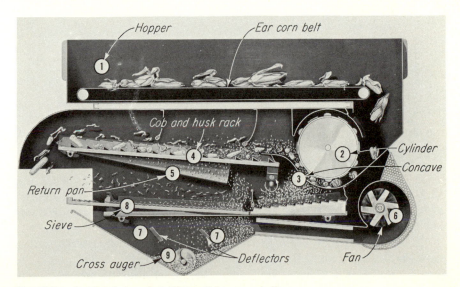

Fig. 18-6   Sectional view of shelling unit for corn sheller.

holes in the screen, while the husks, cobs, and fine trash are ejected and blown out the end of the sheller.

Tests conducted by Pickard indicated that a rasp-bar concave clearance of $\frac{5}{8}$-in (1.6 cm) was more desirable than a $\frac{3}{4}$-in (1.9-cm), that the critical kernel moisture falls between 30 and 25 percent, and that a cylinder speed of 800 r/min (3100 f/min at the periphery) is more desirable than 600 r/min (2350 f/min). Some corn pickers are equipped with a grinder attachment.

Factors Affecting Performance of Corn Pickers

There are a number of factors that will influence the performance of efficiency of corn pickers. Several are listed as follows:

1   Plant characteristics
    a   Variety or hybrid suitable for machine harvesting
    b   Stiff stalks that stand up and do not break over and lodge
    c   Condition of the stalks
    d   Height of ears and stalks
    e   Toughness of the ear shanks
    f   Size of ears—large ears reduce shelling losses
    g   Hard-shelling characteristics reduce shelling losses
    h   Thick, tight husk on ear desirable for snapping but not for husking
2   Mechanical factors
    a   Type of snapping-roll surface
    b   Adjustment of snapping rollers—distance apart
    c   Timing of snapping rollers
    d   Rate of travel
    e   Type of wagon hitch
    f   Adjustment of dividers to pick up stalks which are down
3   Miscellaneous factors
    a   Timeliness of harvest; field loss is less if harvesting is done early
    b   Carefulness of operator
    c   Weather conditions
    d   Cleanliness of fields, that is, freedom from tall weeds and grass
    e   Length of rows
    f   Row spacing suitable for machine

Most of the field losses in snapping ears of corn from the standing plants can be attributed largely to three main factors, namely, (1) down or lodged stalks, (2) timeliness or date of harvest, and (3) rate of travel in harvesting.

Data collected by the author in Texas show that the average percentage of lodged and down stalks in August, September, and October was 12.1, 24.5, and 54.5, respectively. The average percentage of lost shelled corn was 2.9, 7.6, and 12.9, respectively, for August, September, and October. This increased field loss was due largely to the increase in ears lost.

Tests conducted by Burrough and Harbage of Indiana with a picker-sheller show that losses increased for both the snapper unit and the shelling unit as the season became later. They also found that sheller losses decreased as the moisture content of the cob decreased (Fig. 18-7). Cobs with a high moisture

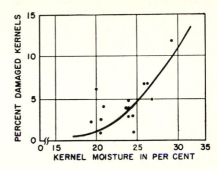
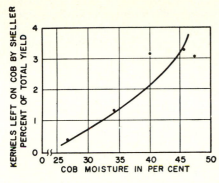

Fig. 18-7   Charts showing how the moisture content of cob and kernel affected sheller loss and kernel damage. (*Agr. Engin., 34(1); 22, 1953.*)

content were spongy and easily broken, and kernels of corn stuck to the whole cob and cob pieces.

Young found that field losses increased as the rate of travel was increased (Fig. 18-8). He recommended that tractors used with corn pickers be operated in first gear.

Methods for Calculating Field Losses

There are several rule-of-thumb methods of quickly calculating the amount of corn being lost by a corn picker. Some are as follows:

    1   Twenty kernels of corn to a hill or $3\frac{1}{2}$ ft (1.1 m) of row equals 1 bushel per acre loss.
    2   One good-size ear in 40 hills, or 133 ft (40.5 m), equals 1 bushel per acre loss.
    3   Shelled-corn loss can be determined by counting the number of

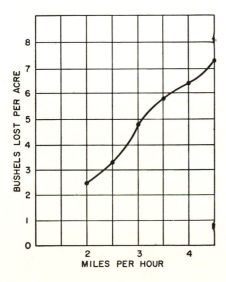

Fig. 18-8   Chart showing the effect of speed of picking on field loss of corn. (*O. L. Young, Illinois State Normal University.*)

kernels/ft^2. There are approximately 74,052 kernels in a bushel of average corn. As there are 43,560 ft^2 in an acre, 1.7 kernels/ft^2 is a bushel loss per acre. If the rows are spaced 42 in, or $3\frac{1}{2}$ ft (1.1 m) apart, and 125 kernels of corn are found in 6 ft of row, or $3\frac{1}{2} \times 6 = 21$ ft^2 of row, $125 \div 21 = 5.95$ kernels/ft^2. Thus $5.95 \div 1.7 = 3.5$ bushels of shelled corn loss per acre.

   4   The ear-corn loss can be estimated by gathering the ears on the ground after harvest in $\frac{1}{100}$ part of an acre and multiplying their weight by 100. It is assumed the plot was gleaned for fallen ears before operating the machine.

Safety Rules for Operating Corn Pickers

Operators of corn pickers have suffered many accidents, probably more than for any other single piece of farm equipment.

The Farm Equipment Institute, working with the National Safety Council, recommends the following safety rules for the safe operation of corn pickers.

   1   Keep the picker on the row.
   2   Keep the power-takeoff shaft covered.
   3   Keep the snapping rolls in good condition.
   4   Be sure the safety clutches are properly adjusted.
   5   Keep riders, especially children, off the picker.
   6   Do not combine hunting with corn picking.
   7   Always shut off the power before dismounting to clean picker if it should clog.
   8   Wear snug-fitting clothing.
   9   Run the picker at the correct speed for the field conditions.
   10   Read the instructions.
   11   Do not use a stick to push stalks through the rolls.
   12   Watch the roll adjustment and the tension of the gathering chains.
   13   Shield the exhaust manifold and the sides of the engine to prevent dry material from collecting and being set on fire.
   14   Use a sediment bulb which is not easily broken.
   15   Keep fire extinguisher available.
   16   Keep the platform of the tractor free of obstructions over which the operator can stumble.
   17   Be especially careful when tired.
   18   Keep untrained operators off the picker.
   19   See that proper lights for night travel are provided.
   20   Do not clean, grease, or adjust the machine while the power is connected.
   21   Always shut off the power when getting off the tractor.

MAINTENANCE AND CARE

   1   Thoroughly inspect the machine before operating. Read the operator's manual and follow instructions.
   2   Check tension on gathering chains and belts.
   3   Check safety shield on power-takeoff shaft on pull-type machines.
   4   Check snapping rolls for proper spacing.

5   Husker and shelling parts should be checked for proper adjustment.
6   At end of season, clean all parts thoroughly.
7   Apply oil to all parts that may rust.
8   Store in a shed or cover with a plastic or canvas sheet. Tie the sheet securely to the machine.

REFERENCES

Adjusting the Corn Picker, *Iowa Agr. Ext. Ser. Pamphlet* 61, 1950.

Barkstrom, Ray: Attachments for Combining Corn, *Agr. Engin.*, **36**(12):799-800, 1955.

Bateman, H. P., G. E. Pickard, and Wendell Bowers: Corn Picker Operation to Save Corn and Hands, *Ill. Agr. Expt. Sta. Cir.* 697, 1952.

Burrough, D. E., and R. P. Harbage: Performance of a Corn Picker-Sheller, *Agr. Engin.*, **34**(1):21-22, 1953.

Gienger, Guy W.: Corn Picker Adjustments, *Maryland Agr. Ext. Ser. Fact Sheet* 360.

Goss, John R., et al.: Field Tests of Combines in Corn, *Agr. Engin.*, **36**(12):794-796, 1955.

Herum, Floyd L., and Kenneth K. Barnes: What's the Best Way to Harvest Corn?, *Iowa SC. Farmer*, **9**(1):7-8, 1954.

Hobson, L. G., and R. H. Wileman: Mechanical Corn Pickers in Indiana, *Ind. Agr. Expt. Sta. Bul.* 362, 1932.

Huber, S. G., and W. E. Stuckey: Efficient and Safe Corn Picker Operation, *Ohio Agr. Col. Ext. Bul.* 325, 1951.

Hurlbut, L. W.: More Efficient Corn Harvesting, *Agr. Engin.*, **36**(12):791-792, 1955.

Johnson, William H., and B. J. Lamp: *Principles, Equipment and Systems of Harvesting Corn*, Agricultural Consultants, Inc., Wooster, Ohio, 1966.

Keller, A. H.: The Engineering Development of a Light, Two-Row Tractor-mounted Corn Picker, *Agr. Engin.*, **29**(5):197-200, 1948.

Morrison, C. S.: Attachments for Combining Corn, *Agr. Engin.*, **36**(12):796-798, 1955.

Pickard, Geo. E.: Laboratory Studies of Corn Combining, *Agr. Engin.*, **36**(12):792-794, 1955.

Scranton, C. J.: Safety and the Mechanical Corn Picker, *Agr. Engin.*, **33**(2):140-142, 1952.

Smith, C. W., W. E. Lyness, and T. A. Kiesselbach: Factors Affecting the Efficiency of the Mechanical Corn Picker, *Nebr. Agr. Expt. Sta. Bul.* 394, 1949.

Smith, H. P., and J. W. Sorenson, Jr.: Mechanical Harvesting of Corn, *Tex. Agr. Expt. Sta. Bul.* 706, 1948.

Young, Orville L.: Pick Slower and Save Corn, *Farm Impl. News*, Sept. 25, 1953.

QUESTIONS AND PROBLEMS

1   Explain the differences between the types of corn pickers.
2   Explain the action of the snapping rolls on a corn picker.
3   Discuss the various factors that affect the performance of corn pickers.
4   Give the rule for calculating the field losses for corn pickers.
5   Give 15 safety rules for operating corn pickers.

Chapter 19

Cotton-harvesting
Equipment

A little more than 100 years passed from the time the first cotton-picking machine was developed until cotton farmers began the uninterrupted use of machines for harvesting the cotton crop. During this time, there were hundreds of cotton harvesting machines patented, some of which would, no doubt, do a satisfactory harvesting job today. There were, however, a number of factors that hindered their acceptance and use at the time they were developed. The cotton farmer had an ample supply of relatively cheap labor. He was looking for a machine that would harvest cotton comparable in cleanliness to that of hand-harvested cotton and, at the same time, leave practically no visible cotton along the row. The gins were not equipped with devices to handle machine-harvested cotton. Generally, the cotton farmer was satisfied with hand harvesting his crop—he was not ready for a change.

The scarcity of labor during the Second World War and new developments in harvesting and ginning equipment played a large part in changing the cotton farmer's viewpoint about harvesting cotton with machines. In 1942, a few bales of cotton were harvested with experimental picking machines. In 1953, it is estimated that there were approximately 15,000 mechanical cotton pickers and 25,000 cotton strippers available. These machines harvested about 25 percent of the 16 billion bales produced. In 1962 it was estimated that 60 percent of the cotton grown in the United States was mechanically harvested. In 1973 practically 100 percent was mechanically harvested.

HISTORY OF DEVELOPMENT

The first patent granted for a mechanical cotton picker was granted to S. S. Rembert and J. Prescott of Memphis, Tennessee, September 10, 1850. Many mechanical, pneumatic, and electrical devices were patented during the century that followed. August Campbell obtained a patent July 16, 1895, on a spindle that has proved to be the basic principle for the successful cotton-picking machine of today. The International Harvester Company acquired the Campbell patents in the early 1920s and spent more than twenty years developing the barbed-spindle machine they placed on the market in 1942. About 1932, John and Mack Rust were granted a patent on a cotton picker which used a smooth, moist spindle.

A machine for the removal of the entire cotton boll from the plant by a stripping action was patented by John Hughes of New Bern, North Carolina, March 28, 1871. Z. B. Sims of Bonham, Texas, obtained a patent September 3, 1872, on a finger-type cotton stripper that combed the cotton bolls from the plants. The machine patented by W. H. Pedrick of Richmond, Indiana, January 27, 1874, was the first in which revolving spiked rolls were used to remove cotton bolls.

The author, in connection with studies for the Texas Agricultural Experiment Station, developed a smooth-roll tractor-mounted cotton stripper in 1930. The principles of this development were incorporated in the Boone machine in 1943. In 1948, the Oklahoma Agricultural Experiment Station developed a stripping roll using longitudinally extending rows of nylon bristles. In 1952, the author developed a stripper roll using longitudinally extending rubber strips that projected radially from a central core (Fig. 19-1). Patents were granted on this type of roll in 1958.

Several commercial horse-drawn cotton strippers were available from 1928 to 1931. The depression period from 1932 to 1942 caused a lag in both farmer and commercial interest. In 1943, a two-row tractor-mounted stripper equipped with two smooth steel rollers was introduced. In 1944, Deere and Company introduced a two-row tractor-mounted stripper equipped with a single steel roller. During the period from 1946 to 1953, at least seven commercial makes of strippers were introduced.

This brief historical sketch shows that there are available two distinct types of cotton harvesters, namely, the *stripper* type, which removes the whole boll from the plant, and the *picker* type, which removes only the locks of seed cotton.

THE COTTON STRIPPER

The first farm-scale use of the mechanical cotton stripper was in northwest Texas about 1914, with a section of a picket fence. Farmers of the area used narrow, boxlike sleds between sorghum rows to collect heads of sorghum grain. An enterprising farmer knocked out the front end of the sled box and nailed some fence pickets to the base so that, as the cotton was combed from the

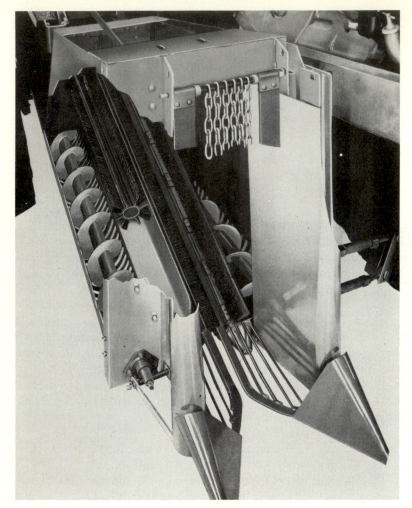

Fig. 19-1   A partially cutaway view of double-roll rubber-strip-brush cotton stripper. Note the auger conveyors. (*International Harvester Company*.)

plants, he could stand in the box and rake the cotton off the fingers. This method was termed *sledding cotton*. Horse-drawn sled strippers were used extensively from 1925 to 1931. Few were used during the depression period from 1932 to 1942 because of an abundance of cheap labor. Their use was discarded shortly after the appearance of the tractor-mounted stripper in 1943.

Types of Cotton Strippers

Cotton strippers are classified largely according to the type of stripping device, such as the double roller with alternate rows of nylon brush bristles and a flexible rubberized strip (Fig. 19-1), and the multiple-finger or comb type (Fig. 19-2). The double-roll cotton stripper may be centrally mounted on the tractor (Figs. 19-3 and 19-4) or it may be self-propelled (Fig. 19-5). The finger- or

Fig. 19-2 Two-row pull-type cotton stripper equipped with long steel bar teeth to comb the cotton bolls from the plants. (*White Farm Equipment.*)

Fig. 19-3 Centrally mounted cotton stripper. Note the green-boll eliminator on the elevator. (*International Harvester Company.*)

Fig. 19-4   Tractor-mounted cotton stripper with basket being dumped. (*Deere & Co.*)

comb-type cotton stripper is mounted centrally on the tractor or is pulled behind the tractor (Fig. 19-2).

Figure 19-6 shows a self-propelled wide-finger-type stripper that will harvest cotton planted broadcast or in narrow rows and in wide rows.

Factors Affecting the Performance of Cotton Strippers

There are many factors, embracing both plant characteristics and mechanical factors, that affect the performance of all types of mechanical cotton strippers.

Plant Characteristics   Results of stripping numerous varieties possessing variable characteristics indicate that the desirable plant type for the mechanical cotton stripper is one which has relatively short-noded fruiting branches 8 to 10 in (20.3 to 25.4 cm) in length, is less than 3 ft (91.4 cm) in height, and has a storm-resistant boll (Fig. 19-7). The locks of a storm-resistant-type cotton are usually not very fluffy and are held tightly in the boll. Fluffy and loosely attached locks (Fig. 19-7) are easily caught and held between limbs and thus are pulled through the stripping space and lost.

Fig. 19-5   Self-propelled cotton stripper. (*Deere & Co.*)

Thickness of Plants in Row   Figure 19-8 shows that plants planted thick and closely spaced in the row are smaller in diameter, have shorter limbs borne higher off the ground, and are not so tall as plants thinly spaced in the row. The performance of the stripper is better with the closely spaced plants. Plant populations of 30,000 to 40,000 plants per acre retard plant growth sufficiently to reduce field losses materially in harvesting with a cotton stripper.

Cultural Practices   When cotton is to be mechanically stripped, the various cultivations should be made with the sweeps set flat and little soil thrown around the base of the plant. At the last cultivation, the soil should be 1 or 2 in (2.5 or 5.1 cm) higher at the base of the plant than in the middle. This permits winds to shift leaves and trash into the depression in the middle between the rows (Fig. 19-8). Every effort should be made to keep the field free of weeds, grass, and vines. Pieces of grass collected with the cotton are hard to remove, and if present in excessive amounts, they will reduce the quality of the cotton lint.

Fig. 19-6   Self-propelled broadcast cotton stripper. The inset shows the finger stripper bars and beater to clear cotton from the fingers. (*Allis-Chalmers Mfg. Co.*)

Pickup Fingers or Limb Lifters   The function of the fingers at the front of the machine (Fig. 19-1) is to slip under and lift the low limbs and bolls and also to strip off the bolls on or near the ground. They should be set rather flat, at not over a 5 to 10° angle with the ground. They should be flexible enough to yield to large-stemmed plants.

Design of the Stripping Unit   The design and type of stripping unit are the most important mechanical factors influencing the performance of a cotton stripper.

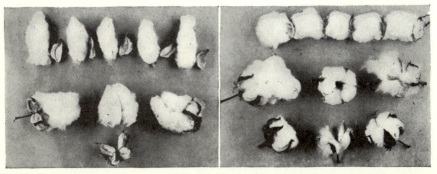

Fig. 19-7   Two types of cotton: (*left*) a typical variety of cotton that is easy to pick and is lacking in storm resistance; (*right*) extra-storm-resistant type of cotton that is hard to remove from the boll.

Fig. 19-8 A slight depression in the middle between the rows of cotton plants permits trash to drift and collect in the depression. (*Texas Agr. Expt. Sta.*)

Generally, cotton-stripper rollers are mounted at an angle of approximately 30° to the ground. The rollers should be long enough so the rear end is 30 to 36 in (76.2 to 91.4 cm) above the ground. This aids in preventing the tops of tall plants from bunching and overlapping at the rear of the rollers. The peripheral or surface speed of the steel rollers should be 25 to 50 percent faster than the travel of the machine. The brush and rubber-paddle stripper rolls are operated faster than the steel rolls, and their speed may range from 400 to 800 r/min.

The fingers on the finger-type stripper are 2 to 3 ft (61.0 to 91.4 cm) in length and set at an angle of about 10 or 15° to the horizontal (Fig. 19.9).

Conveying System There are three methods of conveying the cotton away from the stripping unit. These are (1) finger-beater rolls (2) augers, and (3) air.

The finger-beater rolls are used in connection with the finger-type strippers (Fig. 19-9). They are not suitable for use with the brush and rubber-paddle rolls, because the beaters throw locks of cotton against the downward-traveling outside of the roller and the locks are whipped by the doffer plate onto the ground. The finger beaters beat a high percentage of the dirt and trash out of the cotton as it is conveyed from one beater roll to another.

The auger type of conveyor (Fig. 19-1) is suitable for roller-type strippers but is not used with the finger-type strippers.

The underhousing of all mechanical conveyors should consist of $\frac{1}{2}$- by 3-in (1.3- by 7.6-cm) slots of perforated sheet iron or of heavy $\frac{1}{2}$-in (1.3-cm)-mesh hardware cloth. Much dirt and trash can be screened out of the cotton through the openings in the housing under the conveyors. This is especially true where revolving beater conveyors are used.

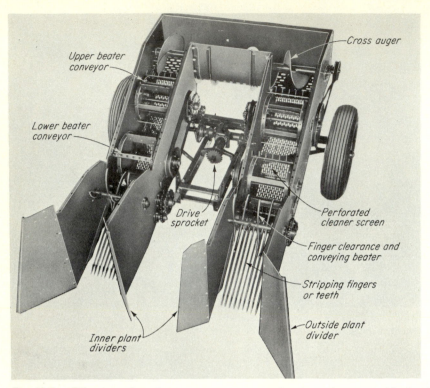

Fig. 19-9   Off-tractor view of two-row tractor-mounted cotton-stripper assembly showing stripping fingers or teeth, beater conveyors, and trash screens. (*White Farm Equipment Co*.)

The elevator shown in Fig. 19-3 has a fan to blow the light cotton into the basket. The velocity of the air stream can be adjusted so that green bolls can be separated from the cotton and dropped into a box.

It is essential that the cotton be conveyed away from the stripping unit as rapidly as possible. Air is a fast method, but the air suction also draws dirt and trash thrown over by the stripping rolls, and some of this dirt may be mixed with the cotton as it is blown into a trailer or basket.

Rate of Travel   At high rates of travel, it is difficult to steer the tractor so that the narrow space between the stripping rolls is always in line with the row of plants. If the yield is high, the volume of cotton may not be adequately handled by the conveyors. The finger-type stripper can be operated at least one tractor-gear speed faster than the roll stripper because precision steering is not required.

THE COTTON PICKER

Although the cotton stripper discussed above is a type of machine for harvesting cotton, its action is so different from the mechanical cotton picker that the two types are discussed separately.

The cotton picker performs the work of the hand picker in that only the locks of seed cotton are removed from the plant.

Types of Pickers

There are four ways of classifying cotton-picking machines:

1 By the method of mounting
2 By the number of rows harvested
3 By the height of the picking drums
4 By the type of spindle used

Cotton-picking units may be mounted on a tractor as a *tractor attachment*, or the machine may be built as a self-contained, *self-propelled* unit. There are *single-row* and *two-row* machines of both the tractor-mounted and the self-propelled types. The two-row self-propelled picker is the most popular type.

The self-propelled picker shown in Fig. 19-10 is a high-drum picker, as it has 20 rows of spindles on each drum. It is suitable for picking cotton from tall plants. The low-drum machine has only 14 rows of spindles and is suitable for low-growing plants.

There are two general types of spindles. The *tapered-tooth* spindle is equipped with three or four rows of machine-cut teeth (Fig. 19-11) designed to catch and hold the cotton fiber while the complete lock is wrapped around the spindle. These spindles are moistened, mainly to keep them free of gum, dirt, and lint.

Fig. 19-10 A two-row self-propelled cotton picker. (*Deere & Co.*)

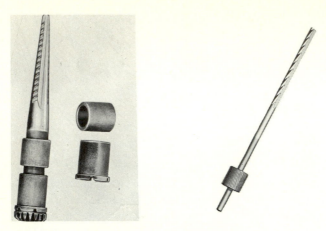

Fig. 19-11   Types of cotton-picking spindles. (*International Harvester Company.*) (*Allis-Chalmers Mfg. Co.*)

The straight spindle may be only slightly roughened, or it may have a row of machine-cut teeth (Fig. 19-11). Both types of spindles are moistened to aid the cotton fiber to adhere to the spindle so that the lock can be wrapped around the spindle as it is removed from the boll.

One manufacturer of mechanical cotton pickers has recently redesigned their machine to travel 10 percent faster along the row or 3.42 mi/h (5.5 km/h). The faster row speed made it necessary to increase the speed of the picker drum and the speed of the tapered barbed spindles. Throwaway doffer units are used.

Methods of Mounting Spindles

There are two general arrangements used for mounting and operating cotton-picker spindles, namely, the *drum* and the *chain-belt*.

The *drum spindle* arrangement is so called because the spindles are arranged in a cylindrical manner, or like a drum set on its end (Fig. 19-12). The spindles are mounted on a bar, the top end of which has a crank arm and a bearing (Fig. 19-13) that travels in a cam track. The cam-actuated picker bars swing the bars around so that the rows of spindles are about $1\frac{1}{2}$ in (3.8 cm) apart as they enter the cotton plants. As the spindles are spaced $1\frac{1}{2}$ in (3.8 cm) on the bars, the spindles are spaced $1\frac{1}{2}$ in (3.8 cm) both vertically and horizontally.

The spindle drums are operated in pairs, one drum on each side of the row, but not directly opposite, as shown in Fig. 19-12. The drum of picking bars and spindles rotates at about the same rate of speed as the tractor as it moves along the row. The spindles can be operated at two speeds to synchronize with the first and second gear speeds of the tractor. The rotation of the spindles ranges from about 2000 r/min for the first tractor gear to about 2700 r/min for the second tractor gear. The complete front drum of one machine rotates at about 60 r/min, and the rear drum rotates at about 79 r/min. Some machines are equipped with spindle bars having 20 spindles per bar, while other machines may have only 12 or 14 spindles per bar. These are termed *high*- and *low*-drum models. The number of spindle bars per drum ranges from 12 to 16.

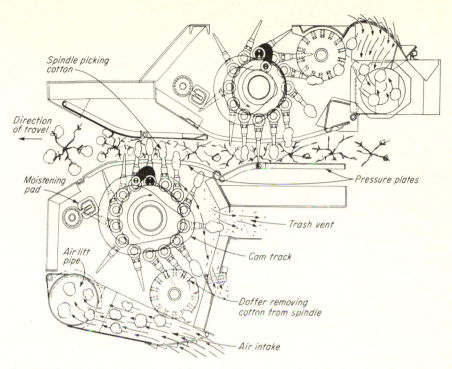

Fig. 19-12   Overhead plan view of the operational action of cotton-picker spindles picking from each side of the row. (*International Harvester Company*.)

The *chain-belt* spindle arrangement is so called because the spindle bars are attached to an endless-chain belt (Fig. 19-14). The belt consists of 80 spindle bars or slats each containing 16 almost smooth spindles. The belt is geared to travel at about 3 mi/h (4.8 km/h), which is also the row travel. Generally, the spindles project into the row of plants only from one side. If two belts of spindles are arranged in tandem on opposite sides of the plant row, spindles project into the plants from each side of the plant row. The straight spindle is friction-driven, while the tapered-tooth spindle is gear-driven.

A moistening agent is used for all types of spindles. Tests have shown that a moistening agent such as textile oil will increase machine efficiency 2 or 3 percent over plain water. The moistening of the spindles makes the cotton adhere to the spindle, keeps the spindle clear of gum, and aids in doffing the cotton from the spindle.

Doffing of the Cotton from Spindles

When the tapered-tooth spindles are withdrawn from the cotton plant with cotton wrapped around them, they rotate about 180° and come in contact with a cylinder of rotating rubber-faced disc doffers which remove the cotton from the spindles. The cotton is dropped at the entrance of the air conveyor system (Fig. 19-15).

The straight spindles are withdrawn from the plants and carried around by

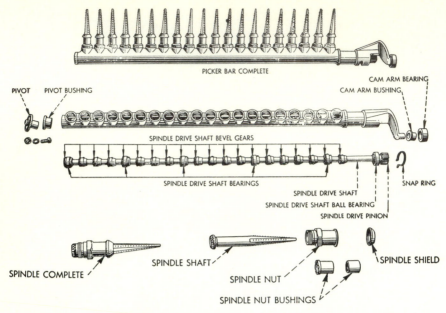

Fig. 19-13   Picker bar complete with spindles. Gearshaft removed, showing drive gears, a complete picker bar, and various parts of a spindle. (*International Harvester Company*.)

the chain belt to the opposite side, where they pass between stripper bars that remove the cotton from the spindles. The cotton is conveyed away either by air or by mechanical elevators.

Elevating or Conveying Systems

All cotton pickers use air for conveying the cotton from the doffing point to a mounted basket. One make has a fan, mounted above and behind the picker units, that sucks the cotton from the doffing point and blows it into the basket (Fig. 19-15). The fan has a tangential bypass which permits the cotton to pass through the fan housing. The fan blades do not come in contact with the cotton, and the seeds are not damaged. At the exhaust point, the air stream is directed against a grid which deflects the cotton into the basket but allows the air to exhaust through the grid. Much dirt and trash are carried out with the exhausting air. Figure 19-16 shows another air conveying system. A fan produces a jet of air in the conveyor duct, and this in turn creates a suction to remove the cotton from the doffing area.

Cotton Baskets   As cotton-picking machines are usually turned almost about-face, it is not feasible to attach trailers to them. Therefore, large baskets are mounted above the picking and power units (Fig. 19-10). The capacity of the baskets ranges from about 500 to 1200 lb (226.8 to 544.3 kg) of seed cotton. When full, the baskets are usually emptied into a wide-bed trailer by tilting the basket with hydraulic power cylinders.

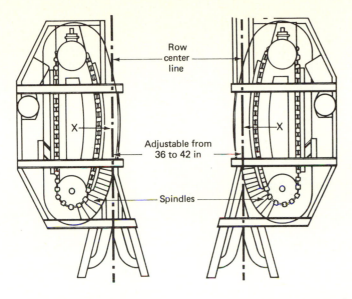

Fig. 19-14 Overhead plan view of cotton-picking spindles mounted on a chain belt. (*Ben Pearson Mfg. Co.*)

Row center line

X

Adjustable from 36 to 42 in

Spindles

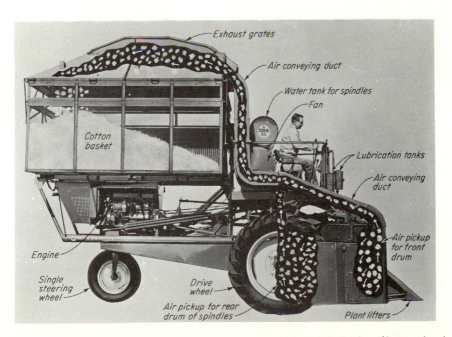

Exhaust grates

Air conveying duct

Water tank for spindles

Fan

Cotton basket

Lubrication tanks

Air conveying duct

Air pickup for front drum

Engine

Single steering wheel

Drive wheel

Air pickup for rear drum of spindles

Plant lifters

Fig. 19-15 The flow of cotton from the picking spindles to the basket. (*International Harvester Company.*)

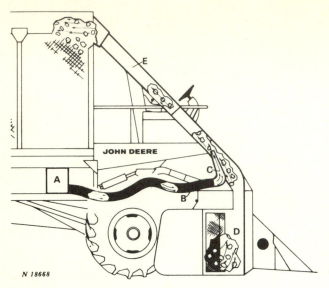

N 18668

Fig. 19-16   A jet air conveying system for cotton picker. (*Deere & Co.*)

How a Cotton Picker Works

The features of the various units of a cotton picker have been discussed above. A better understanding of how a cotton picker works can be obtained if the flow of cotton is traced through the machine, as shown in Fig. 19-15. (1) The plants are folded into the throat of the machine by rounded members, and low limbs and bolls are lifted up by limb-lifting fingers. (2) The revolving drum or belt of rotating spindles passes under moistening pads, and the spindles are thoroughly wetted. (3) The spindles then are projected in among the limbs and bolls to engage the cotton. The plants are pressed against and around the spindles by spring-loaded pressure plates. (4) The spindles rotate and come in contact with the rotating rubber doffers or the stripper-bar doffers. (5) The cotton drops into the air or mechanical conveying system, which conveys the cotton to the basket. (6) Air, dirt, and trash are exhausted through grates which deflect the cotton into the basket. (7) Hydraulic cylinders tilt the basket and dump the cotton into a trailer.

Factors Affecting the Performance of Cotton Pickers

The general classification of factors affecting the performance of mechanical cotton strippers is also applicable to mechanical cotton pickers, that is, plant characteristics, mechanical factors, cultural practices, and miscellaneous factors. The application of these factors to cotton pickers is in most cases different from their application to cotton strippers.

Plant Characteristics   Cotton-picking machines perform best when the cotton plants are of medium size. Medium-sized plants flow through the machine and permit the spindles to engage the cotton better than large plants with many long limbs. As with the stripper, the picker performs better when the lowest limbs are at least 4 in (10.2 cm) above the ground. Cotton-picking machines require a well-opened boll with locks that are fluffy and fiber that is long enough to wrap around the spindle, be held, and be pulled from among the compressed mass of plant limbs.

The chart in Fig. 19-17 shows the performance of a mechanical cotton picker in picking several varieties of cotton having different plant, boll, and fiber characteristics. The compact-lock short-staple Hi-Bred variety gave the poorest machine performance. The locks were so compact and the fiber so short that the spindles could not hold the cotton, and much of it was dropped on the ground.

Defoliation   The collection of green-leaf particles by mechanical cotton harvesters has been the cause of the loss in quality and grade of mechanically harvested cotton ever since the machines were in the experimental and development stage.[1]

Many attempts were made to develop mechanical devices to remove green-leaf particles from mechanically harvested cotton. These were not satisfactory, largely because of the fine pubescent hairs on the leaf particles that

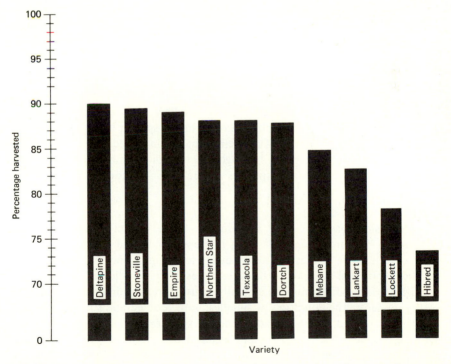

Fig. 19-17   Chart showing the effect of variety of cotton on the performance of a cotton picker. (*Tex. Agr. Expt. Sta.*)

[1] Charles A. Bennett, The Relation of Mechanical Harvesting to the Production of High-grade Cotton, *Agr. Engin.*, **19**(9):386-388, 1938.

caught and hung onto the cotton fiber. Dry-leaf particles were easier to remove, as the hairs became brittle and broke to release the leaf particles.

Weed-killing chemicals were tried, but they were not satisfactory because the foliage was killed quickly. This did not give the plant sufficient time to form abscissa cells between the limb and the leaf stem and cause the leaf to fall from the plant. Then, too, excessive amounts of dry-leaf particles were collected by the mechanical harvester. In some areas this type of chemical is used as a desiccant before stripping.

The first step toward a successful chemical defoliant was made by a farmer who was applying calcium cyanamid dust as a fertilizer. Some of the dust drifted onto some cotton plants. The farmer observed that the leaves of these plants dropped off in two or three days. Cyanamid dust was used extensively until other chemicals were developed.

Now the chemical defoliation of cotton fields is generally a preharvest operation.

Liquid and dust chemical defoliants can be applied by either ground or aircraft equipment.

Bolls of cotton less than 36 days old at the time of defoliation may suffer loss in weight of fiber and seed and in quality of products. The early-set bottom bolls mature and open before the late-set top bolls are 36 days old. A delay in harvesting the early open cotton will result in atmospheric damage to the exposed cotton fiber.

Bottom Defoliation   Early mechanical picking of these bottom bolls without collecting many green-leaf particles can be done if the bottom parts of the cotton plants are defoliated. Liquid defoliants are applied with high-clearance ground sprayers having long drops with the sprayer nozzles low enough to spray only the bottom part of the plant. The picking spindles must operate in the defoliated area of the plant. If a high-drum 20-spindle machine is used, only the bottom 12 or 14 rows of spindles can be used. The top eight to six rows must be removed.

Mechanical Factors   The performance of a cotton picker is greatly affected by the care taken in keeping it properly adjusted and operated. The right amount of water must be kept flowing to the moistener pads to keep the spindles moist and clean. The pressure plates must be adjusted to suit plant conditions. The doffers must be carefully adjusted to the spindles to remove the cotton properly. The condition of spindles, both toothed and smooth, must be frequently checked. Worn-out and damaged spindles should be replaced. The elevating systems should be carefully watched to prevent plugging, stoppages, and loss of cotton. Figure 19-10 shows a cotton picker doing a good harvest job.

Cultural Practices   A slight elevation of the soil at the base of the plants and weed-free fields are as essential for good picker performance as for stripper performance. The row spacing should suit the multiple-row machines. Fairly thick and uniformly spaced plants aid the performance of the mechanical cotton picker.

Cotton-salvaging Machine   The data in Table 19-1 and Fig. 19-18 show that the amount of cotton either left by the mechanical picker on the plant or dropped on the ground varies with the variety of cotton. Some of the cotton on the ground may be the result of natural shedding, and some may be blown out by winds. Under some conditions the amount of cotton on the ground after the mechanical harvester has been used may be sufficient to justify salvaging.

One machine uses a series of slitted belts that are run in contact with the ground to pick up loose cotton lying on the ground. Figure 19-18 shows how the slits in the belt open and close as the belt passes around the pulley. The loose cotton is caught in the closing slits and held until the slits open and release the cotton as the belt passes around the upper pulley. Most of the foreign matter picked up with the cotton is removed as the cotton is passed to a second belt.

There are several other types of cotton-salvaging devices. One uses a series of carding belts to pick the cotton off the ground. Others use rubber finger-wheels and brushes to sweep and flip loose locks of cotton onto the picking spindles.

Miscellaneous Factors   A well-trained and careful operator is essential to obtain the best performance from a cotton-picking machine. The operator should carefully study the instructions in the operator's manual and learn the functions and adjustments of the various parts of the machine. Custom outfits generally have a trained serviceman to take care of machine adjustments, but the operator must handle the machine properly to obtain the best performance.

Table 19-1   Performance of Mechanical Cotton Picker in Harvesting 10 Varieties of Cotton, Brazos River Field Laboratory

Variety	Number of plants per acre	Storm loss before machine picking	Cotton lost by picker — On ground	Cotton lost by picker — On plant	Yield on plants before picking	Cotton harvested by machine	Percentage of cotton on plant harvested by picker
Deltapine D-5	28,880	32	36	77	1034	921	89.1
Empire	36,590	7	18	57	1153	1078	93.5
Texacala	25,483	11	40	129	1051	882	83.9
Northern Star	23,653	10	25	70	859	764	88.9
Lockett 140	41,295	39	51	90	892	751	84.2
Dortch	29,795	88	46	69	951	836	87.9
Lankart 57	21,170	3	26	112	791	653	82.6
Mebane	35,806	17	44	72	815	699	85.8
Hi-Bred	29,534	101	124	124	790	542	68.6
Stoneville	16,988	114	65	41	1054	948	89.9

* All weights shown are pounds of cleaned seed cotton per acre.
Source: H. P. Smith and E. C. Brown, Mechanical Harvesting of Cotton, mimeograph, *Tex. Agr. Expt. Prog. Rpt.* 1527, 1953.

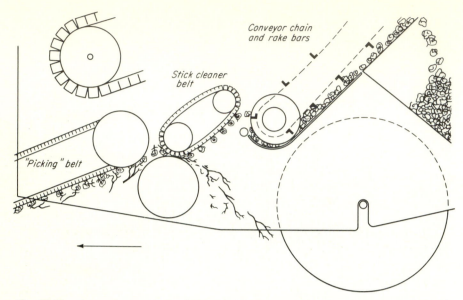

Conveyor chain
and rake bars

Stick cleaner
belt

"Picking" belt

Fig. 19-18 Showing how burs and sticks are removed from cotton picked off the ground by a salvaging machine. The inset shows how the slits in a thick belt open as it passes around a pulley. As the slits close, while in contact with the ground, any cotton will be pinched in the slits and picked up. (*Garland Sales Co.*)

Fig. 19-19 A mechanical rick compactor forming a stack of cotton on the ground in the field. (*Tex. Agr. Expt. Sta.*)

Fig. 19-20 The compaction mechanism of stack or modular builder that forms high-density stacks of cotton on pallets. (*Tex. Agr. Expt. Sta.*)

HANDLING AND STORAGE OF SEED COTTON[1]

Seed cotton has traditionally been transported from the harvesters in the field to the gin with trailers designed for this purpose. The rate at which cotton can be harvested with present-day pickers and strippers is much faster than the rate at which it can be ginned practically and economically. Harvesting should be permitted to continue at an uninterrupted pace to prevent losses due to weathering in the field. The purchase and use of more trailers to accumulate the cotton increases the cost of handling and storage, since the trailers will be making fewer trips to the gin.

Alternate systems have been developed to handle and store the seed cotton in the field or in central storage areas that will receive the cotton at the high harvesting rates and dispense it to the gin at uniform rates. A mechanical ricker as shown in Fig. 19-19 receives the seed cotton from the harvester and forms a long stack in the field. The cotton is loaded into conventional cotton trailers with a front-end loader or lift-truck and transported to the gin.

Another method of handling and storage of seed cotton is referred to as the *modular system*. The module builder, as shown in Fig. 19-20, receives the cotton directly from the harvester and compresses the cotton into a high-density stack

[1] L. H. Wilkes and J. K. Jones, Modular System of Handling Seed Cotton, *ASAE Paper No.* 73-149, presented at annual meeting of American Society of Agricultural Engineers, Lexington, Ky., 1973.

Fig. 19-21   High-density seed cotton stack being off-loaded onto table of the mechanical feeder at the gin. (*Tex. Agr. Expt. Sta.*)

on a pallet. After the stack or module is formed, the tailgate is opened and the machine is moved to the next pallet. The cotton may be stored in the field or transported with a tilting bed trailer to the gin or central storage area. Since the seed cotton is in a uniform shape and lends itself to mechanization, a mechanical feeder has been developed to feed the cotton into the gin with less labor and power requirements than with the conventional pneumatic system (Fig. 19-21).

MAINTENANCE AND CARE

   1   Before operating a cotton-harvesting machine, read the instruction manual and become familiar with all the working parts and their functions.
   2   Check to see that the picking spindles are properly moistened.
   3   Check to see that the doffing device is properly adjusted to the picking spindles.
   4   On cotton strippers, check for the proper spacing of the stripping rolls.
   5   Check the conveying system.
   6   At end of season, thoroughly clean the machine. If it is a picker, steam-clean, if possible.
   7   After cleaning a picker, operate the picker drums and spray a light oil on the spindles. Just a speck of rust on a spindle can cause trouble in doffing.
   8   Thoroughly clean the bristles and rubberized strips on cotton stripper rolls.
   9   Store machines in a shed or cover with a plastic or canvas sheet. Tie the sheet securely to the machine. Why leave a $20,000 machine exposed to the weather for 10 or 11 months of the year?

REFERENCES

Capstick, Daniel F.: Economics of Mechanical Cotton Harvesting, *Ark. Agr. Expt. Sta. Bul.* 622, 1960.
Colwick, Rex F., et al.: Mechanization of Cotton Production, *So. Coop. Series Bul.* 33, 1953.

—— et al.: Mechanized Harvesting of Cotton, *So. Coop. Series Bul.* 100, 1965.

Corley, Tom E.: Correlation of Mechanical Harvesting with Cotton Plant Characteristics, *Trans. of the ASAE,* **13**(6):768, 1970.

Crowe, Grady B., and Harry R. Carns: The Economics of Cotton Defoliation, *Miss. Agr. Expt. Sta. Bul.* 552, 1955.

Hagen, C. R.: Twenty-five Years of Cotton Picker Development, *Agr. Engin.,* **32**(11):593-596, 1951.

Hall, W. C., G. B. Truchelut, and H. C. Lane: Chemical Defoliation and Regrowth Inhibition in Cotton, *Tex. Agr. Expt. Bul.* 759, 1953.

Harrison, George J.: Breeding and Adapting Cotton to Mechanization, *Agr. Engin.,* **32**(9):486-488, 1951.

Harvest-Aid Chemicals for Cotton—Defoliants, Desiccants and Second-growth Inhibitors, *ARS Special Rept., U. S. Dept. Agr.,* 1960.

Horne, Roman L., and Eugene G. McKibben: Mechanical Cotton Picker Works, *Progr. Administration Rept.* A-2, August 1937.

Johnston, E. A.: The Evolution of the Mechanical Cotton Picker, *Agr. Engin.,* **19**(9):383-388, 1938.

Oates, W. J., R. H. Witt, and W. S. Wood: The Development of a Brush-type Cotton Harvester, *Agr. Engin.,* **33**(3):135-136, 1952.

Smith, H. P.: Harvesting Cotton, chap. 7 in *Cotton Production, Marketing, and Utilization,* edited and published by William Baker Andrews, State College, Miss., 1950.

—— and E. C. Brown: Mechanical Harvesting of Cotton, mimeograph, *Tex. Agr. Expt. Sta. Prog. Rpt.* 1527, 1953.

—— and D. L. Jones: Mechanized Production of Cotton in Texas, *Tex. Agr. Expt. Sta. Bul.* 704, 1948.

——, ——, and H. F. Miller, Jr.: The Cleaning of Mechanically Harvested Cotton, *Tex. Agr. Expt. Sta. Bul.* 720, 1950.

—— et al.: Mechanical Harvesting of Cotton, *Tex. Agr. Expt. Sta. Bul.* 452, 1932.

—— et al.: Mechanical Harvesting of Cotton as Affected by Varietal Characteristics and Other Factors, *Tex. Agr. Expt. Sta. Bul.* 580, 1939.

—— et al.: Comparison of Different Methods of Harvesting Cotton, *Tex. Agr. Expt. Sta. Bul.* 683, 1946.

Tavernetti, J. R., C. G. Leonard, and L. M. Carter: Spindle Moistening Agents, *Cotton Gin and Oil Mill Press,* Nov. 30, 1957.

—— and H. F. Miller, Jr.: Mechanized Cotton Growing, *Calif. Agr.,* May 1953.

Tharp, W. H., et al.: Effect of Cotton Defoliation on Yield and Quality, *ARS, U.S. Dept. Agr. Prod. Res. Rpt.* 46, 1961.

Watson, Leonard J.: The Effect of Mechanical Harvesting on Quality, *Proc. Fifth Ann. Cotton Mechanization Conf.,* Chickasha, Okla., 1951.

Wilkes, Lambert H., and Billy J. Cochran: Bottom Defoliation and Harvesting of Bottom Defoliated Cotton, Paper Presented to Southwestern Section of the American Society of Agricultural Engineers, 1961.

Williamson, E. B., and J. A. Riley: The Interrelated Effects of Defoliation, Weather, and Mechanical Picking on Cotton Quality, *Amer. Soc. Agr. Engin. Paper* 60-641, 1960.

——, O. B. Wooten, Jr., and F. E. Fulgham: Factors Affecting the Efficiency of Mechanical Cotton Pickers in the Yazoo-Mississippi Delta, *Miss. Agr. Expt. Sta. Bul.* 515, 1954.

Williamson, M. N., Q. M. Morgan, and Ralph Rogers: Economics of Mechanical
Cotton Harvesting in the High Plains Cotton Area of Texas, *Tex. Agr. Expt.
Sta. Prog. Rpt.* 735, 1951.

QUESTIONS AND PROBLEMS

1   Explain why mechanical cotton-harvesting machines were not adopted by cotton farmers before 1943.
2   Explain the difference in principle of operation between a cotton-stripping and a cotton-picking machine.
3   Explain the differences between the different types of cotton strippers.
4   Discuss the various factors that may affect performance of a cotton stripper.
5   Explain the different methods of classifying cotton-picking machines.
6   Explain the structural difference between a tapered and a straight spindle, (*a*) the method of mounting, (*b*) the method of doffing, and (*c*) the method of spindle drive.
7   Explain the need for moistening the spindles of cotton pickers.
8   Discuss the factors that affect the performance of a cotton picker.
9   What is meant by defoliation?
10   Why is bottom defoliation practiced?
11   Describe the action of the slitted belt-type cotton-salvaging machine.
12   Describe the module method of handling seed cotton.

Chapter 20

Root-harvesting Equipment

Several field crops, such as potatoes, beets, peanuts, and sweet potatoes, form tubers or roots below the surface of the ground. Specially designed machines are required to dig and separate them from the soil.

POTATO HARVESTERS

In general, potato harvesters (diggers) can be classed as one-row and two-row tractor-drawn machines. The two-row machines can be further divided into three types, namely, the *two-unit* or *separate apron*; the *two-row, open-throat, connected apron*; and the *combination digger-cleaner-sacker,* or *loader*. The last can be called a potato combine.

One-Row Harvester

The one-row and two-row potato harvesters are drawn behind the tractor, but the elevating and shaking apron is power-takeoff-driven (Fig. 20-1). A long power shaft extends from the power takeoff to an enclosed gearbox on the harvester. This gearbox usually consists of a three-speed transmission which permits the elevator to be operated at speeds suited to the field condition.

Several types of digging blades are available. The blade is run deep enough to scoop up the buried potatoes. The complete mass of soil and potatoes is

Fig. 20-1   A one-row power-takeoff-driven potato digger. (*Lockwood Corp.*)

delivered to a traveling elevator chain which is agitated up and down by elongated sprockets on each side of the chain (Fig. 20-2). This agitation shifts loose soil through the links of the chain, which are usually offset alternately; that is, one link is high, while the next link is set low. The chain belt may be constructed as a continuous belt from front to rear, or it may be divided into two sections (Fig. 20-2).

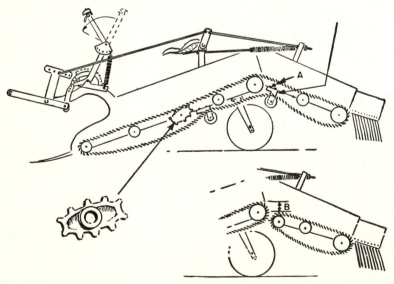

Fig. 20-2   Types of chain-belt elevator aprons used on potato harvesters: (*A*) the continous apron; (*B*) the divided apron. The inset shows the elliptical sprocket for agitating the belt.

Two-Row Harvesters

The two-row potato harvester consists of two sets of digger blades and conveyor shaker belts. The whole unit is usually power-takeoff-driven, but some machines may be equipped with an auxiliary engine. Some two-row machines drop the potatoes on the ground while others deliver the potatoes into a tank on the machine or to a truck driven along beside the harvester. The depth of the digging blades is controlled by hydraulic cylinders.

Most potato growers use chemicals to kill the tops or power-operated flail-beaters to dispose of the vines and weeds before operating the potato harvester.

SWEET POTATO HARVESTERS

The sweet potato has long, tangled vines, which makes mechanical harvesting difficult. It is necessary that these vines be cut and removed from the bed before any machine can be successfully used to uproot the potatoes. A number of vine-cutting devices or deviners have been tried with varied results. A successful devining machine must remove the vines and about 1 in (2.5 cm) of soil from the top of the bed row. The flail forage harvester can be used if the flails are made of rubber and cut to fit the contours of the bed row. The rotary stalk cutter or pasture clipper equipped with gage wheels does a good job of shaving off the tops of the bed row.

Much research work has been done in an effort to develop a sweet potato harvester. Tools such as the ordinary moldboard plow or middlebreaker with rods (Fig. 20-3) have been extensively used. Several types of tractor-drawn and -operated harvesters are now being manufactured and used.

Figure 20-4 shows a one-row power-takeoff-driven sweet potato digger-harvester. This machine is used in harvesting sweet potatoes grown for the market. It consists of a digger blade and a rod-link belt which is agitated to shift

Fig. 20-3 Tractor-mounted sweet potato digger.

Fig. 20-4   A one-row pull-type power-takeoff-driven sweet potato harvester. Note the trailer bed for the graders and racks for crates. (*Johnson Manufacturing Company.*)

the soil through the belt. A rubber belt extends over the center of a trailer bed. As many as eight men stand by the belt and pick, grade, and crate or pallet the potatoes.

SUGAR-BEET HARVESTERS

Sugar beets are grown under a wide variety of soil and climatic conditions. These factors cause the roots and tops to develop differently in different areas, making it difficult to adapt machines to the varied types of growth, soil, and weed conditions and to meet the desires of the growers.

Patents were granted on mechanical sugar-beet harvesters as early as 1898. Machines were used to some extent in the 1930s, but it took the scarcity of labor during the Second World War to bring farmers to accept and use the machines available. Large-scale use of beet harvesters began in 1943, and by 1953 probably 80 percent of the beets grown were mechanically harvested. In 1962 almost all the sugar beet crop was mechanically harvested.

Beet harvesters may be either tractor-mounted, trailed behind the tractor, or self-propelled (Fig. 20-5). The mechanism in all three types is power-takeoff-driven. One-, two-, three-, and six-row sizes are available. The harvesters are operated from $2\frac{1}{4}$ to 5 mi/h (3.6 to 8.0 km/h) and can harvest from 25 to 30 acres (10.1 to 12.1 hectares or more) per day.

Topping the Beets

Beet leaves can be removed by letting herds of sheep graze them off, by beating them with rotary beaters, or by topping devices. Most beets are topped before they are lifted from the ground. Figure 20-6 shows a typical topping device mounted on the harvester. It consists of a power-driven, notched-edged disk set flat to scoop off the beets slightly below or above the lowest leaf scar, depending on the size of the beets. The position of the topping disk is gaged by a *finder*,

Fig. 20-5   A pull-type power-takeoff-driven beet harvester. (*Deere & Co.*)

which may be a sliding shoe, a power-driven wheel roller, or a belt. The pressure spring should not exert a force in excess of 60 lb (27.2 kg). The tops are thrown to one side by a power-driven *top flinger*. Beets can be topped before the digger is operated by tractor-mounted scalping knives.

Lifting the Beets

The lifting unit is mounted just back of the topper. The soil can be loosened around the beets by two large, notched rolling colters set a few inches apart and at an angle.

There are four types of beet lifters. The oldest and simplest consists of two helical-shaped blades that straddle the row. The second type consists of two spade- or solid-rim-type wheels tilted or angled so the rim closes under and lifts the beets from the soil. The third method of lifting beets consists of a large

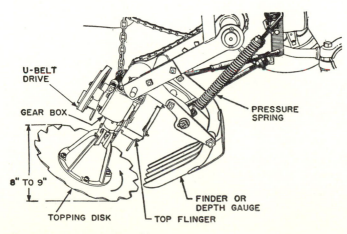

Fig. 20-6   Enlarged view of beet topper.

spiked wheel some 30 in (76.2 cm) in diameter (Fig. 20-7). The spikes are forced into the roots as the wheel rolls over them. As the wheel turns, the roots are lifted from the soil. The fourth device for removing beet roots from the soil consists of a pair of inclined spring-loaded belts that grasp the tops and lift them as the belts move up the incline. The soil is loosened around the roots by a pair of digger blades.

The Cleaning Mechanism

The soil that is lifted with the beet roots is removed as the beets pass over several rows of rotating and overlapping star-shaped kicker wheels by some types of beet harvesters. The soil drops out between the star wheels, but the beets are flipped from one row of wheels to another until they are delivered to the elevator.

Beet Combines

Some machines are provided with a sorting belt so that rocks and large, hard clods of dirt can be picked out manually.

Some beet harvesters have elevators that dump the beets directly into a truck box, while others deposit the beets in a tank on the machine. When the tank is filled, it is unloaded by a chain flight elevator (Fig. 20-5). This type of beet harvester can be termed a *beet combine*.

PEANUT HARVESTING

The harvesting of peanuts is generally a three-stage operation: (1) the tap root is cut and the soil loosened around the peanuts; (2) the vines and nuts are lifted

Fig. 20-7   Spike-studded wheel to lift beets out of the soil and carry them under a topping device above the wheel. (*Blackwelder Mfg. Co.*)

from the soil, passed over a shaker to shake off loose soil, and collected in windrows; and (3) the windrows of peanuts and vines are picked up and passed through a picker or thresher to separate the nuts from the vines.

Stages 1 and 2 are usually done as a single operation, as equipment to dig and shake the vines is operated by the same tractor. The vines are left in the windrow from 3 to 10 days before stage 3, or threshing, is done. Peanuts have been dug and threshed in a single operation, but it required longer to dry the nuts and the quality was below standard.

Peanut Diggers

The peanut plant has a central tap root. Some varieties have vines or runners radiating out 8 to 12 in (20.3 to 30.5 cm) from the tap root, and there may be nuts under the entire plant. Other varieties do not have runner vines, and the nuts are bunched on fibrous roots near the tap root.

Digging of peanuts requires a long knife set fairly flat with the cutting edge extending backward at an angle of about 30° (Fig. 20-8). The knife is set to run

Fig. 20-8   A peanut digger-shaker-windrower and inverter. (*Lilliston Corp.*)

about 2 in (5.1 cm) deep underneath the plants to cut the tap roots and loosen the soil around the nuts. The knives for digging two rows are usually mounted centrally on the tractor. Some growers use two long-bladed cultivator half-sweeps mounted on a cultivator frame to dig peanuts. The knives and sweep blades are set so that they extend toward each other to aid in partially windrowing the vines. Rods can be attached to the knives and knife standards to aid in windrowing the vines from the two rows.

Peanut Shakers

Peanuts for combine or stationary threshing should be as free of dirt as possible. For combine threshing, four to six rows are windrowed together. The side-delivery hay rake has been extensively used to windrow and shake peanuts. This type of work is heavy for a side-delivery rake and causes excessive wear and breakage of parts. The peanut vines are tangled, and the windrow is too compact for rapid drying.

Special equipment has been developed which lifts, shakes, and places the vines in relatively light and untangled windrows. There are a number of machines available for lifting and shaking peanuts. The machine shown in Fig. 20-8 digs, shakes, and delivers the peanut vines in inverted (upside-down) windrows. Tests in Alabama indicate that a peanut shaker should be approximately 54 in (137.2 cm) wide to handle runner peanuts where rows are spaced 34 to 36 in (86.4 to 91.4 cm) apart. The shaker should be designed to raise or elevate the peanuts about 48 in (121.9 cm) in order to shake and windrow the peanuts effectively.

Peanut Threshers and Pickers

There are generally two types of machines for separating the peanuts from the vines. They are classified according to the type of teeth used on the cylinder and are termed *threshers* and *pickers*. The thresher has the regular straight teeth similar to those used on a grain thresher, except that they are spaced farther apart on the cylinder and concave bars. The picker has spring teeth on both the cylinder and concave bars (Fig. 20-9).

Peanut Combines

The commercial peanut combine embodies all the features of a peanut thresher with the addition of a combination pickup-feeder chute (Fig. 20-10).

Stokes and Reed[1] state that

> The two most critical phases of combining peanuts are lifting the peanuts off the ground and into the picking unit, and getting the nuts out of the hay. The speed of the pickup unit was found to have a decided effect on the peanuts lost from cured windrows. When the peripheral speed of the ends of the teeth on the pickup was greater than the forward speed of the combine, there was a tendency to either pull the windrows apart and cause

---

[1] C. M. Stokes and I. F. Reed, Mechanization of Peanut Harvesting in Alabama, *Agr. Engin.*, **31**(4):175, 1950.

Fig. 20-9 Tractor-drawn power-takeoff-driven peanut combine with peanut tank. (*Lilliston Corp.*)

loose nuts to drop off or for the teeth to tear through the windrow knocking off peanuts. Reducing the speed of the pickup unit on one machine from 87 to 40 r.p.m. reduced the loss of peanuts by approximately 10 per cent. This made the peripheral speed of the pickup cylinder approximately equal to the forward travel of the combine. Separating the peanuts from the hay was not a particular problem in the regular picker

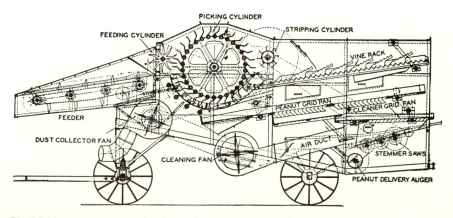

Fig. 20-10 Peanut picker-thresher with spring teeth on cylinder and concaves.

units used as combines, if they were equipped with a hood over the hay discharge to keep tail winds from affecting their operation. It was found, however, that the hay tended to bunch on the racks of some of the converted grain combines. Peanuts were carried over the racks in bunches. The amount of foreign matter left in the peanuts varied with adjustments and could be kept within desirable limits on all three types of machines.

A self-propelled peanut combine is available for picking up windrows of peanuts. It is equipped with bagging attachments and has an industrial gasoline engine for power.

Combined peanuts have a high moisture content and must be dried before storage.

MAINTENANCE AND CARE

1 Potato, beet, and peanut harvesters should be operated and checked to determine if all parts are functioning properly.

2 Two-row harvesters should be checked for proper row spacing.

3 Read and follow instructions in operator's manual.

4 At end of season thoroughly clean equipment, especially peanut threshers.

5 Coat digging blades with heavy oil to prevent rust. Apply oil to other parts that may rust while in storage.

6 Store in a shed or cover with a plastic or canvas sheet. Tie sheet securely to machine.

REFERENCES

Burkhardt, G. J., et al.: Mechanical Harvesting and Handling of Sweet Potatoes, *Trans. of the ASAE*, **14**(3):516, 1971.

Duke, G. B.: Mechanized Harvesting of Virginia Peanut Crop, *Amer. Soc. Agr. Engin. Trans.*, **3**(2):138-139, 1960.

Hammerle, James R.: The Design of Sweet Potato Machinery, *Trans. of the ASAE*, **13**(3):281, 1970.

A Handbook of Peanut Growing in the Southwest, *Tex. Agr. Expt. Sta. Bul.* 727 and *Okla. Agr. Expt. Sta. Bul.* B-361, 1950, joint publication.

Harlan, Reginald: Costs, Returns and Efficiency of Potato Production in Maine, *Maine Agr. Expt. Sta. Bul.* 701, 1973.

Langley, B. C., J. W. Sorenson, Jr., W. E. McCullough, and H. P. Smith: Harvesting and Drying Peanuts in Texas, mimeograph, *Tex. Agr. Expt. Prog. Rpt.* 1124, 1948.

Martin, J. W., and E. W. Humphrey: Development of Idaho Potato Harvesters, *Agr. Engin.*, **32**(5):261-263, 1951.

Mills, William T.: New Method of Harvesting Virginia Bunch Peanuts, *Trans. of the ASAE*, **4**(1):26-27, 1961.

—— and J. W. Dickens: Peanuts Harvesting and Curing the Windrow Way, *N.C. Agr. Expt. Sta. Bul.* 405, 1958.

Park, Joseph K.: New Sweet Potato Equipment: Sweep and Digger, *Agr. Engin.*, **30**(7):330-335, 1949.

——, M. R. Powers, and O. B. Garrison: Machinery for Growing and Harvesting Sweet Potatoes, *S.C. Agr. Expt. Sta. Bul.* 404, January 1953.

Pierce, Walter H., and William T. Mills: An Evaluation of a Mechanized System of Peanut Production in North Carolina, *N.C. Agr. Expt. Sta. Bul.* 413, 1961.

Poole, Wiley D.: Harvesting Sweet Potatoes in Louisiana, *La. Agr. Expt. Sta. Bul.* 568, 1963.

——: Mechanical Harvesting of Sweet Potatoes in Louisiana, Paper presented at the Conference of Collaborators from Southern Agr. Expt. Stations, March 1970.

Schrumpf, William F.: Practices, Costs, and Tuber Bruising in Digging Potatoes in Aroostook County, Maine, *Maine Agr. Expt. Sta. Bul.* 472, 1949.

Stokes, C. M., and I. F. Reed: Mechanization of Peanut Harvesting in Alabama, *Agr. Engin.*, **31**(4):175-177, 1950.

Stout, Myron, S. W. McBirney, and Charles A. Fort: Developments in Handling Sugar Beets, *U.S. Dept. Agr. Yearbook*, pp. 300-307, 1950-1951.

QUESTIONS AND PROBLEMS

1 Make a list of the root crops.
2 Explain the operation of a potato digger.
3 Explain the differences between harvesting Irish and sweet potatoes.
4 Trace the flow of beets through a beet-digging machine, explaining the action of the different sections of the machine. What is a beet combine?
5 Explain the action of (*a*) a peanut shaker, (*b*) a peanut thresher-picker, (*c*) a peanut combine.

Special
Crop-harvesting
Equipment

Machines are being developed to harvest mechanically almost every kind of farm product grown for market. Sugar cane and castor crops can now be harvested with special self-propelled machines. As new crops gain in importance, special machines are designed, developed, and adapted to their harvesting.

SUGAR-CANE HARVESTERS

The mechanical harvesting of sugar cane requires a machine that must perform several operations simultaneously (Fig. 21-1). (1) The green watery tops must be severed and discarded. (2) The tremendous amount of leaf and trash must be removed from the stalks. (3) The stalks must be cut as close to the ground as possible. (4) The stalks must be either placed in piles for loading or put directly into wagons (Figs. 21-2, 21-3, and 21-4).

Sugar cane is grown in bedded rows spaced about 6 ft (1.8 m) apart. The plants grow to a height of 12 to 15 ft (3.7 to 4.6 meters), with a diameter of about 1 in (2.5 cm). The plants may be so thick as to average about 1 in (2.5 cm) apart in the row. In Louisiana, Texas, and Florida, the harvesting period ranges from early October to late December. Sugar cane is grown in most tropical countries. The world production of sugar from sugar cane averages approximately 24 million tons (21.8 million metric tons) annually. It may

Fig. 21-1   A self-propelled sugar-cane harvester. (*J. & L. Engineering Co.*)

require 10 tons (9.1 metric tons) of sugar cane to produce 1 ton (0.9 metric ton) of sugar.

A self-propelled sugar-cane harvester is shown in Fig. 21-1. The dead leaves are usually burned out before harvesting is done. The stalks are cut at the top and at the ground with adjustable power-operated knives or disc blades. The severed stalks are conveyed through the machine in a vertical position. It is claimed that under favorable conditions a sugar-cane harvester can harvest as much as 1000 tons per 10-hour day. The cost may range around $1.65 per ton.

A self-propelled cane harvester will weigh about 16,000 lb (7.3 metric tons). It is powered by two 50-hp diesel engines. The whole machine is mounted on rubber-tired wheels. The two drive wheels are generally equipped with 13-38 high-lug cane-field tires.

The sugar-cane stalks are collected in piles or windrows on the ground from which special hydraulic-actuated loaders (Fig. 21-2) are available to load the

Fig. 21-2   A self-propelled sugar cane loader. (*J. & L. Engineering Co.*)

Fig. 21-3   Loading sugar cane from windrow to trailer. (*J. & L. Engineering Co.*)

cane onto specially built wagons. The front tongs of the grapple fork are used to drag a bundle of stalks to the open hopper. Then the tongs are closed by hydraulic power and grip the stalks while the boom is lifted and swung over a special cane wagon (Fig. 21-3).

Where cane must be transported several miles to the mill, the cane is transferred from the field wagon to large transport trailers or trucks (Fig. 21-4).

CASTOR HARVESTERS

The castor was formerly known as the *castor bean.* In 1969 the crop terminology committee of Crop Science of America suggested that, since the crop was not a true bean, the better term would be *castor* and not *castor bean.*

The acreage of the castor crop has declined during the past decade. Therefore, no special machines are being manufactured to harvest the crop. A special header attachment for self-propelled combines is now the main harvesting equipment for the castor.

Cultural equipment for cotton and corn can be used in the production of castor.

Plant types can be divided into three groups: (1) dwarf or semidwarf, (2) those of intermediate plant height, and (3) the tall-plant group. The plants of group 1 are the most suitable for mechanical harvesting.

The *racemes,* or clusters of seed, are distributed over the plant somewhat like the bolls of cotton plants. Another similarity to cotton is the characteristic lack of uniform maturity of the seed clusters. At frost, some seed clusters will be mature while others will be in various stages of development. The well-matured seed clusters have a tendency to shatter easily.

Fig. 21-4   A transfer loader transferring sugar cane from field trailer or stack to large transport truck. (*J. & L. Engineering Co.*)

KENAF FIELD HARVESTER

Kenaf is a soft-bast, long-fiber plant.[1] The seed are planted with a grain drill. The stalks vary in size from $\frac{1}{2}$ to $\frac{3}{4}$ in (1.3 to 1.9 cm) in diameter and grow to a height of 8 to 12 ft (2.4 to 3.7 meters). Kenaf has some characteristics of the jute plant grown in India and Bangladesh.

The experimental harvesting machine developed by the Florida Field Station of the U.S. Department of Agriculture is a combined harvester and ribboner. A 36-in (91.4-cm) swath is cut at the ground, separated from the standing plants, and 12 to 18 in (30.5 to 45.7 cm) of the top is cut off. The severed stalks are delivered to a feed table from which they are fed into the crushing unit. Large drums crush and break the stalks and separate the woody core from the fiber. The ribbons of fiber are delivered to the rear of the machine. The fiber is then retted and processed.

SESAME HARVESTING

This is an oilseed crop and is relatively new to farmers of the United States, though it is an old crop in Asia. The first strains introduced and grown in the United States were dehiscent (shattering) types. When the seed pod became mature and dry, it popped open and the seeds were lost. Plant researchers have developed indehiscent (nonshattering) varieties. Average yields range from 600 to 1500 lb (272.2 to 680.4 kg) of seed per acre.

[1] Leonard G. Schoenleber, Machines for New Crops, *U.S. Dept. Agr. Yearbook,* p. 434, 1960.

The dehiscent types must be harvested with a row binder and the bundles placed in shocks for a period of 10 to 14 days, after which the bundles are threshed by feeding them into a combine. The combine cylinder speed is reduced and adjustments of the screens and sieves are necessary. If the seeds are cracked, free fatty acids are produced which are unfit for human consumption.

FLAX HARVESTING

Flax is a plant native to Asia. In the United States and Canada flax is grown as a commercial crop for the oil-bearing seed, even though high-grade linen cloth is made from the flax fiber. Linseed oil is expressed or squeezed from the seed. The crop is grown in the northern Midwestern states and in southern Texas. The seeds are harvested with a combine. As the seeds are flat and coated with a film of oil, special screens and adjustments on the combine are necessary.

EDIBLE-BEAN HARVESTING

Generally, the beans are cut and windrowed, then threshed with a combine fitted with a pickup attachment. The cylinder speed of the combine is reduced, and the screens adjusted or changed. Plant scientists are working to develop varieties of beans that grow upright and can be cut and threshed in one operation with the combine.

TREE HARVESTING EQUIPMENT

Mechanical devices are being developed for the harvesting of fruits and nuts. Pickup machines are used to pick up prunes and figs from the ground. Self-propelled catchers and conveyors are used to slip around and under plum and apricot trees so the fruit can be mechanically shaken from the tree onto the catcher-conveyor.

Similar equipment is used to harvest walnuts and pecans.

The catcher consists of a large butterflylike frame covered with canvas. Long, tractor-mounted booms contact the tree limbs. A mechanical or hydraulically actuated knocker agitates and shakes the limbs to loosen the nuts.

Tung nuts are grown along the eastern coastal area of the Gulf of Mexico. The nuts are allowed to mature and fall on the ground, to be picked up by hand. Researchers are working to develop a tung-nut rake and pickup machine. A reel of rubber fingers set to operate somewhat like a reel-type side-delivery hay rake is being tested to rake the nuts from under the trees and deliver them to an elevator-cleaner.

Figure 21-5 shows a view of a self-propelled orchard lift. A man in the basket can control the lift to move the basket up or down and slightly sidewise.

Fig. 21-5 A self-propelled hydraulic orchard lift. The lift is controlled by the man in the basket. (*Edwards Equipment Co.*)

VEGETABLE OR TRUCK HARVESTERS

A self-propelled conveyor-elevator-loader is used in harvesting lettuce, cabbage, and pineapple. A long conveyor 20 to 30 ft (6.1 to 9.1 m) in length extends over the rows so that the hand-harvested heads of lettuce, cabbage, or pineapple can be placed on a belt that conveys them to an elevator that deposits them into a truck.

Machines are being developed to harvest radishes, carrots, cucumbers, tomatoes, grapes, berries, and other vegetable and truck crops.

GRASS-SPRIG HARVESTER

"Grass farmers" of the South have been handicapped in digging sprigs of grasses, such as Coastal Bermuda, to develop better pastures. A sprig digger now available plows up the sod of the well-established grass. The slices of sod are passed under a toothed drum which breaks up dirt and clods, pulls the sod apart, and separates the sprigs into about 6-in (15.2 cm) lengths. The drum throws the sprigs onto a separating conveyor which removes excess dirt and discharges the clean sprigs into a trailer. The sprigs are set in the soil by a special sprig planter.

GRASS-SEED HARVESTERS

Machines are available for harvesting grass seed, such as bluestem, side oats grama, and Bermuda. One homemade machine uses long naillike teeth on a 9-ft (2.7-m) drum. The teeth comb off the seed as they come against a steel pan.

Other grass-seed harvesters use large suction fans to pull air through a narrow-slatted hood set close to the ground. Drag fingers ahead of the suction nozzle loosen the seed so they can be drawn into the nozzle. The air discharges the trashy seed into a trailer box.

MAINTENANCE AND CARE

1   Before special crop harvesting equipment is used, it should be operated to determine if all parts are functioning properly.

2   Understand the operation of sugar-cane harvesters.

3   Check for proper screens and adjusters of equipment for seed harvesters.

4   At end of season thoroughly clean equipment and treat parts that may rust with oil.

5   Store in a shed or cover with a plastic or canvas sheet tied securely to the machine.

REFERENCES

Arms, Milo F., and Lloyd W. Hulburt: An Experimental Harvester for Castor Seed, *Agr. Engin.,* **33**(12):784-786, 1952.

Blanchi, R. H., and Arthur G. Keller: Clean, Fresh Cane; How Much Is It Worth? *La. Eng. Expt. Sta. Bul.* 28, 1952.

Porterfield, J. G., and F. J. Oppel. Jr.: An Experimental Castor Bean Huller, *Agr. Engin.,* **33**(11):713-715, 1952.

Powers, J. B.: The Development of a New Sugar Beet Harvester, *Agr. Engin.,* **29**(8):347-351, 1944.

Schoenleber, L. G., and W. M. Hurst: Performance of Castor Bean Hulling Plants, *Agr. Engin.,* **33**(11):708-710, 1952.

—— and W. E. Taylor: A Two-Row Castor Bean Harvester, *Okla. Agr. Expt. Sta. Bul.* B-395, April 1953.

Schroeder, E. W., and I. F. Reed: Developing Tractor-mounted Castor Bean Harvesters, *Agr. Engin.,* **33**(12):775-776, 1952.

Thomson, B. C.: Sugar Cane Harvesting, part I, *Farm Impl. News,* **73**(24):38, December 1952.

——: Sugar Cane Harvesting, part II, *Farm Impl. News,* **74**(1):86, Jan. 10, 1953.

Walz, Claude W.: The Mechanization of Sugar Beet Harvesting, *Agr. Engin.,* **27**(12):549, 1946.

QUESTIONS AND PROBLEMS

1   Name several special or minor crops that can be mechanically harvested.
2   Describe the action of a sugar-cane harvester.
3   Describe the action of a sugar-cane loader.
4   Explain how the castor crop can be harvested with a combine.
5   What are some of the new crops that are becoming important farm crops?
6   Describe how tree crops can be mechanically harvested.
7   How does a grass-sprig harvester operate?

Chapter 22

Crop-processing Equipment

Crop-processing equipment includes machines that are used to dispose of the crop residue after harvest and machines to process harvested material and put it into a more usable form. Machines that perform such treatments include stalk-cutter-shredders, shellers, feed grinders, and crop dryers.

CROP-RESIDUE DISPOSAL EQUIPMENT

The plants of cotton and corn are left in place in the field after the bolls and ears have been harvested. The stubble and straw of wheat, rice, and sorghum remain in the field. This residue of one crop must be disposed of in some manner before seedbed preparation begins for the next crop. Burning of the residue has long been a thorough way of disposing of crop residue. This method is now in disfavor because soil-improving practices show the need of returning all crop residue to the soil.

Types of Machines for Crop Residue Disposal

Crop-residue disposal machines can be divided into two general types: (1) the power-operated flail beater, and (2) the power-operated horizontal-rotating-knife cutter-shredder.

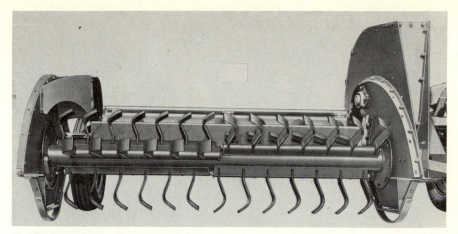

Fig. 22-1 Cutaway view with housing removed to show construction of the flail cylinder and knives. (*Gehl Bros. Mfg. Co.*)

Power-operated Flail Beaters The beater-type residue-disposal machines generally consist of a horizontal solid or fabricated drum to which are attached rubber flails, steel bars, or steel chains (Fig. 22-1). Hammer heads may be attached to the end of the bars or chains. The hammer heads or shredding knives may be flat bars or L-, V-, or T-shaped (Fig. 22-2). In some cases the heads are made to be removed. Power is transmitted from the tractor power takeoff through a propeller shaft to a centrally located gearbox. Cross-shafts extend from the gearbox to either one or both ends. The power is transmitted from the cross-shafts to the rotor drum by means of sprocket and roller chain or V belts. The ratio of the sheave size is arranged to increase the revolutions per minute of the rotor or beater drum to give a peripheral speed to the hammers of about

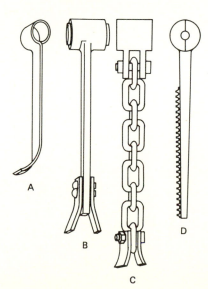

Fig. 22-2 Types of flail bar and heads: (*A*) bar with curved ends; (*B*) bar with V-shaped head; (*C*) chain with V-shaped head; (*D*) rubber rasp-bar flail.

Fig. 22-3 Flail-type stalk cutter-shredder with disk harrow to work the shredded vegetation into the soil. (*E. L. Caldwell & Sons.*)

6000 ft (1829 m)/min. Hammer arms of different lengths can be arranged on the rotor to conform to the contour of bedded land.

Potato tops can be beaten off on the top and sides of beds, and winter cover crops planted in the furrow can be disposed of with the beater. Rubber flails are used where the vegetation does not contain much fiber, such as sweet potato vines. More power is required if the hammers strike the vegetation when they are moving in the direction of travel than if the hammers strike the material rotating opposite to the direction of travel. In the latter case, the material is quickly discharged and the pressure of the hammers against the vegetation has a tendency to push the tractor forward. Power requirements will vary with the type and density of the vegetation. Figure 22-3 shows a stalk cutter-shredder with disk harrow to work the shredded vegetation into the soil.

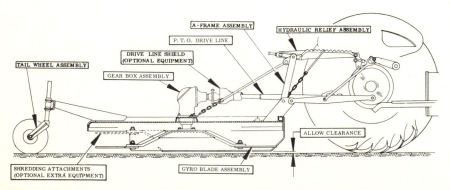

Fig. 22-4 Rotary cutter with three-point hitch. (*Servis Equipment Co.*)

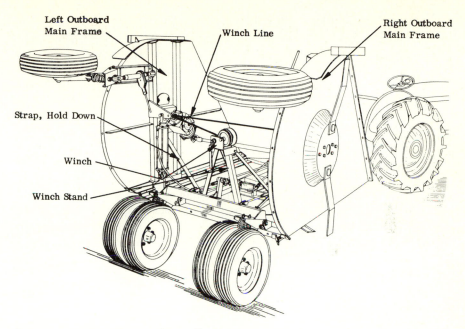

Left Outboard
Main Frame

Winch Line

Right Outboard
Main Frame

Strap, Hold Down

Winch

Winch Stand

Fig. 22-5 Folded side wings. (*Servis Equipment Co.*)

Horizontal-rotating-Knife Cutter Mowers The knives of this type of cutter rotate in a horizontal plane close to the ground surface (Fig. 22-4). The number of driveheads per machine ranges from one to three. The number of knives per drivehead may be either one or two. Side wings are also used on some machines (Fig. 22-5). The width of swath cut, or the diameter of the cutting circle, on the single-head machines is about 57 in (144.8 cm). The width of swath on the larger machines ranges from 10 to 12 ft (3.0 to 3.7 m).

Power is transmitted from the power takeoff to a gearbox on the cutter. When the machine has two or three sets of knives, V belts transmit the power from the gearbox to sheaves for the knife driveheads. Free-swinging, hammerlike knives can be used on the ends of the knife bars, which may be either straight or offset as shown in Fig. 22-6.

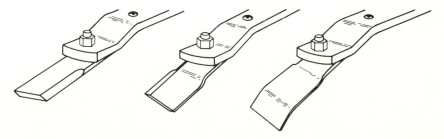

Fig. 22-6 Three types of cutting knives for rotary cutters: (*left*) flat; (*center*) suction; (*right*) pickup. (*Deere & Co.*)

Rotary cutter-shredder-mowers may be trailed behind (Fig. 22-7), rear-mounted (Fig. 22-4), or suspended under the center of the tractor. The trailing types are supported by wheels on each side of the machine. The mounted types can be held at adjustable heights by skids or side- or rear-attached wheels.

In addition to the disposal of crop residues, this type machine can be used to mow weeds in pastures and to mow highway shoulders, and the heavy-duty types can be used to cut and shred brush.

CORN SHELLERS

Most corn is shelled as a part of the harvesting operation. Farmers who grow both small grain and corn use a corn-harvesting attachment on the combine and let the threshing unit shell the corn (Chap. 17). Corn farmers are using a shelling attachment in combination with the corn picker (Chap. 18). Where corn is harvested with the husk left on the ear or the husk removed, corn shellers are required to prepare the corn for the shelled-corn market.

Types of Corn Shellers

There are two types of corn shellers: the *spring* and the *cylinder.*

The spring sheller has a plate under spring pressure to hold the ears against a rotating disc which loosens and separates the kernels from the cobs (Fig. 22-8). The kernels pass downward through the cleaning unit, while the cobs are ejected by conveyors.

There are several types of cylinder corn shellers, listed as follows:

Fig. 22-7   Four-row pull-type rotary cutter with three horizontal rotating blades. (*Deere & Co.*)

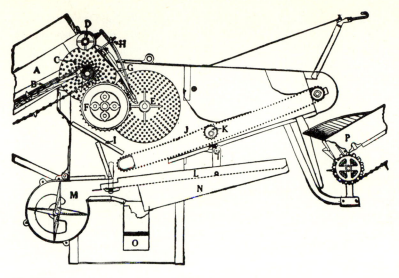

Fig. 22-8   Sectional view of two-hole spring corn sheller.

1   The stationary large industrial and the small farm units. The industrial sizes are used by corn buyers, while the smaller are used on farms where a large quantity of corn is shelled.

2   Truck-mounted portable auxiliary engine-driven types can be used for custom shelling from farm to farm.

3   Two-wheel trailing portable power-takeoff-driven types can be moved to different locations and used for custom work.

4   The tractor-mounted power-takeoff-driven type has the same general uses as the two-wheel trailing type.

The power requirements of a corn sheller will vary from 10 to 35 hp, the cylinder speed ranges from 600 to 1000 r/min, and the capacity is influenced by the percentage of husk on the ears, the moisture content of the kernels, the rate of feeding, and the size of the cylinder. The large stationary units will shell 40 to 50 tons (36.3 to 45.4 metric tons)/h, while the smaller portable units will shell from 100 to 250 bushels per hour.

Some corn shellers are equipped with blowers to handle both the shelled corn and the cobs separately. Sacking attachments are available for some of the portable units.

FEED GRINDERS AND MILLS

In the feeding of livestock, it has been found that more animal nutrition and food constituents can be assimilated and put into flesh on an animal if the feed is ground rather than left whole. Every farmer who has any livestock to feed, therefore, would find it advantageous to secure a small feed grinder to grind the feed before it is fed to the stock. Small feed grinders can be operated by gasoline engines or small electric motors. These grinders can be divided into three types, depending upon the method of grinding, namely, the *burr,* the *hammer*, and the rarely used combination *grain-roughage.*

Burr Grinders

Most of the burr feed grinders are equipped with flat, roughened, chilled-iron plates which are often called *burrs*; hence the name *burr grinders.* Burr feed-grinder mills are generally adaptable to grinding husked corn, the small grains, and material of low fiber content.

Hammer Mills

The hammer mill differs from the burr mill in that, instead of flat disc plates for grinding, there are hammerlike projections mounted on a cylinder (Fig. 22-9). This cylinder of hammers revolves at a high rate of speed and grinds the material by beating it to pieces. It is claimed that this type of mill will grind almost any material that is used for feed.

Krueger gives the following advantages of a hammer mill:

1   It is not dulled by running empty.
2   Foreign material in the feed will not ordinarily injure it.
3   There is greater range in fineness.
4   Replacements are fewer.
5   Wear does not impair its efficiency.

Sizes   Hammer-mill feed grinders vary in size from the small, compact mill with a direct-connected 1-hp electric motor to the large mill requiring 75 to 100 hp to operate it. When a feed hopper is attached to the mill, it will automatically feed itself and requires no attention. Automatic grinding out of a grain bin is practical for farm installations, as shown for roller mills in Fig. 22-10.

Hammers   The hammers are fastened on a cylinder and may be rigid or swinging. The free-swinging hammer is hinged, but the rigid hammer is fastened to a rotor shaft or cylinder by jam nuts. The shape of the hammer's cutting edge varies according to the ideas of the designers. The hammer should, however, be made of high-grade hardened steel to prevent excessive wear.

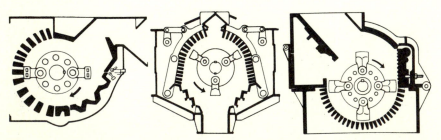

Fig. 22-9   Cross section of hammer mill showing arrangement of two, three, and four sets of hammers. (*Abex Corporation.*)

Fig. 22-10   Automatic roller feed mill equipped with electronic controls: (*left, bottom*) overhead view of four auger feeds; (*left, top*) the rate of feed can be set on the scale. (*Crimpomatic Mill & Equipment Co.*)

Screens   In most machines, the lower half of the cylinder is enclosed by a screen, usually of one piece. It consists of holes punched through sheet steel. Various-sized holes are used, depending upon the fineness of grinding desired. The size of the holes ranges from $\frac{5}{64}$ in to 2 in (0.2 cm to 5.1 cm). The smaller holes are used when grinding grains, while the larger sizes are used when grinding roughage, such as sorghum stalks, cornstalks, or hay.

Grinding Process   The material to be ground is fed directly into the compartment where the hammers are revolving. The hammers strike the material with such violent force that it is practically exploded. The material is retained on the screen until it is beaten fine enough to pass through the perforations.

Capacity and Power   The capacity of a hammer mill depends on many factors, such as the rate of feeding, speed of hammers, power available, kind of material being used, fineness of grinding as determined by size of opening in screen, and size of mill.

Elevating Attachments   The ground feed is removed from under the mill by suction and blown into the large collector hopper, equipped with either a bagging attachment or a swivel spout for delivery into a truck or trailer. The ground material can also be blown directly into the bin. This eliminates most of the dust resulting from the grinding of the feed.

Portable Hammer Mills   When a farmer wishes to do custom grinding, he can mount a hammer feed grinder on a truck so that it can be easily transported from farm to farm with little lost time. Portable mills can be mounted on a trailer cart and operated by the power takeoff of the tractor.

Roller Feed Mills

Roller feed mills are equipped with two grooved rollers about 10 in (25.4 cm) in diameter. One roller is under spring pressure which can be adjusted to the desired degree of crushing. The grain passes between the rollers, which crush and crimp the grain. Figure 22-10 shows an automatic roller mill which is electronically controlled. It has four feed augers, so that a complete ration of two to four ingredients can be crushed and crimped simultaneously. Figure 22-11 shows how different kinds of feed are conveyed from storage bins to the mill and by a conveyor from the mill to the finished-feed bin. The operation is completely automatic. If any one of the ingredients runs out, the mill is automatically stopped.

FEED MIXERS

When farmers wish to mix two or more ground feeds of known feeding value to obtain a balanced-ration feed, a feed-mixing machine is needed (Figs. 22-12 and 22-13).

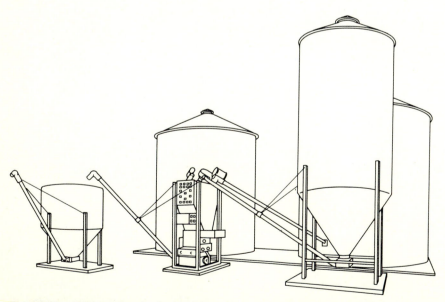

Fig. 22-11   Diagram showing how the automatic feed mill conveying system is arranged. (*Crimpomatic Mill & Equipment Co.*)

Fig. 22-12 Portable power-takeoff-driven feed mill that grinds and mixes small grain, hay, and shelled or ear corn with concentrates. (*Gehl Bros. Mfg. Co.*)

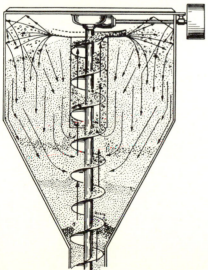

Fig. 22-13 Sectional view of feed mixer: (*top*) overhead view; (*bottom*) side view.

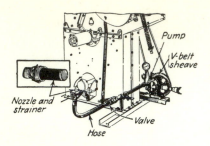

Fig. 22-14 Molasses pump to mix molasses with ground or chopped green feed.

Adding of Molasses to Feed

If molasses is to be mixed with the feed, a special molasses pump is required to pump the molasses from a tank or drum and inject it into the conveyor system so that the molasses is automatically mixed with the feed as it is blown from the housing of the mill (Fig. 22-14).

Pellet and Wafer Feeds

Many commercial feed plants have equipment where the feed is ground, steamed, and squeezed through holes in a plate to make pelleted feed. Field wafering machines were discussed in Chap. 16.

CROP DRYERS

A crop dryer may be any structure or contrivance which includes the facilities for the drying or curing of agricultural products with forced, heated or unheated air.

The drying of fruits by exposure to the sunshine is an ancient practice. The damp climatic condition of England prompted the people of that country to investigate artificial drying methods many years ago. Tobacco growers of Virginia and the Carolinas attempted the use of heated air to color tobacco as early as 1830. Heated air for the curing of sweet potatoes was used in the early 1930s. Forced air drying of hay came into use in the late 1930s and early 1940s.

The drying of farm products on the farm and in commercial plants is now an essential phase of agricultural storage and marketing requirements. Many farm

Table 22-1   Moisture Contents for Safe Storage

Crop	Moisture, %
Shelled corn in the bin	13
Ear corn in the crib	16-18
Oats in the bin	13
Wheat in the bin	13
Barley in the bin	13
Sorghum in the bin	12-13
Rice in the bin	12-13
Soybeans in the bin	10-11
Peanuts	7-9

products, when harvested, contain too much moisture for safe storage. Table 22-1 shows the percentage of moisture for safe storage of several crop products.

Types of Crop Dryers

There are many types of farm dryers, as shown by the outline below.

I   Farm-constructed dryers
    A   Duct-lateral systems
      1   Central duct
      2   Side duct
    B   Duct-slatted floor system
    C   Underfloor plenum-chamber sack
    D   Trailer-wagon
    E   Bin dryer
II   Manufactured crop dryers
    A   Continuous-flow
      1   Column
      2   Horizontal
    B   Batch
      1   Column
      2   Bin

Farm-constructed Dryers   The dryers for both hay and grain can be constructed on the farm. The duct-lateral and slatted systems (Fig. 22-15) are used primarily for the drying and finishing of hay. The hay dryer or finisher consists of a system of air ducts on the floor of the haymow or barn on which the hay is placed to a depth of 6 to 10 ft (1.8 to 3.0 m), depending upon the moisture content of the hay. Air is forced through the air ducts and up through the hay by a centrifugal or propeller-type fan driven by an electric motor. With this system, the hay is allowed to cure partially in the field to a moisture content of 45 to 50 percent before it is placed on the dryer. Unheated air forced through the system will ordinarily dry the hay to 20 percent in 7 to 14 days, depending upon the outside atmospheric conditions. Alfalfa hay will usually dry in the swath in midsummer to 50 percent within 3 to 4 hours after it is cut; therefore, it is possible to get the hay into the barn on the same day that it is cut.

Other types of dryers constructed on the farm are the duct in a trailer or wagon box, crib and bin dryers, and specially constructed column batch dryers.

Manufactured Crop Dryers   The manufactured crop dryers can be divided into two classes, namely, the *continuous-flow* and the *batch*. The continuous-flow types are available as tower dryers (Fig. 22-16) and as horizontal dryers.

Heat Exchangers or Furnaces

A heat exchanger is a unit which draws in cold air and heats it as it passes through the unit into the drying section. In a sense, cold air is exchanged for heated air. The unit has a furnace for heating the air and a fan to move the air. The heat exchangers shown in Fig. 22-17 are generally termed *crop dryers*, but they really are only a part of the complete dryer. The product to be dried is in a

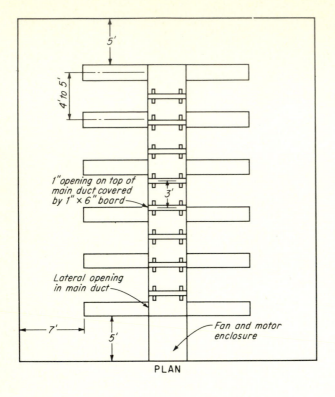

5'

4' to 5'

1" opening on top of
main duct covered
by 1" × 6" board

3'

Lateral opening
in main duct

7'

5'

Fan and motor
enclosure

PLAN

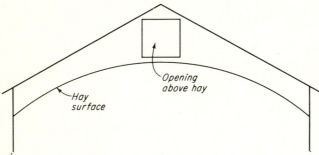

Opening
above hay

Hay
surface

Fig. 22-15   Arrangement of a central duct with laterals for drying loose hay. (*Illinois Agr. Ext. Ser. Cir. 757.*)

separate unit, having ducts, chambers, or passages to conduct and direct the air through the product.

There are two general types of heat exchangers, the *direct* and *indirect* The direct type draws the air past the burner and flame and blows heated air and burned gases directly through the product to be dried. In the indirect type, the flame and gases flow around air tubes and pass out through a vent. The hot air blown into the hay or grain is drawn through the tubes. Some drying outfits have the flame from the burner directed into the fan intake or into an intake tube extending out at least two diameters of the intake.

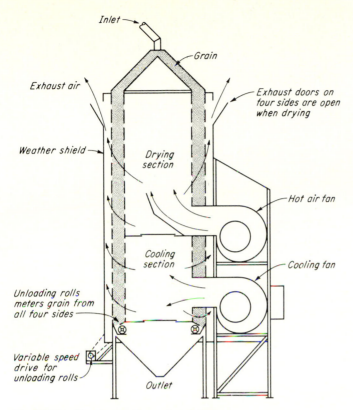

Fig. 22-16   Continuous-flow column dryer.

Burners   Oil and gas burners are used in heat exchangers. There are two types of oil burners, the *pressure-atomizing* and the *pot*. In the *pressure-atomizing* burner, the oil is pumped under pressure and the rate of burning controlled by a nozzle of the desired size. In the *pot* burner, the oil flows into a round pot. The draft of the regular fan or an auxiliary fan aids in atomizing the fuel.

Natural or LP (liquid petroleum) gas burners are low-pressure burners and operate with 1 to 30 lb of gas pressure. Some types may operate with pressures as low as 4 oz.

Fuels   The type of fuel used for a burner or drying system is recommended by the manufacturer. Different types of fuel have different Btu ratings as shown in Table 22-2.

Types of Fans

The *centrifugal*, or *multivane*, type of fan is designed to deliver large volumes of air over a wide range of pressures. It is made in *straight, forward-curve,* and *back-curve* types, depending upon the shape of the impellers, or blades (Fig. 22-18). The forward-curve fans operate at relatively low speeds but vary widely

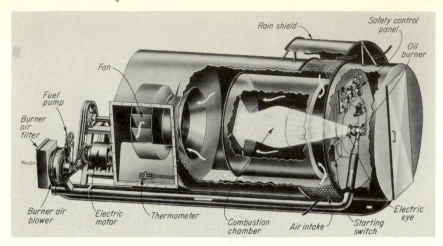

Fig. 22-17   Sectional view of a heat exchanger. (*Campbell Dryer Co.*)

in volume of air delivered and power required against varying static pressures.[1] The back-curve fans operate at approximately twice the speed of the forward-curve but give a more uniform air delivery and power demand against widely varying static pressures. The straight radial-blade fan operates at a speed between the forward- and backward-curve fans. The number of blades in a backward-curve fan may range from 14 to 24, the number in a forward-curve fan may range from 32 to 66, while the straight radial-exhaust fan may have from 5 to 12 blades. Pressure-blower and supercharger types may have up to 22 blades.

Centrifugal fans are made in both single- and double-inlet types, depending upon whether the air enters one or both sides of the fan housing. Most fan manufacturers have developed lightweight, light-duty centrifugal fans in sizes up to 24 to 30 in (61.0 to 76.2 cm) (size refers to diameter of rotor or fan blades), especially designed for use in air-conditioning systems where static pressures rarely exceed $1\frac{1}{2}$ to 2 in (3.8 to 5.1 cm) water column. These are available in

Table 22-2   Heat Content of Various Fuels

Fuel	Btu/gal	Btu/ft^3
Natural gas	. . . .	1000
Butane	104,000	3400
Propane	92,000*	2570
Kerosene	144,000	
Gasoline	117,000	
Fuel oil	130,000	
Distillate	135,000	

* Vapor per gallon of liquid at 60°F and 30 in mercury = 36.4 ft^3.

[1] *Static pressure* is the force exerted by the fan in forcing air through a duct system. It is usually expressed in *inches water column,* or the distance the pressure will depress a free column of water in a U tube.

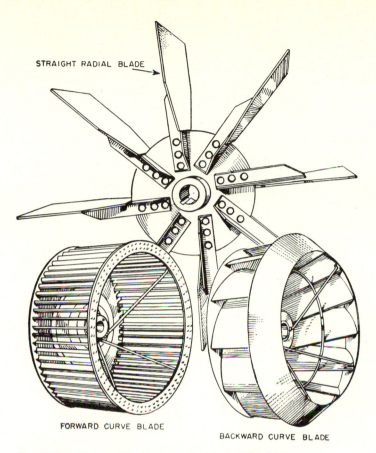

STRAIGHT RADIAL BLADE

FORWARD CURVE BLADE

BACKWARD CURVE BLADE

Fig. 22-18 Types of fan blades used with centrifugal fans.

either single- or double-inlet types. In some cases, two or more fan units are mounted on a single shaft for increased capacity.

The heavy-duty centrifugal fans are made in both forward- and backward-curve types and in a wide range of sizes from about 12 in (30.5 cm) up to 180 in (457.2 cm) or more. The heavy-duty fans are ordinarily used when the amount of air required exceeds the capacity of the larger-sized light-duty fans.

Propeller-type or *axial-flow* fans are designed to deliver large volumes of air against relatively low static pressures (Fig. 22-19). They operate at relatively high speeds and are somewhat noisy. Most propeller fans are designed for operation against static pressures of $\frac{3}{4}$ in (1.9 cm) water column or less, although some special types will operate against pressures of 3 in (7.6 cm) or more. The ordinary *attic ventilator* type of propeller fan designed to exhaust into open air is not satisfactory for use with hay or grain dryers.

The fan selected should deliver an air flow under the existing static pressure sufficient to carry away the moisture from the product being dried. Different crop products of hay and seeds offer different resistance to air flow as shown in Table 22-3 and Fig. 22-20.

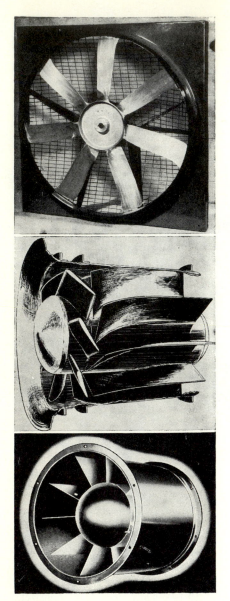

Fig. 22-19   Axial-flow fans.

Fan Laws   The following general fan laws apply to all types of fans.

1   The air capacity in ft^3/min varies as the fan speed. (Twice the speed results in twice as much air.)

2   The developed pressure varies as the square of the fan speed. (Twice the speed results in four times the pressure.)

3   The required horsepower varies as the cube of the fan speed. (Twice the speed will result in a requirement of eight times as much power.)

Table 22-3   Resistance to Air Flow of Grains and Seeds

Kind and condition of grain*	Loose fill Lb/ft³	Loose fill Pressure ratio†	Packed fill Lb/ft³	Packed fill Pressure ratio†
Ear corn (clean)		1.00		
Ear corn (94.6% clean)		1.00		
Peanuts in shell (farmer's stock)	14.0	1.00	14.9	1.54
Fescue grass seed (clean)	11.9	1.00	13.7	1.65
Soybeans (clean)	47.4	1.00	49.9	1.41
Shelled corn (Yellow Dent) (clean)	45.6	1.00	47.7	1.34
Barley (clean)	37.3	1.00	39.9	1.46
Oats (clean)	33.0	1.05	36.1	1.73
Rough rice (clean)	38.2	1.00	40.6	1.32
Grain sorghum (clean)	47.4	1.00	50.0	1.41
Wheat (clean)	49.8	1.00	51.8	1.30
Crimson clover seed (clean)	47.4	1.00	51.3	1.44
Crimson clover seed (unclean)	29.5	0.60	32.1	1.01
Flax seed (clean)	42.4	1.00	44.4	1.27
Alfalfa seed (clean)	50.2	1.00	53.6	1.46
Sericea Lespedeza seed (clean)	49.1	1.00	51.8	1.50
Sericea Lespedeza seed (unclean)	44.0	1.03	46.6	1.55
Red clover seed (clean)	50.1	1.00	52.2	1.42
Alsike clover seed (clean)	50.2	1.00	53.4	1.49

* Arranged in order of resistance as in Fig. 22-20.
† To find pressure drop per foot depth of grain, read the pressure from Fig. 22-20 and multiply it by this ratio.
Source: C. K. Shedd, Resistance of Grains and Seeds to Air Flow, *Agr. Engin.,* 34(9):618, 1953.

Example: A fan running at a speed of 473 r/min delivers 14,850 ft³/min against a resistance of ¾ in water gauge and requires 3.18 hp.

If 16,850 ft³/min were required, what would be the speed, the pressure, and the horsepower?

Use the first law to find the required speed:

$$\text{Speed} = \frac{16,850}{14,850} \times 473 = 537 \text{ r/min}$$

Use the second law to determine the pressure:

$$\text{Pressure} = \left(\frac{537}{473}\right)^2 \times 0.75 = 0.97 \text{ in water gage}$$

Use the third law to determine the horsepower:

$$\text{Hp} = \left(\frac{537}{473}\right)^3 \times 3.18 = 4.65$$

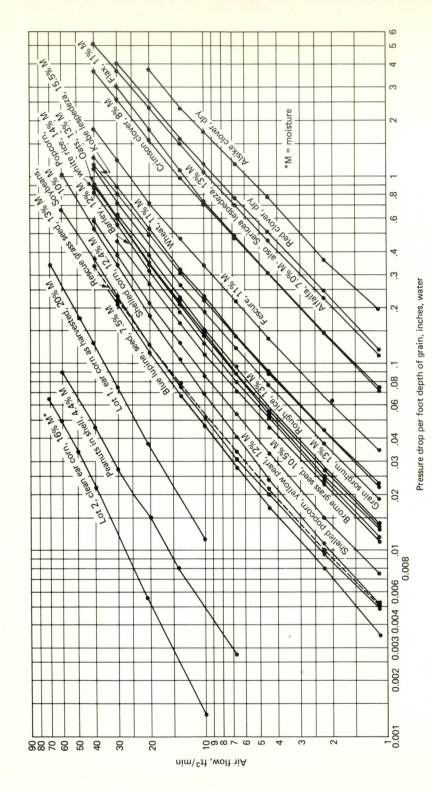

Fig. 22-20 Chart showing resistance to air flow of grain and seeds. (*Agr. Engin. 34(9): 617, 1953.*)

   4   If the air handled by the fan is of a different density from standard air, due to temperature, barometric pressure, or altitude, it becomes necessary to convert the conditions to standard air conditions in order to select the fan from the performance tables.

Effect of Temperature on Rate of Drying Grain Sorghum

Tests conducted in Texas on the drying of grain sorghum indicate that the drying cycle consists of at least two stages. The first stage is the evaporation of the surface moisture. The second stage is the diffusion of the moisture from the inside of the kernel to the surface, where it is absorbed by the air moving through the grain column. The velocity of the air is an important consideration during the first stage, as it governs the rate of drying. The drying rate during the second stage depends on the rate of diffusion. With the higher air temperatures, the rate of diffusion is high at first, but as the drying advances, this process becomes slower, with the drying rate falling off rapidly near the end. The ultimate time of drying is, therefore, governed by the rate of both surface moisture removal and diffusion. The charts in Figs. 22-21 and 22-22 show the effect of four ranges of temperatures on the time required for drying sorghum grain and the moisture content of the grain during the drying period.

   In column-type dryers, the temperature used depends on the type of dryer.

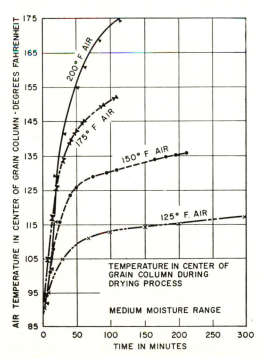

Fig. 22-21   Temperature in sorghum grain and time required for drying at four temperatures. (*Tex. Agr. Expt. Sta. Bull. 710.*)

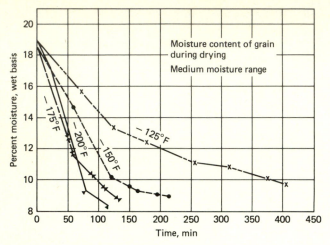

Fig. 22-22 Moisture content of sorghum grain and time required to dry at four temperature ranges. (*Tex. Agr. Expt. Sta.*)

Water Removed in Drying Hay   The number of pounds of water removed per ton in drying hay is shown in Table 22-4.

Distribution of Air Essential for Uniform Drying

Data obtained in tests with many crop materials dried in various types of dryers indicate that the distribution of the air through the product is an important factor in uniform drying throughout the product. The design of the duct laterals for hay as well as the design of the overall structure must be considered. Open-mesh prefabricated ducts have given excellent results in drying both hay and grain.

Table 22-4   Water Removed in Drying Hay

Moisture content, percent (wet basis)		Water removed per
Beginning	End	ton of dry hay, lb.
80	20	6000
65	20	2570
60	20	2000
50	20	1200
45	20	910
80	10	7000
65	10	3140
60	10	2500
50	10	1600
45	10	1270

Source: P. T. Montfort, Supplemental Heat in Barn Hay Curing, *Agr. Engin.*, **28**(3):95-97, 1947.

Drying with Heated and Unheated Air

The use of heated air increases the rate of drying, as heated air has a greater moisture-carrying capacity. The heat causes the moisture from inside the hay stem or grain kernel to diffuse to the surface, where it can be absorbed by the heated air flowing through the mass. The use of heated air, however, requires air-heating equipment, fuels, safety controls, and close attention.

The unheated air is used extensively in connection with the barn drying of hay by means of the duct-lateral and slatted floor systems. Grain, rice, and other crop products have been successfully dried with unheated air. The time required is much longer than that required when heated air is used. The humidity of the natural air must be fairly low to permit drying of the product.

Economics of Drying

The data in Table 22-5 show how drying affects the value of farm products.

Table 22-5   Economics of Drying

(1)	(2)		(3)	(4)	(5)	(6)
	Lb water/bu*					Bu required to break
Moisture content, %			Equivalent value, $/bu	Discount price, $/bu‡	Gain if dried, $/bu†	even, $2\frac{1}{2}$¢ total /bu §
	Actual	Excess				
12	6.5			1.000		
13	7.1			1.000		
14	7.7			1.000		
15.5	8.7	Standard	1.000	1.000		
16	9.0	0.3	0.994	0.985	0.009	
17	9.7	1.0	0.982	0.955	0.027	
18	10.4	1.7	0.970	0.925	0.045	33,250
19	11.1	2.4	0.959	0.895	0.064	17,051
20	11.8	3.1	0.947	0.865	0.082	11,667
21	12.5	3.8	0.935	0.825	0.110	7824
22	13.3	4.6	0.923	0.785	0.138	5885
23	14.2	5.5	0.911	0.745	0.166	4716
24	14.9	6.1	0.899	0.695	0.204	3715
25	15.8	7.1	0.888	0.645	0.243	3050
26	16.6	7.9	0.876	0.595	0.281	2598
28	18.4	9.7	0.852	0.495	0.357	2003
30	20.2	11.5	0.828	0.395	0.433	1630

* Based on 47.32 lb dry matter per bushel at $15\frac{1}{2}$ percent moisture.
† Value of dry matter content = market price × [(100 − initial %)/(100 − final %)].
‡ Typical discounts: No discount for up to $15\frac{1}{2}$ percent moisture, $1\frac{1}{2}$ cents per bushel for each $\frac{1}{2}$ point up to 20 percent, 2 cents per bushel for each $\frac{1}{2}$ point between 20 and 23 percent, $2\frac{1}{2}$ cents per bushel for each $\frac{1}{2}$ point over 23 percent.
§ Typical annual fixed cost of ownership of $665 includes amortization over 10 years, interest, taxes, and insurance. Number of bushels to break even (column 6) = [annual fixed cost ($665 typical)]/[gain (column 4) minus drying cost per bushel ($2\frac{1}{2}$ cents typical)].

MAINTENANCE AND CARE

1   Read and follow instructions in operator's manual before using equipment.

2   Check to see that all safety shields are in place.

3   Check machines for proper adjustments before using.

4   Keep knives on crop-residue disposal equipment sharp.

5   Practice regular inspection of long-season usage equipment, such as feed mills, feed mixers, and crop dryers.

6   Store field equipment in a shed or cover with a plastic or canvas sheet tied securely to the machine.

7   Clean field equipment and treat parts that may rust with oil.

REFERENCES

Baker, Vernon H.: The Oil-burning Crop Drying Unit, *Agr. Engin.*, **32**(12): 657-660, 1951.

Collins, Tappan: Flow Patterns of Air through Grain during Drying, *Agr. Engin.*, **34**(11):759-760, 1953.

Forth, M. W., R. W. Mowery, and L. S. Foote: Automatic Feed Grinding and Handling, *Agr., Engin.*, **32**(11):601-605, 1951.

Foster, George H.: Minimum Air Flow Requirements for Drying Grain with Unheated Air, *Agr. Engin.*, **34**(10):681-684, 1953.

Hukill, W. V.: Types and Performance of Farm Grain Driers, *Agr. Engin.*, **29**(2):53-54, 1948.

——: Grain Drying with Supplemental Heat, *Agr. Engin.*, **42**(9), 488-489, 1961.

Madison, D. Richard: *Fan Engineering*, 5th ed., Buffalo Forge Company, Buffalo, N.Y., 1949.

Montfort, P. T.: Supplemental Heat in Barn Hay Curing, *Agr. Engin.*, **28**(3):95-97, 1947.

Olson, E. A., G. M. Peterson, and F. D. Young: *Nebr. Agr. Col. Ext. Bul.* E. C. 735, 1951.

Pedersen, T. T., and W. F. Buchele: Drying Rate of Alfalfa Hay, *Agr. Engin.*, **41**(2):86-89, 1960.

Ramser, J. H., F. W. Andrew, and R. W. Kleis: Better Hay by Forced-Air Drying, *Ill. Agr. Ext. Ser. Cir.* 757, 1956.

Rickey, C. B., Paul Jacobson, and Carl W. Hall: Drying Farm Crops, *Agricultural Engineers' Handbook*, pp. 646-671, McGraw-Hill Book Company, New York, 1961.

Shedd, Claude K.: Some New Data on Resistance of Grains to Air Flow, *Agr. Engin.*, **32**(9):493-495, 1951.

——: Resistance of Grains and Seeds to Air Flow, *Agr. Engin.*, **34**(9):616-619, 1953.

Simons, J. W.: How to Dry and Store Grain and Seed on Georgia Farms, *Ga. Agr. Expt. Sta. N.S.* 33, 1956.

Smith, H. P., and H. F. Miller: Performance of Stalk Cutter-shredders, mimeograph, *Tex. Agr. Expt. Sta. Prog. Rpt.* 1444, March 1952.

Sorenson, J. W., Jr., et al.: Drying and Its Effects on the Milling Characteristics of Sorghum Grain, *Tex. Agr. Expt. Sta. Bul.* 710, 1949.

——— and M. G. Davenport: Drying and Storing of Flaxseed in Texas, *Agr. Engin.,* **32**(7):379-382, 1951.

——— and L. E. Crane: Drying Rough Rice in Storage, *Tex. Agr. Expt. Sta. Bul.* 952, 1960.

——— et al.: Research on Farm Drying and Storage of Sorghum Grain, *Tex. Agr. Expt. Sta. Bul.* 885, 1957.

Steiner, Kalman: *Fuels and Fuel Burners,* McGraw-Hill Book Company, New York, 1946.

Turner, C. C.: *The Bottled Gas Manual,* Jenkins Publications, Inc., Los Angeles, Calif., 1946.

Vutz, William: Some Observations on Hammer Type Feed Grinders, *Agr. Engin.,* **12**(7):271-274, 1931.

Weaver, John W., Jr., Norman C. Teter, and Sidney H. Usry: The Development of a Farm Crop Drier, *Agr. Engin.,* **30**(10):475-476, 1949.

QUESTIONS AND PROBLEMS

1 Explain the need for crop-residue disposal.
2 Explain the differences in design and operation of the different types of crop-residue disposal machines.
3 Give the advantages of hammer feed mills, and explain the grinding operation.
4 Define crop dryers.
5 Make an outline showing the various types of crop dryers.
6 Explain the function of a heat exchanger.
7 Explain the design differences between fans used with crop dryers.
8 Explain the effect of temperature and air flow on the rate and uniformity of drying crop products.
9 List the fan laws.

Special Farm and Ranch Equipment

The handling of farm products such as wheat, oats, milo, shelled and ear corns, bales of hay, and manure is a tiresome job if done with manual labor. Power-operated equipment is now available whereby manual labor is often reduced to the operation of a power unit such as a tractor. Laborsaving equipment for the farm and ranch consists of elevators, power hoists, power loaders, wagon unloaders, transport mixer-feeders, post-hole diggers, and brush saws. In addition to these, there are laborsaving gadgets of various kinds too numerous to mention here.

ELEVATORS

In general the elevators used on the farm may be classified as *portable elevators* and *stationary elevators*.

Portable Elevators

The portable elevator makes the farmer's life easier and farm work faster and helps to solve labor shortages. The portable elevator is designed so that it can be moved easily from one location to another. Plans for building homemade portable elevators can be obtained from the extension services of many states. Many different sizes and types are being manufactured. There are three types of portable elevators: the *chain drag-flight* type, the *auger* type, and the *blower* type.

Chain Drag-Flight Elevator   The *chain drag-flight* type (Figs. 23-1 and 23-2) is available in lengths ranging from 16 to 50 ft (4.9 to 15.2 m). The trough may be narrow and V-shaped, using a single chain, or it may be as wide as 20 in (50.8 cm) with a chain on each side of the chute to support each end of the drag flights. The wide types have a great range of applications. They can be used to elevate various types of loose grain, ear corn, and bales of hay. The elevator chute should be well braced and trussed to prevent sagging and twisting when long lengths are used.

Two types of hoppers are available for use with the drag-flight elevator: the *trapezoidal-shaped* hopper, which requires that the grain be scooped into it, and the trailer-wagon box, which has a narrow opening to let the grain pour into the hopper. The long *rectangular-folding*-type hopper is suitable for handling crop products that are unloaded on a wide-bed scale or where the entire endgate is removed. The front of the truck or trailer can be tilted to let the crop product flow into the hopper. A spout is provided at the top end of the elevator to guide grain into the bin.

Where long, heavy portable elevators are used, a special derrick lifting arrangement (Fig. 23-2) enables one man to raise the long chute to any height desired. The derrick is mounted on pneumatic-tired wheels. When the chute is lowered, the complete elevator can be trailed behind a tractor or truck. The average angle for satisfactory operation varies from approximately 20 to 45°. Greater angles can be used, but they reduce the capacity of the elevator.

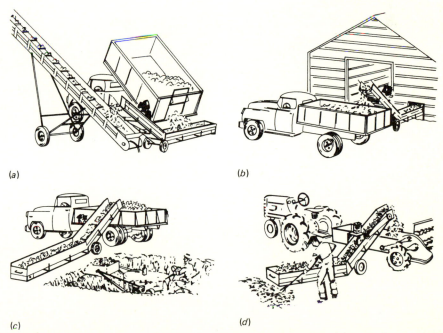

(a)   (b)

(c)   (d)

Fig. 23-1   Four applications of chain drag-flight portable elevators: (*a*) conveying ear corn from trailer to barn; (*b*) conveying material from bin to truck; (*c*) loading truck; (*d*) loading manure spreader.

Fig. 23-2 Portable chain-drag elevator. (*Portable Elevator Div., Dynamics Corp. of America.*)

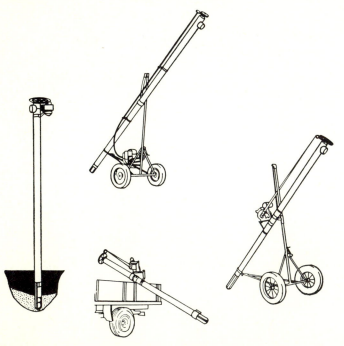

Fig. 23-3 The auger-type portable elevator can be operated at almost any angle. The auger is driven from the top by a gasoline engine or electric motor that can be mounted at various locations as shown. (*Wyatt Mfg. Co.*)

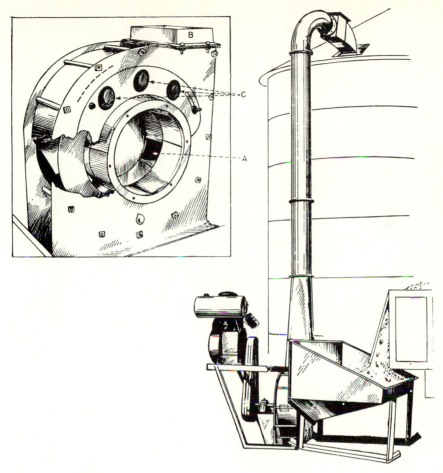

Fig. 23-4 Blower-type elevator unloading grain from truck to bin. (*Inset*) Fan housing with auxiliary air ports.

Auger Elevators The *auger*-type portable elevator is simple in construction, as it consists of a long enclosed section of a screw conveyor (Fig. 23-3). The lower end is not enclosed; when the elevator is inserted into a pile of grain, the exposed part of the revolving auger automatically picks up the grain and conveys it to the other end. The auger conveyor, however, is suitable only for elevating and conveying grains and seeds.

Both the chain drag-flight and the auger-type portable elevators can be operated by small electric motors, by gasoline engines, or from the tractor power takeoff.

Air Elevators The blower-type elevator is useful where large quantities of grain are handled. Blowers have been designed so that the grain passes through the fan housing on a cushion of air without being hit and cracked by the fan blades. Figure 23-4 shows a blower elevator. The grain is dumped from the

Fig. 23-5 Portable suction-blower elevators can be used either to unload or to load trucks and trailer wagons.

truck into the blower hopper, from which it is fed into the stream of air. Small blowers are attached to the truck body and driven by a special power takeoff. Grain blowers will elevate from 300 to 1200 bushels of grain per hour to heights of 25 to 30 ft (7.6 to 9.1 m). A 5-hp electric motor has sufficient power to operate the average-sized blower. Figure 23-5 shows a tractor-mounted suction-blower elevator that can be used for unloading cotton from trailers or for the loading of trailers from piles of cotton. Silage blowers were described under "Forage Harvesting Equipment" (Chap. 16).

Stationary Elevators

When a farmer has a barn with bins for the storage of small grain and corn, he may wish to install a permanent bucket-type elevator such as that shown in Fig. 23-6. Several types can be obtained, but this is typical. The elevator chute is set vertical or almost vertical, so that cups attached to chains can be used to elevate the grain.

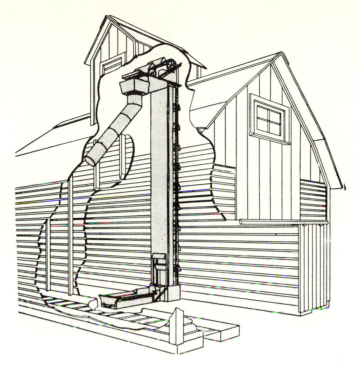

Fig. 23-6   Bucket elevator for small grain installed in driveway of barn. (*King and Hamilton Co.*)

TRUCK AND WAGON HOISTS

When trucks or trailer-wagon loads of material are to be unloaded into elevator hoppers, time and labor can be saved by the use of hoists for tilting the vehicle. Mechanical hand, power-operated, and hydraulically operated hoists are available. Hydraulic cylinders may be used to tilt either the body of the truck only (Fig. 23-7) or the entire truck.

Fig. 23-7   A hydraulic hoist is used to tilt trailer or truck box so material will flow out by gravity. The hydraulic remote-control cylinder is operated from the tractor. (*Deere & Co.*)

TRACTOR-HITCH UTILITY CARRIER

Figure 23-8 shows a carrier rack, or platform, attached to a three-point tractor hitch. The platform can be lowered to the ground and raised to truck- or trailer-bed height. Barrels, rolls of wire, and many heavy machines and tools can thus be lifted and transported on the farm without much manual lifting.

POWER LOADERS

The tractor-mounted power loaders are also called *manure loaders*. They received this name, no doubt, because this type of machine was developed largely for the loading of manure. It has, however, many other uses. It can be used to load onto trucks bales of cotton, baled hay, dirt, gravel, sand, and many other materials (Fig. 23-9). If manure is to be loaded in a barn where the ceiling is low, a loader should be selected with the lifting beams arranged so they do not rise up and hit the ceiling before the scoop has been lifted high enough to dump into the manure spreader. The loader shown has a frame linkage that permits lifting the load above the highest part of the machine.

Most power loaders use hydraulic power for lifting the scoop, but a few use mechanical winches or a block and tackle. Figure 23-10 shows several methods of mounting power loaders on tractors.

Fig. 23-8   Carrier racks or platform lifts can be attached to the three-point tractor hitches for lifting heavy articles.

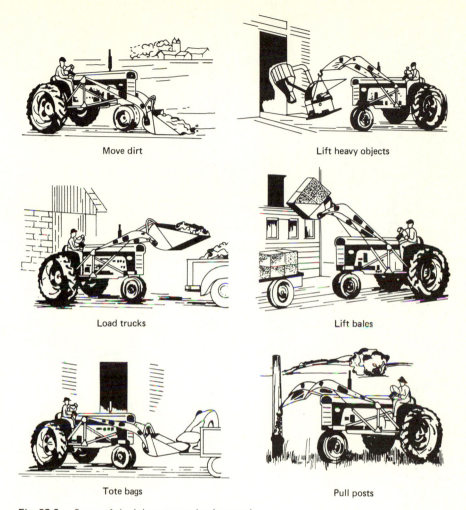

Move dirt   Lift heavy objects

Load trucks   Lift bales

Tote bags   Pull posts

Fig. 23-9   Some of the jobs a power loader can do.

The scoop or bucket can be removed and a special platform attached to the boom frame and the lift then used to harvest fruit and do many other aboveground tasks that otherwise would require a ladder. Different types of scoops, bulldozer blades, hay forks, and other attachments are available for power loaders.

Figure 23-11 shows a combination loader and ditcher. This type of machine is commonly called a *backhoe-loader*. The ditcher unit has many uses on the farm.

TRUCK, WAGON, AND BIN UNLOADERS

Power-operated truck and trailer-wagon unloaders save time and labor in unloading chopped silage into the blower at the silo. Generally, five methods can

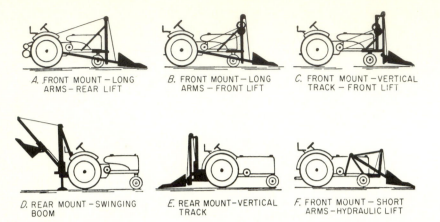

A. FRONT MOUNT – LONG   *B.* FRONT MOUNT – LONG   *C.* FRONT MOUNT – VERTICAL
      ARMS – REAR LIFT   ARMS – FRONT LIFT   TRACK – FRONT LIFT

D. REAR MOUNT – SWINGING   *E.* REAR MOUNT – VERTICAL   *F.* FRONT MOUNT – SHORT
      BOOM   TRACK   ARMS – HYDRAULIC LIFT

Fig. 23-10   Methods of mounting loaders on tractors. (*South Dakota Agr. Expt. Sta. Bull. 378.*)

be used to unload vehicles, namely, *fence wire, movable front endgate, movable bottom, tilt body,* and *air.*

The *fence-wire* method consists of laying a section of wire fencing on the floor of the trailer box and extending it up over the front end of the trailer, where it is attached to a wood or steel bar. To unload, a cable is attached to the bar which extends back over the load to a truck or tractor. As the bar with the wire attached is pulled backward, the chopped material is rolled out the rear of the box. The disadvantage of this method is that the front part of the load will roll up on top of the rear part of the load and fall in amounts too large for the blower to handle. A man with a fork is required to distribute the material.

The *movable-front-endgate* unloading method consists of a false front endgate which is drawn backward as cables attached to it are wound around a power-driven shaft at the rear of the box. To slide green, heavy silage over the floor of a trailer box requires considerable power.

Fig. 23-11   A combination loader and ditcher. It is called a *backhoe.* (*International Harvester Company.*)

Conveyor Unloaders

A canvas or metal link conveyor belt is installed over the solid bottom of the wagon box (Fig. 23-12). As the belt moves back, the load is moved with it. Figure 23-13 shows a conveyor web that consists of two steel chains with steel angle crossbars. The load is moved by the crossbars as the conveyor moves to the rear of the box. This is the same type of conveyor that is used on manure spreaders (Chap. 14).

The conveyor is driven by a gear-reduction box powered by an electric motor, gasoline engine, or tractor power takeoff.

The *tilt-body* method consists of lifting the front end of the box or entire vehicle. When the rear endgate is removed, the material slides out by gravity (Fig. 23-7). Dry grain and ear corn will readily flow when the box is at 40 to 45° to the horizontal. Wet material, such as silage, requires a greater angle for the body to clear itself.

The *air* unloading method shown in Fig. 23-5 sucks the material out of the box and blows it into the silo or bin.

Bin Unloaders

A chain drag-flight type of bin unloader is shown in section 2 of Figure 23-1. The long horizontal unit can be projected into the bin, or it can be set under a

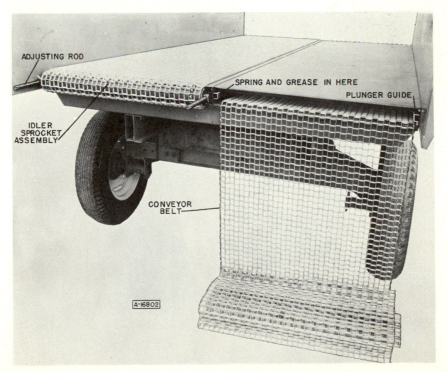

Fig. 23-12   Power-driven steel link belt that moves the material out the rear end of the wagon.

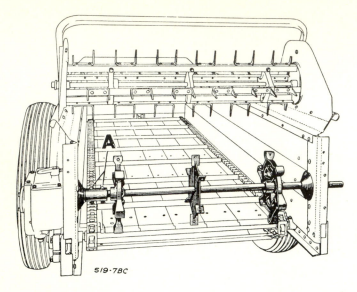

519-7BC

Fig. 23-13   Power-driven web apron conveyor for unloading manure from box.

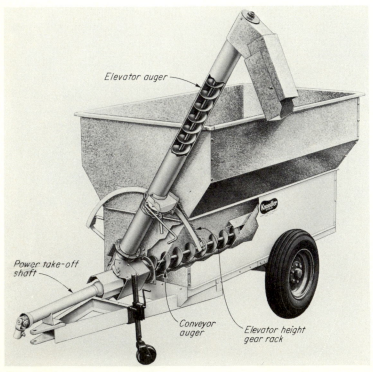

Fig. 23-14   Trailer-type bulk-feed mixer wagon operated by tractor power takeoff. (*Knoedler Manufacturers, Inc.*)

spout on the side of the bin. Small and medium-sized auger portable elevators can be projected into the bin and the grain unloaded, or an extension section can be used to unload bins.

TRANSPORT MIXER-FEEDERS

Where large numbers of livestock are being fed, the use of a truck-mounted or trailer-type transport mixer-feeder will save countless hours of scoop labor in handling the feed. Chopped roughages and ground grains are mechanically loaded into a specially constructed, inverted semi-trapezoidal box equipped with a power-driven agitator and a mechanical power-driven unloader elevator (Fig. 23-14). The feed is mixed in transit, and as the outfit is driven along the feed troughs, the feed is mechanically elevated and delivered directly into the troughs. The agitator and elevator on the trailer mixer-feeders are usually driven from the power takeoff of the tractor that tows the machine.

POWER POST-HOLE DIGGERS

The digging of post holes to build fences about the farm and pasture is slow, hard work when done by hand. Power-driven post-hole diggers make the job easier and many times faster. A power-driven post-hole digger with power pressure and lift can dig a standard post hole in a few seconds. The majority of post-hole-digging attachments are mounted on the rear of the tractor so that the power can be obtained from the power takeoff (Fig. 23-15). The best power

Fig. 23-15   Tractor-mounted power-takeoff-driven post-hole digger. (*Danuser Machine Co.*)

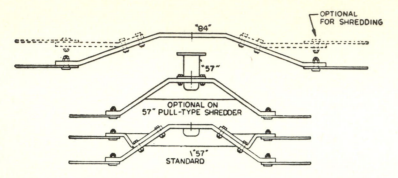

Fig. 23-16   Types of horizontal rotating knives used for medium-size brush.

Fig. 23-17   Chain saw suitable for cutting trees and brush. (*McCulloch Corp.*)

Fig. 23-18   A small air compressor that can be operated by either a gasoline engine or an electric motor. (*Champion Pneumatic Machinery Co.*)

post-hole diggers have an arrangement whereby pressure can be applied to force the auger into the soil and a power-lifting device used to lift the auger from the hole.

BRUSH-CUTTING EQUIPMENT

The destruction of small to medium-sized brush can be accomplished with heavy-duty rotary beaters, rolling cutters, and the horizontal-rotating-knife-type cutter-shredders (Fig. 23-16). These machines were described under "Crop-processing Equipment" (Chap. 22).

Circular saws mounted on tractors and on two-wheel carts are suitable for sawing down larger brush and trees. A small gasoline engine furnishes power for operating the saw. Hand-held chain saws are useful in cutting timber (Fig. 23-17).

AIR COMPRESSORS

As most farm equipment and tractors are equipped with pneumatic tires, every farm should have a small air compressor (Fig. 23-18). Tires can be easily inflated, and with an air-jet tip, radiators can be cleaned and dirt and dust blown out of hard-to-get-to places.

SPECIAL TRAILERS AND WAGONS

Trailers and wagons have been developed to handle special farm products such as grain, chopped forage, machine-harvested cotton, and livestock. Figure 23-19 shows a trailer wagon equipped with conveyor and beaters that are driven by the tractor power take-off. The box unloads itself.

Almost every farm and ranch that has livestock of any kind has some type of truck or trailer for transporting the animals. Many are made at local shops, and design varies with the fancy and whims of the individual.

Fig. 23-19   A self-unloading chopped-forage trailer wagon.

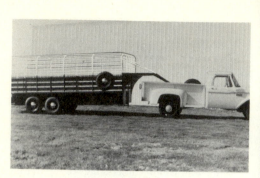

Fig. 23-20 (*Left*) Goose-neck trailer hitched to pickup truck; (*right*) hitch plate in bed of truck and trailer hitch unit in hook-up position. (*Hanover Mfg. Co.*)

Figure 23-20 shows a trailer with a *gooseneck*-type trailer hitch. A hitch plate is installed in the pickup truck bed over the pickup rear axle. The frame of the trailer curves over the endgate of the pickup truck. The curve of the frame is high enough so that a right-angle turn can be made (Fig. 23-21). This type of trailer is available in a number of sizes to suit the requirements of the job. Livestock trailers, enclosed vans, and flatbed types are made.

Trailers with the gooseneck type of hitch are popular with farmers and ranchers.

DETASSELING AND TOPPING EQUIPMENT

Figure 23-22 shows a high-clearance machine equipped with knives for detasseling corn. Similar machines are available for topping tobacco.

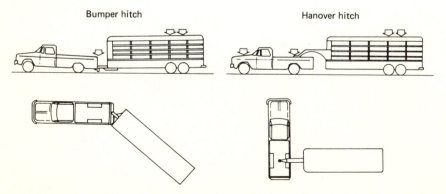

Fig. 23-21 Comparison of the bumper and goose-neck trailer hitches in maneuverability and load distribution. (*Hanover Mfg. Co.*)

Fig. 23-22 A high-clearance machine equipped with corn detasseling knives. (*W. F. Larson, Inc.*)

MAINTENANCE AND CARE

1   The operator of any type of laborsaving equipment should be familiar with the machine to be efficient.

2   Safety measures should be carefully followed in operating the machines.

3   Check hydraulic lifts regularly.

4   Keep the machines clean and protect them from weather damage.

REFERENCE

Anonymous Push-button Farming, *Farm Impl. News,* **74**(11):40-41, June 25, 1953.

QUESTIONS AND PROBLEMS

1   Explain the differences between the various types of elevators.

2   Give several uses of power loaders.

3   Describe the various types of unloading equipment for trucks and trailer wagons.

4   What are the advantages of using a transport mixer-feeder?

5   Describe the gooseneck type of trailer hitch.

Chapter 24

Selection of
Farm Machinery

After studying the various individual farm implements and their construction features, it is well to consider some of the important items that apply to all implements in general. These points or qualities that a machine may have or that it may lack are abstract in a way, yet they are fundamental in their bearing on the quality of the machine. They are factors that will enable the student to judge a machine better; they will call to the student's attention the points to look for and may have an important influence on the selection of one machine in preference to another.

Trademark

The standard definition of a trademark is as follows: *Trademark* is a distinguishing mark, device, or symbol affixed by a manufacturer, merchant, or trader to his goods in order to identify them as his goods and distinguish them from the goods manufactured, sold, or dealt in by others. Most countries give special statutory protection to such trademarks when they are registered according to law. This registration gives the owner of the trademark exclusive right to its use.

The importance of the trademark in the selection of farm machinery lies in what it stands for. A manufacturer spends many years and much money in building a reputation and establishing the company's trademark. After the reputation and trademark are thoroughly established and well known among the trading world, the maker will continue to try to maintain the same standards. It is not possible to judge a machine by its appearance or to determine whether good materials are used in its construction simply by looking at it. Therefore, if it bears the trademark of a firm that has a good reputation, it is fairly certain that the manufacturer of such an implement will stand behind that particular piece of machinery. If any defect occurs within a reasonable length of time, the firm will make it good. In other words, then, we may say that the trademark of a machine is a guarantee of reliability and of what lies beneath the paint. Look well to the builder of your machine when you are judging and preparing to invest.

Trade Name

The trade name is the name by which an article is called or the name given by a manufacturer to an article to distinguish it as one produced by that company. This name is also called a *brand name*. It is entirely different from the trademark. A trade or brand name can be registered in the U.S. Patent Office and thus have statutory protection in a manner similar to a trademark. A trade name will be found only on one particular type of equipment, such as Farmall for a line of row-crop tractors. A well-known brand name is valuable from a sales standpoint. The use of brand names in the farm-equipment industry is being abandoned in favor of model designations.

Models

Models in farm equipment may indicate a type of machine, a size, an improvement or new design of an old machine, a special-purpose machine, or a combination of one or more of these features. Model designations may be by a series of numbers or letters or a combination of both numbers and letters. For example, a combine model may be given as SP-12, which indicates that this particular machine is self-propelled and cuts a 12-ft (3.7-m) swath. In another case, a tractor-mounted planter is listed as a 34-400 two-row planter. The 400 indicates the size of the tractor. In the number 34, the 3 indicates a two-row planter or cultivator while the 4 indicates a four-row planter or cultivator. Each manufacturer has his own special system of designating models. Farm-equipment models are not dated on an annual basis as are automobiles; consequently, when a machine is needed, there is no necessity to wait for next year's model.

Repairs

Before considering the purchase of any machine, it is well to look into the source of repairs. Can repairs be made near at hand, or will it be necessary to send several hundred miles away? No farm implement has yet reached the stage of perfection where it will not break, wear out, or meet with accidents; therefore, it will need repairs. Many times the saving of a crop depends upon the speed with which repairs can be completed. If breakdowns occur in the midst of plowing, planting, or harvesting, they may cause so much delay that the crops will be lost. The larger implement companies maintain repair supplies at many points so that they can render quick service to every part of the country. The machine should be examined to see whether the various parts are accessible for making repairs when needed. Provision should be made in all implements for taking up the wear of bearings and gears. Look well to the source of supplies before buying a machine.

In making up the order for repair parts that are needed, be sure to secure the following information:

1   The name and address of the manufacturer.
2   Brand name, model number, and year made or purchased.
3   Number of the part wanted.
4   If the number of the part cannot be determined, then get the numbers of the parts with which it works.

Keep the pamphlets that are furnished with the machine, especially the one containing the repair parts list. When repair parts are needed, the part numbers and their description will be found in the list of parts. From the standpoint of repairs, it is economical to standardize on a single line of equipment.

Design

Design is the arrangement of the parts to show the difference of make-up in machines of the same type. Manufacturers may put out the same line of implements, but the machines will not be exactly alike. It is this difference in the arrangement of the component parts that makes up the design of the machine. In studying the general construction of the machine, keep in mind the number of castings, gears, points of wear, bearings, and ease of lubrication and adjustment. Wherever possible, gears should run in a sealed oil bath. Provision of safety devices should be carefully considered. Be sure there are shields over power-takeoff drive shafts. In general, does the machine have a finished appearance and style without sacrificing strength and performance? Environmental conditions must also be considered in farm machinery design.

Ease of Operation

Many implements that look well are found to require an unnecessary amount of power and labor to make them operate successfully. Of course, it is not always feasible to have the machine demonstrated to see if it will operate easily; nevertheless, such things should be considered in the selection of the machine. The ease of operation may simply depend upon the correct adjustment. It is not an uncommon thing for a farmer to purchase an implement, take it home, and, after attempting to use it, condemn the machine because of its hard operation. The user may go so far as to take it back to the dealer and ask to have the purchase money back. A dealer who is a good one will usually take the machine out, have the farmer go along, make the necessary adjustments, and see that the machine is running perfectly before turning it over to the farmer.

Power and hydraulic lifts have taken the place of the manually operated levers. When the machine has once been properly adjusted, little effort is required of the operator other than the steering and turning of the machine.

Ease of Adjustment

In the selection of farm equipment, careful study should be made of the methods for adjusting the various parts. Devices designed to simplify adjusting the equipment are time and labor savers. The owner's or operator's manual should be studied thoroughly to understand the method of adjusting the equipment as planned by the designer and the test engineer. Many operators of farm equipment are not inclined to take sufficient time to make needed slight adjustments. The author suggested some adjustments to a farmer who was operating a harvesting machine. His reply was, "I know it will do better work if I make the changes, but I'm in a hurry to get my crop out." His crop losses were, no doubt, many times greater than the value of the thirty minutes' time it would have taken to make the adjustments. Too many operators take the "I'm in a hurry" attitude about adjustments.

Adaptability to Work and Conditions

There are many implements on the market which are not adaptable to every condition. A machine may work in one locality and be an absolute failure in another because it is adapted to certain soil conditions or types of crops grown. To take an example: Tools built for the Southeastern and Gulf Coast states are not suitable for use in the Southwestern states, New Mexico, or Arizona because of the difference in climate, which influences the methods of preparing the seedbed, planting, and cultivating.

Quick Change of Units

The time and labor required to dismount one unit and mount another are important considerations in selecting farm equipment. Some lines of equipment are built in unit packages and are designed so that changes can be made in a few minutes from a plow to planter, from planter to cultivator, or from one unit to another for all the units supplied. There are other lines that require one or more days to change from planter to cultivator and the same time to change back to planter.

Most integral-tractor-mounted equipment is designed for a certain make of tractor and cannot be used on any other make.

Maneuverability

As a general rule, tractor-mounted equipment is provided with power and hydraulic lifts. The units can be lifted and the tractor maneuvered almost as though no equipment was attached. When trailing equipment is attached to the drawbar of a tractor, turns cannot be made so sharply as with mounted equipment. Extended and swinging drawbars are an aid for short turning. Maneuverability problems are increased when a trailer is drawn behind a trailing-type harvester. The maneuverability is greatly reduced when machines are mounted in front of row-crop tractors. The small wheels sink into loose soil, drop into shallow ditches and furrows, and the tractor is difficult to turn.

Comfort

As the operator of power equipment must spend days upon days riding upon it, the comfort and safety of the seat should be considered. A comfortable seat should be supported with shock-absorbing devices. The seat should be stable and adjustable to suit different-sized individuals. Cabs give protection from cold, heat, dust, and excessive noises.

Other Factors

Other factors to keep in mind in the selection of farm equipment are the power requirements, cost of operation, initial cost, years of service expected, and whether the purchase of the equipment is economical in relation to the size of the farm and the work to be performed by the equipment.

FARM-EQUIPMENT SAFETY

The operation of farm equipment has many hazards, and accidents can be reduced if the safety precautions shown in Fig. 24-1 are carefully followed.

"Safety first" is a slogan that should be kept in mind at all times. Many of the accidents that occur in operating farm equipment are the direct result of thoughtlessness and carelessness. Accidents are more likely to occur with power-operated equipment that has moving parts to perform its operating function. The tractor perhaps is involved in more farm accidents than any other type of equipment, largely because of its versatility and extensive use. The corn picker is considered a hazardous machine because of the number of accidents that happen to those who operate it. The major number of accidents result from attempts to unchoke the downward revolving snapping rolls without disengaging the power and stopping them. Power-takeoff shafts should be provided with shields. Slip clutches should be installed to protect the machine in case of plugging.

Design engineers add protective features around moving parts, but if the operator of the machine fails to use these safety features, he is likely to be involved in an accident that will cause serious injury or perhaps death.

All the national societies and associations, such as the American Society of Agricultural Engineers, The Society of Automotive Engineers, The Farm Equipment Institute, The Manufacturers Association, and Dealers Association, have safety committees. The National Safety Council gives assistance to the 45 permanently organized State Farm Safety Committees. These organizations promote safety programs, publish safe farm practice leaflets, and supply safety information to both adult and youth rural organizations.

Stop a minute; think and plan for safety. The operator who says, "I'm in a hurry" is likely to have an accident.

CABS FOR FARM EQUIPMENT

Cabs on farm equipment serve as safety devices on tractors, so the cab should be reinforced with a roll-bar. A strong cab protects the operator from being crushed in case a tractor rolls over. Cabs on combines, cotton harvesters, and self-propelled equipment protect the operator from dust, noise, heat, and cold. A big advantage of a cab is that it greatly reduces noise made by the tractor engine and other equipment. This in turn prevents to a large extent damage to the operator's hearing. Protection from dust, heat, and cold provides the operator with comfort. In fact, it may be so comfortable that the operator may be reluctant to leave the cab and check and adjust machine troubles. However, specialized control systems—operations monitors—can let the operator know when there is machine trouble.

When a cab is used, a check should be made to be sure that the exhaust pipe from the engine is extended far enough away from the cab to prevent exhaust fumes from entering the cab.

All tractors not equipped with strong cabs should be provided with safety roll bars.

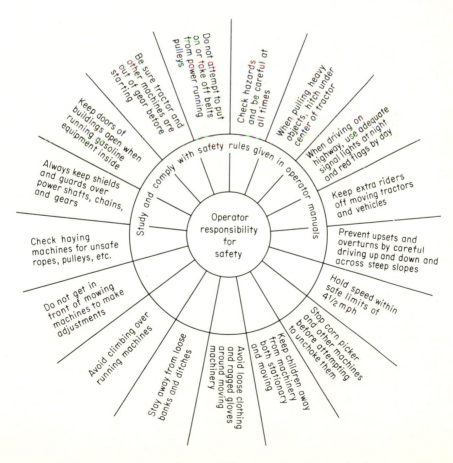

Fig. 24-1   Chart showing operator responsibility for various factors of safety in operating farm equipment.

SAFETY EMBLEM

The National Safety Council recommends that all slow-moving vehicles operated on highways have a safety emblem attached to the equipment in a manner so that the emblem can be easily seen. Figure 24-2 shows the standard safety emblem. The emblem is triangular in shape. The central portion is fluorescent orange while the 4-in (10.2-cm) border is reflective dark red. If a slow-moving vehicle is moved at night, it should be equipped with amber-colored flashing lights.

Fig. 24-2 Safety emblem for slow-moving vehicles when operated on highways. The emblem is fluorescent orange in the center and reflector red on the borders.

REFERENCES

ASAE Standard: Safety for Agricultural Equipment, *ASAE Yearbook*, 1973.
Hunt, Donell: A Fortran Program for Selecting Farm Machinery, *Agr. Engin.*, 48(6):332, 1967.
Kirby, Edwin: Safety Programs for Agriculture, *Agr. Engin.*, 55(10):17, 1974.
National Safety Council literature.
Sullivan, Herbert: Safety Standards in Farm Equipment Design, *Agr. Engin.*, 55(4):29, 1974.

QUESTIONS AND PROBLEMS

1   Explain the difference between a trademark and a trade name.
2   Discuss methods of designating models in farm equipment.
3   Make up an order for repair parts for a plow, a planter, a combine.
4   Discuss the importance of design in selecting farm equipment.
5   Discuss the need for considering maneuverability and ease of operation and adjustment in selecting farm equipment.
6   Explain the value of having equipment in which the units can be interchanged quickly.
7   Enumerate and discuss the various safety rules to be followed in operating farm power equipment.
8   Why are cabs useful on farm equipment?
9   Describe the safety emblem for slow-moving vehicles on the highway.

Chapter 25

Economics and Management of Farm Equipment

Generally, the size and type of farm equipment used by a farmer are closely related to the size of the farm, in acres, and the kind of crop grown.

The USDA and the Census Bureau has recently changed the definition of a farm. The definition now reads as follows: "A farm is now any establishment from which $1000 or more of agricultural products is sold or would normally be sold during a year." The number of acres involved is not specified.

This definition of a farm places farming in the United States on a productive business-operation basis rather than the old idea of farming for a living.

Butz[1] states that American agriculture has had three great revolutions and lists them as follows:

The first great revolution came in the middle of the 19th Century, when we began to substitute animal energy for human energy.

The second great revolution began in the 1920's with the substitution of mechanical energy for animal energy.

The third revolution is the undergirding of agricultural production and marketing with vast amounts of science, technology, and business management.

[1] *U.S. Dept. Agr. Yearbook,* p. 380, 1960.

The modern commercial farm resembles a manufacturing plant in many respects. The large amount of equipment in use on the farm represents a substitution of capital and machinery for labor.

This development obviously calls for a high level managerial capacity. It is more difficult to manage the modern commercial farm conglomerate successfully than it is to manage the family-sized manufacturing concern, grocery store, or foundry shop in the city.

The successful commercial farmer must determine what farm equipment meets his needs according to his acreage and crop. There are many factors to consider.

Matching Equipment to Farm Needs

Before buying farm equipment, a farmer must decide which make, size, and type of machine will be the most efficient for both the farm and the equipment. It is a difficult job to match equipment to meet the farm needs. Farmers must decide whether or not their acreage, production, and especially income are sufficient to justify the purchase of expensive equipment.

He must decide whether it is more economical to own the equipment and furnish the labor and supplies for its operation or to hire the equivalent services through custom work.

Ownership Essentials

When a farmer buys equipment for his farm, he must assume a number of expense items. These can be divided into fixed and variable costs.

The fixed costs include:

1	Original purchase	5	Repairs
2	Depreciation	6	Insurance
3	Interest on investment	7	Shelter
4	Taxes		

The variable costs include:

1	Fuel (for self-powered units)	3	Labor
2	Lubricants	4	Tractor, fuel, oil, and grease

Figure 25-1 shows the relative importance of the various total fixed and variable costs when plowing with a four-plow tractor and four 16-in bottoms.

Purchasing Equipment   Equipment for single-crop farming, such as cotton, corn, or small grain, is less expensive than that required for mixed-crop farming of cotton and grain sorghum, wheat and corn, and combinations of other crops.

A cotton farm requires tractor power, plows, harrows, planters, weed- and insect-control equipment, and harvesters. If corn is included in the enterprise organization, additional harvesting equipment is required. Small-grain crops

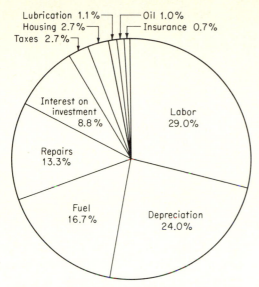

Lubrication 1.1% — — Oil 1.0%
Housing 2.7% — — Insurance 0.7%
Taxes 2.7%

Interest on
investment
8.8%

Labor
29.0%

Repairs
13.3%

Fuel
16.7%

Depreciation
24.0%

Fig. 25-1 The relative importance of each item of cost when plowing with a four-plow tractor and four 16-in (40.6-cm) mounted bottoms. (*Kans. Agr. Expt. Sta. Bul. 417.*)

require special seeding and harvesting equipment that is not needed for the production of cotton. Some equipment is a common requirement for most crops, such as the power unit, plows, and harrows. The combine with adjustments and attachments can be used to harvest a number of crops. The cotton picker and the corn picker are limited in use to single crops.

The retail prices of farm equipment are not given here because the prices vary from one section to another as the result of transportation charges and state sales taxes.

Table 25-1, however, shows the cost per hour of using various types of farm machines. The table also shows the approximate years and hours of service for the basic types of equipment used on farms.

COST OF CROP MACHINES

The depreciation cost for the machines shown in Table 25-1 has been calculated on a straight-line basis, using a life based on whichever occurs first, wearing out or obsolescence. For instance, a grain drill which will be obsolete in 20 years or worn out in 1200 hours will depreciate one-twentieth of its new cost per year until it is used over 60 hours per year, when it will depreciate one twelve-hundredth of its new cost per hour. Depreciation cost can be estimated for a particular machine by applying the life data in Table 25-1 to the chart in Fig. 25-2.

Repair cost has been calculated as a constant cost per hour based on the *life repair cost* divided by the *hours of life*. The above grain drill, with a *life repair cost* of 25 percent of the new cost, would have an hourly repair cost of one forty-eight-hundredth of the new cost. Repair cost can be estimated for a

Table 25-1 Cost per Hour of Using Farm Machines

Machine	Years until obsolete	Hours to wear out	Total repair cost, % of new cost	Cost per hour of use per $100 of new cost*						
				20 h per year	40 h per year	60 h per year	100 h per year	150 h per year	250 h per year	350 h per year
Tractor plow	15	2000	80	$0.600	$0.319	$0.226	$0.152	$0.120	$0.108	$0.103
Tractor disk harrow	15	2000	30	0.575	0.295	0.202	0.127	0.095	0.083	0.078
Spring-tooth harrow	20	2000	40	0.495	0.253	0.179	0.115	0.100	0.088	0.083
Spike-tooth harrow	20	2500	30	0.487	0.250	0.171	0.107	0.082	0.070	0.065
Roller	25	1500	10	0.432	0.220	0.149	0.118	0.103	0.091	0.086
Soil pulverizer	20	2000	15	0.483	0.245	0.167	0.103	0.088	0.076	0.071
Endgate seeder	20	800	30	0.512	0.275	0.238	0.208	0.193	0.181	0.176
Grain drill	20	1200	25	0.496	0.258	0.179	0.149	0.135	0.122	0.117
Corn planter	20	1200	30	0.500	0.263	0.184	0.153	0.138	0.126	0.121
Field sprayer	10	1500	30	0.745	0.383	0.262	0.165	0.117	0.105	0.100
Rotary hoe	15	1500	20	0.573	0.293	0.200	0.125	0.110	0.098	0.093
Tractor cultivator	12	2500	40	0.662	0.341	0.234	0.148	0.106	0.078	0.073
Rotary cutter	12	2000	35	0.659	0.338	0.232	0.146	0.104	0.086	0.081
Tractor mower	12	2000	75	0.679	0.358	0.252	0.166	0.124	0.106	0.101
Dump rake	10	1500	25	0.742	0.379	0.259	0.162	0.114	0.102	0.097
Side-delivery rake	15	1500	50	0.591	0.312	0.219	0.145	0.130	0.118	0.113
Tractor buck rake	12	1500	25	0.657	0.337	0.231	0.145	0.114	0.102	0.097
Hay loader	10	1200	25	0.746	0.383	0.263	0.166	0.134	0.122	0.117
Forage harvester†	12	2000	60	0.671	0.350	0.244	0.158	0.116	0.098	0.093
Forage blower	12	2500	25	0.651	0.330	0.224	0.138	0.096	0.068	0.063
Pickup baler (auto tie)†	12	2500	40	0.657	0.336	0.230	0.144	0.102	0.074	0.069

Table 25-1 Cost per Hour of Using Farm Machines (Continued)

Machine	Years until obsolete	Hours to wear out	Total repair cost, % of new cost	Cost per hour of use per $100 of new cost*						
				20 h per year	40 h per year	60 h per year	100 h per year	150 h per year	250 h per year	350 h per year
Swather	12	1200	25	0.662	0.341	0.235	0.149	0.134	0.122	0.117
Combine†	10	2000	40	0.745	0.383	0.262	0.165	0.117	0.088	0.083
Corn binder	10	1000	40	0.765	0.402	0.282	0.185	0.170	0.168	0.153
Stationary silage cutter	10	1200	30	0.750	0.387	0.267	0.170	0.138	0.126	0.121
Husker-shredder	10	2500	25	0.735	0.372	0.252	0.155	0.107	0.068	0.063
Corn picker	10	1500	30	0.745	0.383	0.262	0.165	0.117	0.105	0.091
Spindle cotton picker	10	2000	55	0.753	0.390	0.270	0.173	0.125	0.096	0.091
Manure loader	10	2000	25	0.738	0.375	0.255	0.158	0.110	0.086	0.076
Manure spreader	15	2500	25	0.568	0.289	0.195	0.122	0.084	0.068	0.063
Feed grinder	15	2000	25	0.571	0.292	0.198	0.125	0.093	0.081	0.076
Portable elevator	15	1500	15	0.568	0.289	0.195	0.122	0.107	0.095	0.090
				60 h per year	150 h per year	300 h per year	500 h per year	750 h per year	1000 h per year	1500 h per year
Wagon gear box	15	5000	50	0.196	0.084	0.047	0.039	0.036	0.035	0.033
Tractor	15	12,000	120	0.196	0.084	0.047	0.032	0.025	0.023	0.021

* Based on 4½ percent of new cost as total annual charge for interest, housing, taxes, and insurance.
† Operating costs such as fuel, oil, grease, wire, twine, etc., not included.
Source: Agr. Engin. Yearbook, 1962.
Note: since these data were calculated, the cost of new farm machinery has increased an estimated 25 percent.

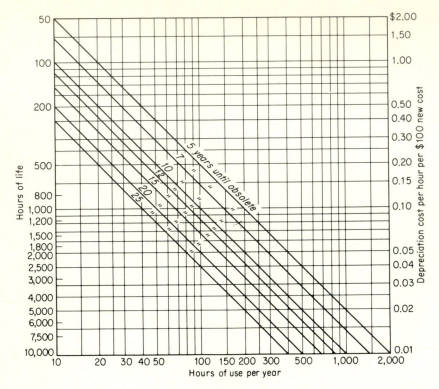

Fig. 25-2 Depreciation cost of farm machines. To find depreciation cost per hour per $100 new cost: 1, locate intersection of hours of use per year with hours of life; 2, locate intersection of hours of use per year with years until obsolete; 3, from upper of these intersections move to right and read. (*ASAE Yearbook.*)

particular machine by applying the life and repair data in Table 25-1 to the chart in Fig. 25-3.

Interest, housing, taxes, and insurance can be grouped together, since the cost of each can be expressed by an annual charge. Annual costs for interest, housing, taxes, and insurance can usually be determined best by considering the specific case. If better estimates are not available, the following costs can be used as typical for farm machines in general:

Interest	$6.50 per year for each $100 value
Housing	$2.08 per year for each $100 value
Taxes	$2.60 per year for each $100 value
Insurance	$0.52 per year for each $100 value

CAPACITY AND EFFICIENCY OF FIELD EQUIPMENT

The capacity of a farm machine is the rate at which it can cover a field while performing its intended function or useful work. This is usually figured as the acres per hour a machine will cover. The factors involved are the width of useful work and the speed of travel with an allowance for lost time in turning and servicing the machine.

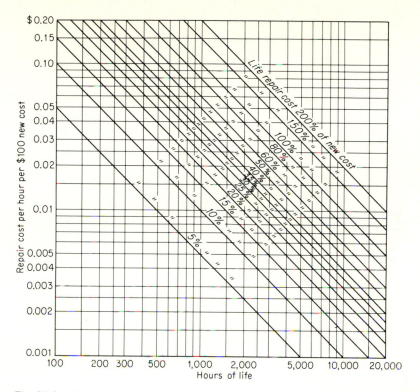

Fig. 25-3 Repair cost of farm machines. To find repair cost per hour, locate intersection of hours of life with life repair cost percentage, move to left, and read. (*ASAE Yearbook.*)

The *theoretical field capacity* of an implement is the rate at which it will perform its intended function if it is operated continuously at its rated width. It is the actual acres covered per hour. There is no allowance for loss of time and servicing. The theoretical field capacity is calculated by multiplying the forward speed in miles per hour by the operating width and dividing the product by the factor 8.25. The factor is found by dividing the number of square feet in an acre by the number of feet in a mile, or $43,560 \div 5280 = 8.25$. This gives a dimensionally consistent equation.[1]

The *effective field capacity* of an implement is the average rate at which it covers a field expressed in acres per hour. This includes an allowance for loss of time in turning and servicing. Figure 25-4 shows a chart from which the effective capacity of machines can be read.

The effective field capacity of a machine in acres per hour is determined by dividing the product of the speed, in miles per hour, and the rated width by 8.25 and multiplying the result by the field efficiency expressed as a decimal fraction.

The *field efficiency* of a machine is effective field capacity divided by the theoretical field capacity and multiplied by 100 to express it as a percentage.

[1] Kenneth K. Barnes and Paul E. Strickler, Management of Machines, *U.S. Dept. Agr. Yearbook,* pp. 346-354, 1960.

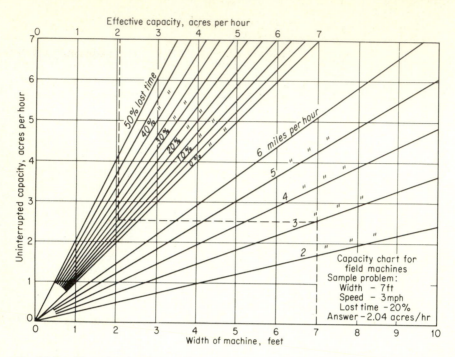

Fig. 25-4   Chart showing the capacity of field equipment for various widths, speeds, and percentage of lost time, without making calculations. (*ASAE Yearbook*).

This field efficiency of a machine is always less than 100 and is influenced by various factors involved in the effective capacity. The field efficiency of a plow may have a range of 74 to 84; of a disk harrow, 77 to 90; of a grain drill, 60 to 78; of a mower, 77 to 85; of a combine, 63 to 70; of a cotton picker and corn picker, 55 to 70.

Doubling the size of a machine does not double the theoretical or effective capacity. It may decrease the field efficiency, as shown in the data in Table 25-2. The data show that to double the theoretical capacity of a four-row planter, a nine-row planter would be required.

Machine efficiency is applied largely to harvesting equipment, such as combines, corn pickers, and cotton pickers and strippers. The efficiency of a machine is the percentage of the crop yield harvested by the machine. It is determined by dividing the yield per acre harvested by the machine by the total yield per acre and multiplying the result by 100. The total yield includes the amount harvested by the machine minus the amount of storm loss before the machine was operated minus the amount lost by the machine.

The family farm operator who succeeds today combines knowledge of genetics, land and water use, conservation, chemistry, and physics with business ability. He combines modern science and ancient art with machine power. The result is greater efficiency in the production of food, clothing, and other materials for home and industry.

Table 25-2   Idealized Capacity Performance of Row-Crop Planters*

| Activity | Distribution of time, min/acre | | | |
| | Field observations | | Predictions | |
	4-row	6-row	9-row	12-row
Planting	8.22	5.50	4.11	2.75
Turning	0.50	0.50	0.50	0.50
Filling hoppers	1.40	1.40	1.40	1.40
Adjusting, checking, cleaning	0.54	0.54	0.54	0.54
Miscellaneous	0.75	0.63	0.56	0.50
Total	11.41	8.57	7.11	5.69
Capacity summary				
Theoretical field capacity, acres/h	7.3	10.9	14.6	21.8
Effective field capacity, acres/h	5.3	7.0	8.4	10.5
Field efficiency, %	73	64	58	49
Relative effective field capacity	100	1.32	1.58	1.98

* Assuming 40-in rows, 1200 ft long, and a speed of 4.5 mi/h

Source: K. K. Barnes, Field Capacity of Row-Crop Planters, *Implement & Tractor,* 76(5):50, 1961. Reproduced with special permission from *Implement & Tractor.*

CUSTOM USE OF FARM EQUIPMENT

The custom use of farm equipment began, no doubt, in ancient times with borrowing whatever type of tool was available or swapping work for its use. The old-time threshing ring was probably one of the first large-scale types of custom hiring of farm equipment in America as a business operation. Now almost all types of equipment used on the farm can be hired. Table 25-3 shows the prevailing custom rates for six agricultural regions of the United States in 1972. Custom work is defined by Nelson and Thompson[1] as "the hiring of men with machines for performance of a specified machine task." The most common types of equipment custom-hired are the harvesting machines. Aircraft for seeding and for the application of fertilizers, herbicides, and insecticides are practically always custom-hired. Aircraft and many of the harvesting machines require highly trained and skilled operators, beyond the qualifications of most farmers.

Many farmers do not have sufficient crop acreage to justify economically the investment in equipment. Some farmers, however, may be able to own and operate equipment to handle a large crop acreage up to harvest, then, to reduce the effects of weather and other hazards, to hire additional harvest equipment.

Many agricultural economists have studied the advantages and disadvantages of custom service of farm equipment. These are listed as follows:

[1] Ted Nelson and Dale Thompson, Farm Custom Rates Paid in Nebraska in 1959, *Nebr. Ext. Ser. and U.S. Dept. Agr. Ext. Ser. E.C.* 60-806, 1959.

Advantages and Disadvantages of Custom Hiring of Farm Equipment

Advantages	Disadvantages

1 Costs of ownership are eliminated.

2 Capital required to own equipment can be invested in other enterprises.

3 Some labor is furnished with hired equipment.

4 Less power and associated equipment must be obtained by the farm operator.

5 Farmers may benefit from newer machinery and techniques and skilled operation.

6 Repairs, maintenance, and securing of materials are the custom operator's responsibility.

7 Farm operator with smaller jobs can gain benefits of large machines.

8 Risk of premature equipment obsolescence is eliminated.

1 Service may not be available when job is ready.

2 Irresponsible custom operators may do poor work and lose quantity and quality of products.

3 Greater risk of crop loss and/or quality because of time delays.

4, Risk of carrying noxious weeds and diseases from farm to farm.

5 For large jobs, total cost may be higher than in owning equipment.

6 Custom operators prefer large jobs, may refuse or postpone small jobs.

7 Farmers may not be able to realize return on labor released by hiring custom work done.

OWN OR HIRE FARM EQUIPMENT?

When a farmer needs a machine for a particular phase of his farm operation, he must, from a business standpoint, determine whether it is more economical to buy and own the machine or to hire men with machines to do the work.

Ownership essentials and cost of use of crop-farm equipment have been discussed above. The prevailing custom rates for various field operations are given in Table 25-3. The farmer must determine by figuring the comparative costs whether it is more profitable to own or hire the equipment.

Ownership   Anderson and Tefertiller have developed a simple method of figuring the cost of machine ownership as shown in Table 25-4 (see also Table 25-5). The estimated ownership costs in the table are for a 14-ft self-propelled combine that costs $7100 new.[1] The combine has an estimated life of 10 years or 2000 hours and a 10 percent trade-in value.[2]

---

[1] A part of unpublished manuscript, *Farm Machinery Decisions*, by C. G. Anderson and K. R. Tefertiller, respectively, Vocational Agriculture, Farm Management Specialist and Assistant Professor, Department of Agricultural Economics and Sociology, A. & M. College of Texas, College Station, Tex.

[2] Reproduced by special permission.

Table 25-3   Machinery Custom Rates in Six Agricultural Regions of the United States, 1972

Job	Basis of charge	North Central	North-east	South	Plains	Mountain & Pacific
		Plowing and cultivating				
Plowing, moldboard plow	acre	$4.85	$6.00	$4.70	$4.50	$5.00
chisel plow—8 to 12 in deep	acre	4.45	5.00	4.40	3.00	4.60
One way	acre	2.50	–	–	2.30	4.00
Disking, tandem	acre	2.10	3.30	2.80	1.80	2.50
bush and bog	acre	4.00	4.40	4.00	2.40	4.25
offset	acre	4.00	3.80	3.50	2.40	3.20
Harrowing, spike-tooth	acre	1.20	2.50	2.10	1.20	1.40
spring-tooth	acre	1.50	2.30	1.90	1.20	1.70
Cultivating						
sweep cultivator	acre	1.70	2.90	2.10	1.50	2.25
rotary hoe	acre	1.30	1.90	2.00	1.20	1.70
Stalk cutter, power takeoff	acre	2.10	3.00	2.50	1.75	2.25
Summer fallowing						
rod weeder	acre	–	–	–	1.40	1.40
graham hoe me	acre	–	–	–	1.65	2.40
duckfoot	acre	–	–	–	1.60	2.00
Subsoiling, 2 ft or deeper	acre	5.50	4.00	7.50	4.25	6.50

443

Table 25-3 Machinery Custom Rates in Six Agricultural Regions of the United States, 1972 (*Continued*)

Job	Basis of charge	North Central	North-east	South	Plains	Mountain & Pacific
		Planting and drilling				
Corn planting,						
with fertilizer attachments	acre	$2.60	$3.40	$3.30	$2.50	$3.00
without fertilizer attachments	acre	2.10	2.90	2.70	2.00	2.60
Till-plant	acre	5.00	6.00	5.80	4.75	–
Drilling small grain	acre	1.90	2.90	2.50	1.90	2.00
Planting soybeans	acre	2.00	2.75	2.60	2.00	–
Planting cotton	acre	–	–	2.60	2.10	2.10
Planting potatoes	acre	5.00	5.60	5.50	5.80	7.50
Seeding alfalfa, clover, etc.	acre	1.85	2.60	2.40	1.80	1.90
Airplane, seeding legumes	acre	2.00	–	1.30	1.40	1.60
seeding small grain	acre	2.00	–	1.90	1.30	1.60
seeding rice	acre	–	–	2.25	–	1.80

Table 25-3   Machinery Custom Rates in Six Agricultural Regions of the United States, 1972 (Continued)

		Chemical application (excluding materials)				
Spraying, aerial						
vegetables	acre	$2.00	$2.75	$1.60	$1.50	$1.80
tree fruit	acre	—	2.50	2.00	1.50	—
cotton	acre	—	—	1.40	1.20	1.70
row crops	acre	1.75	2.50	1.40	1.40	1.70
small grain	acre	1.40	2.50	1.40	1.30	1.70
Spraying, ground						
vegetables	acre	1.80	2.00	—	—	2.00
cotton	acre	—	—	1.50	1.40	2.00
row crops	acre	1.45	2.40	1.50	1.20	2.00
small grain	acre	1.30	2.35	1.50	1.71	1.80
Spraying fence rows	hour	7.00	—	8.00	6.50	10.00
Spraying buildings (excluding material),						
with insecticide	hour	9.00	—	10.00	5.40	8.00
with paint	hour	7.00	—	7.00	5.90	—
Spraying cattle (no material)	head	.40	.50	.32	.23	.28
Spreading fertilizer (excluding material costs),						
bulk dry	acre	1.00	1.80	1.50	.70	1.50
liquid	acre	1.20	2.00	1.25	1.00	1.50
side-dressing	acre	1.50	2.30	2.00	1.30	2.30
airplane	acre	2.20	—	1.50	1.30	1.80
anhydrous ammonia	acre	1.85	2.00	1.95	1.60	1.85
aqua ammonia	acre	1.50	1.75	—	1.60	2.00

Table 25-3 Machinery Custom Rates in Six Agricultural Regions of the United States, 1972 (Continued)

Job	Basis of charge	North Central	North-east	South	Plains	Mountain & Pacific
		Silage and hay making				
Field chop furnishing chopper and blower with:						
1 man, 2 wagons, 1 tractor	hour	$16.90	$15.00	$16.70	$15.00	$15.00
1 man, 2 wagons, 1 tractor	acre	11.20	12.00	–	8.00	–
2 men, 2 wagons, 2 tractors	hour	22.50	20.00	20.50	23.00	25.00
2 men, 2 wagons, 2 tractors	acre	–	18.00	–	13.00	–
Silo filling						
upright	ton	1.80	2.00	2.60	2.00	2.50
trench	ton	1.90	1.90	2.60	2.20	2.00
Mowing hay	acre	1.90	2.90	3.00	1.70	3.00
Raking hay	acre	1.20	2.00	2.00	1.00	1.70
Mowing & conditioning hay	acre	2.90	3.90	4.00	2.90	4.50
S P Windrower	acre	3.00	4.30	2.75	2.50	3.50
Pickup baling						
twine	bale	.12	.13	.17	.13	.13
wire	bale	.16	.16	.19	.17	.17
Baling with loader or pull wagon	bale	.14	.15	.20	.15	.13
Haul bales to barn and store	bale	.11	.08	.12	.13	.10
Haying, cut, rake, stack	ton	–	10.00	8.50	10.00	10.00
Haying, cut, rake, bale, store	bale	.34	.40	.39	.37	.40

Table 25-3 Machinery Custom Rates in Six Agricultural Regions of the United States, 1972 (*Continued*)

		Harvesting				
Corn picking—2-row	acre	$6.10	$ 9.50	$ 8.70	$6.00	$12.00
Corn combining	acre	8.10	12.00	10.10	7.50	12.50
Combining, soybeans	acre	6.40	10.00	8.70	5.80	—
seed crops, alfalfa, etc.	acre	6.60	9.50	8.20	5.10	12.50
grain sorghum	acre	6.50	9.00	8.40	4.70	7.50
wheat, oats, barley	acre	6.00	9.00	7.80	4.35	7.50
rice	acre	—	—	10.00	—	—
Crop handling						
dry corn	bu	.08	.11	.08	.06	.10
dry sorghum	cwt	.12	—	.13	.10	.10
shell corn	bu	.04	.09	.07	.05	.10
Cotton harvesting						
machine-pick	cwt	—	—	2.10	2.00	3.00
machine-strip	cwt	—	—	1.75	1.00	—
Potato harvesting	ton	—	4.00	—	—	.60
Tobacco cutting	hour	2.10	3.00	2.20	—	—
Tobacco stripping	hour	1.40	1.70	1.35	—	—

Table 25-3 Machinery Custom Rates in Six Agricultural Regions of the United States, 1972 (*Continued*)

Job	Basis of charge	North Central	North-east	South	Plains	Mountain & Pacific
		Miscellaneous				
Grinding feed						
corn	cwt	$0.16	$0.16	$0.28	$0.16	$0.21
corn and cobs	cwt	0.16	0.16	0.30	0.20	0.21
oats	cwt	0.15	0.18	0.30	0.16	0.22
sorghum	cwt	0.15	0.18	0.31	0.15	0.20
barley	cwt	0.15	0.15	0.29	0.17	0.24
Mixing feed	cwt	0.10	0.13	0.15	0.12	0.20
Bulldozing	hour	19.00	18.00	17.60	17.00	16.60
Machine tiling, no tile	rod	3.00	3.10	4.00	—	—
Cleaning septic tanks	job	32.00	32.00	33.00	29.00	30.00
Sawing wood, chain saw	hour	5.50	5.00	4.00	4.25	5.00
Post-hole digging	hole	0.21	0.35	0.21	0.20	0.20
Mowing weeds	hour	5.80	6.00	5.40	6.50	5.60

Table 25-3 Machinery Custom Rates in Six Agricultural Regions of the United States, 1972 (Continued)

		Rental rates (no labor or fuel)				
Truck						
1 ton or smaller	hour	$4.50	$3.50	$4.00	$3.00	$3.00
1½ ton or larger	hour	5.10	4.20	6.00	4.00	4.00
Tractors, wheel-type						
6-bottom	hour	6.50	7.80	7.00	5.00	7.25
5-bottom	hour	5.50	6.10	6.00	4.60	5.75
3-bottom	hour	3.90	4.00	5.10	3.90	4.75
Tractor and loader	hour	6.75	6.10	9.00	6.00	9.30
Dry fertilizer spreader	acre	0.50	0.70	0.95	0.45	0.60
Anhydrous ammonia applicator	acre	0.50	1.25	0.95	0.50	0.55
Self-propelled combine with:						
platform head	acre	5.50	8.50	6.00	5.50	6.50
corn head	acre	6.40	10.00	8.50	7.50	—
Hay conditioner	acre	1.30	1.50	2.20	—	—
Stalk shredder, pto	acre	1.40	—	2.25	1.30	—
Trailer sprayer	acre	0.70	1.25	0.80	0.80	—
Self-propelled sprayer	acre	1.10	—	—	1.25	—
Grain elevator or auger	bu	0.02	0.02	—	0.02	0.03
Steam cleaner	day	25.00	—	—	12.50	20.00

Doane's Agricultural Report—Copyright 1972 by Doane Agricultural Service, Inc. Material is based on factual information believed to be accurate but not guaranteed.
Data reproduced by special permission of Doane Agricultural Service, Inc.

**Table 25-4   Form for Estimating Machinery Ownership Costs;
Machine: 14-ft (4.27-m) Self-propelled Combine**

| | Annual fixed costs | |
| | | Estimated |
Item	Explanation	costs
Depreciation	Original cost ($7100) minus trade-in value at end of useful life ($710) divided by expected life of machine (10 years)	$639.00
Interest	Original cost ($7100) plus trade-in value ($710) divided by 2 x interest rate (6%)	234.30
Taxes	Expected annual taxes to be paid on machine (estimated at 0.50% of original cost)	35.50
Insurance	Liability, fire, theft, windstorm, etc. (estimated at 0.25% of original cost)	17.75
Shelter	Portion of depreciation, interest, and maintenance of shelter used by this machine (1% of original cost)	71.00
Total annual fixed costs		$997.55

Notes: Fixed costs per unit: Total annual fixed costs ($997.55) divided by units of work done per year (800 acres) = $1.25 per acre.

The method can be used with present costs.

Hire or Buy Formula   When can a farmer economically justify owning a particular machine? The following formula can be used to estimate the approximate "break-even" point between owning a machine and hiring a custom operator.

$$\frac{\text{Total annual fixed costs}}{\text{custom rate minus operating or variable costs per unit}} = \frac{\$997.55}{\$3.00 - \$0.88\ (\$2.12)} = 471 \text{ acres } (190.4 \text{ hectares) or break-even point}$$

Table 25-5   Operating or Variable Costs per Unit

| | Cost |
Item	per hour
Fuel: 4.0 gal/h @ $0.18 gal	$0.72
Oil: 0.5 gal/day @ $1.00 gal ÷ 10-h day	0.05
Lubricant: $7100 x 0.2%/year ÷ 200 h/year	0.07
Repair and maintenance: 40% of original cost ($2840) ÷ 2000 h	1.42
Operators' labor	1.25
Total operating or variable costs per hour	$3.51

Operating costs per acre: total operating costs per hour divided by acres per hour ($3.51 ÷ 4.0) = $0.88 per acre

Total costs per acre = fixed costs per acre ($1.25) plus total operating costs per acre ($0.88) = $2.13 per acre

Note: Local cost figures can be substituted for present conditions.

With the estimated cost figures used in this example, a combine owner must harvest approximately 470 acres (190.4 hectares) per year economically to justify owning the machine. It should be understood that these cost figures do vary from farm to farm and are useful only as long as they are reasonably accurate and applicable to the individual situation. Cost figures should be used from local farm records.

There are other factors which the above comparison does not include. What is the value of the farmer's labor? What are timeliness and convenience worth to the farmer? Some farmers can better afford individual ownership. The final decision as to owning or hiring a machine is likely to vary among farmers.

LEASING EQUIPMENT

Many big-acreage farmers are leasing their farm equipment from the manufacturer's dealers. As equipment costs more and more, leasing becomes more and more attractive as a sound management practice. Most major farm machinery manufacturers offer leasing plans in one form or another. Consult with the equipment dealer for details on leasing.

REFERENCES

Anonymous: How to Pin Down Your Machinery Costs, *Progressive Farmer*, **90**(3):74, 1975.

Barnes, Kenneth K.: Field Capacity of Row Crop Planters, *Implement & Tractor*, **76**(5):50, 1961.

——— and Paul E. Strickler: Management of Machines, *U.S. Dept. Agr. Yearbook*, pp. 346-354, 1960.

Berge, Orvin I., and Glen C. Pulver: Costs of Farm Machinery, *Wis. Ext. Ser. Cir.* 589, 1960.

Capstick, Daniel F.: Costs of Harvesting with Grain Combines, *Ark. Agr. Expt. Sta. Bul.* 630, 1960.

Changes in Farm Production and Efficiency—A Summary Report: *U.S. Dept. Agr. Stat. Bul.* 233, 1961.

Fairbanks, Gustave E., et al.: Cost of Using Farm Machinery, *Trans. of the ASAE*, **14**(1):98, 1971.

Heady, Earl O., Dean E. McKee, and C. B. Haver: Farm Size Adjustments in Iowa and Cost Economics in Crop Production for Farms of Different Sizes, *Iowa Agr. Expt. Sta. Res. Bul.* 428, 1955.

Hoover, Leo M.: Farm Machinery—To Buy or Not to Buy, *Kans. Agr. Expt. Sta. Bul.* 379, 1956.

Hunt, Donnell: Farm Equipment Operating Costs, *Implement & Tractor*, **76**:30-32, Apr. 15, 1961.

———: Choosing Implements of Efficient Widths, *Implement & Tractor*, **76**:23-25, May 1, 1961.

———: Timeliness and Efficient Sizes, *Implement & Tractor*, **76**:20-24, June 1, 1961.

Larson, G. H., G. E. Fairbanks and F. C. Fenton: What It Costs to Use Farm Machinery, *Kans. Agr. Expt. Sta. Bul.* 417, 1960.

Loftsgard, Laurel D., Dale O. Anderson, and Marvin T. Nordbo: Owning and
    Operating Costs for Farm Machinery, *N. Dak. Agr. Expt. Sta. Bul.* 436,
    1961.
Nelson, Ted, and Dale Thompson: Farm Custom Rates Paid in Nebraska in
    1959, *Nebr. Ext. Ser. and U.S. Dept. Agr. E.C.* 60-806, 1959.
Nordbo, Marvin T., LeRoy Schaffner, and Sigurd Stangeland: Decision Making
    Processes in Farm Machinery Selections, *N. Dak. Agr. Expt. Sta. Bul.* 410,
    1957.
Oregon Agricultural Experiment Station: Farm Machinery—Own or Hire, *Oreg.
    Agr. Expt. Sta. Cir. of Inf.* 604, 1961.
Parker, Cecil A., and Willie, L. Ulich: Reduce Farm Machinery Costs, *Tex. Agr.
    Ext. Ser.* MP-281, 1958.
Parsons, Merton S.: Farm Machinery: A Survey of Ownership and Custom Work,
    U.S. Dept. Agr. Res. Ser. Bul. 279, 1961.
——, Frank H. Robinson, and Paul E. Strickler: Farm Machinery: Use,
    Depreciation, and Replacement, *U.S. Dept. Agr. Res. Ser. Stat. Bul.* 269,
    1960.
Renoll, E. S.: A Method for Predicting Field-Machinery Efficiency and Capacity,
    Trans. of the ASAE, **13**(4):448, 1970.
Ryland, D. W., and P. K. Turnquist: Effect of Cab Soundproofing and
    Exhaust-Control at Operator's Site, *Trans. of the ASAE,* **13**(1):149, 1970.

QUESTIONS AND PROBLEMS

1   Give the U.S. Department of Agriculture definition of a farm.
2   What and when were the three great revolutions in American agriculture?
3   Discuss the importance of matching equipment to farm needs.
4   List the various fixed and variable costs involved in the ownership of farm
    equipment.
5   Find the annual repair cost for a two-row self-propelled cotton picker
    costing $15,000.
6   Define the following: (*a*) theoretical field capacity; (*b*) effective field
    capacity; (*c*) field efficiency; (*d*) machine efficiency.
7   Calculate the effective field capacity of a corn planter planting six rows
    spaced 42 in (106.7 cm) apart and operated at a speed of 4.5 mi/h
    (7.2 km/h).
8   Define farm custom work.
9   List five advantages and five disadvantages of custom hiring of farm
    equipment.
10 Give the formula for determining the break-even point between owning a
    machine and hiring custom work.

Appendix

CONTENTS

453

Table 1   Equivalents

1 acre =
43,560 square feet
160 square rods
4,046.87 square meters
0.40469 hectare

1 atmosphere, standard
29.9212 inches of mercury
33.9006 feet of water
14.6969 pounds per squre inch
2,116.35 pounds per square foot
1.03329 kilograms per square centimeter

1 British thermal unit
778.104 foot-pounds
0.000393 horsepower-hour
0.0003984 horsepower-hour, metric
1,054.9 joules
1,052 watt-seconds
0.000293 kilowatt-hour
107.577 kilogram-meters
0.252 calorie

1 British thermal unit per second
778.104 foot-pounds per second
1.41474 horsepower
1.43436 horsepower, metric
1,054.9 watts
1.0549 kilowatts
107.577 kilogram-meters per second
0.252 calorie per second

1 bushel
2,150.42 cubic inches
1.24446 cubic feet
32 dry quarts
8 dry gallons
35.2393 cubic decimeters

1 calorie
3,087.77 foot-pounds
3.96832 British thermal units
0.001559 horsepower-hour
0.001581 horsepower-hour, metric
4,186.17 joules
426.9 kilogram-meters
0.001163 kilowatt-hour

1 calorie per second
3,087.77 foot-pounds per second
5.61412 horsepower
5.692 horsepower, metric
3.96832 British thermal units per second
4,186.17 watts

4.18617 kilowatts
426.9 kilogram-meters per second

1 centimeter
0.3937 inch
0.0328 foot
10 millimeters

1 chain
792 inches
66 feet
0.0125 mile
20.1168 meters
0.02012 kilometer

1 decimeter, cubic
61.0234 cubic inches
0.03531 cubic foot
1.0567 liquid quarts
0.02838 bushel

1 degree centigrade
1.8 degrees Fahrenheit
0.8 degree Reaumur

1 degree Fahrenheit
0.5555 degree centigrade
0.4444 degree Reaumur

1 degree Reaumur
1.25 degrees centigrade
2.25 degrees Fahrenheit

1 foot
12 inches
0.3333 yard
0.06061 rod
0.01515 chain
0.0001894 mile
304.8 millimeters
30.48 centimeters
0.3048 meter

1 foot, square
144 square inches
929.03 square centimeters
0.0929 square meter

1 foot, cubic
1,728 cubic inches
29.9221 liquid quarts
7.48055 liquid gallons
0.80356 bushel

Table 1   Equivalents (*Continued*)

1 foot per second
0.68182 mile per hour
0.3048 meter per second
1.09728 kilometers per hour

1 foot per second per second
0.68182 mile per hour per second
0.3048 meter per second per second
1.09728 kilometers per hour per second

1 foot-pound
0.0003239 British thermal unit
0.0000005051 horsepower-hour
0.0000005121 horsepower-hour, metric
1.35573 joules
0.13826 kilogram-meter
0.0003239 calorie
0.0000003766 kilowatt-hour

1 foot-pound per second
0.001285 British thermal unit per
   second
0.001818 horsepower
0.001843 horsepower, metric
0.13826 kilogram-meter per second
0.0003237 calorie per second
1.35573 watts
0.001356 kilowatt

1 gallon, liquid
231 cubic inches
0.13368 cubic foot
4 liquid quarts
0.8327 British Imperial gallon
3.78543 cubic decimeters

1 gallon, dry
268.803 cubic inches
0.15556 cubic foot
4 dry quarts
0.96817 British Imperial gallon
0.125 bushel
4.40492 cubic decimeters

1 grain
0.002083 ounce, troy
0.002285 ounce, avoirdupois
0.0001736 pound, troy
0.0001428 pound, avoirdupois
0.0000648 kilogram

1 gram per centimeter
39.1983 grains per inch
0.0056 pound per inch
0.0672 pound per foot

0.10 ton, metric, per kilometer
0.10 kilogram per meter

1 gram per cubic centimeter
0.03613 pound per cubic inch
62.4283 pounds per cubic foot
1000 kilograms per cubic meter
100 kilograms per hectoliter

1 gravity
32.1717 feet per second per second

1 hectare
107,639 square feet
2.47104 acres
0.003861 square mile
10,000 square meters
0.01 square kilometer

1 horsepower
550 foot-pounds per second
33,000 foot-pounds per minute
0.70685 British thermal unit per second
0.17812 calorie per second
76.0404 kilogram-meters per second
1.01387 horsepower, metric
745.65 watts
0.74565 kilowatt

1 horsepower-hour
1,980,000 foot-pounds
2,544.65 British thermal units
641.24 calories
1.01387 horsepower-hours, metric
2,684,340 joules
273,745 kilogram-meters
0.74565 kilowatt-hour

1 horsepower, metric
542.475 foot-pounds per second
0.69718 British thermal unit per second
0.98632 horsepower
0.17569 calorie per second
75 kilogram-meters per second
735.448 watts
0.73545 kilowatt

1 horsepower-hour, metric
1,952,910 foot-pounds
2,509.83 horsepower-hour
2,647,610 joules
270,000 kilogram-meters
632.467 calories
0.73545 kilowatt-hour

Table 1   Equivalents (*Continued*)

1 inch
0.08333 foot
25.4 millimeters
0.0254 meter

1 inch of mercury
0.49119 pound per square inch
13.58 inches of water

1 inch of water
0.0361 pound per square inch
0.0735 inch of mercury

1 inch, square
0.006944 square foot
6.4516 square centimeters
0.0006453 square meter

1 inch, cubic
0.0005787 cubic foot
0.01732 liquid quart
0.0004329 liquid gallon
0.000465 bushel
16.39 cubic centimeters

1 joule
0.73761 foot-pound
0.000948 British thermal unit
0.0000003725 horsepower-hour
0.0000003777 horsepower-hour, metric
0.0002389 calorie
0.10198 kilogram-meter
0.0000002778 kilowatt-hour

1 kilogram
15,432.4 grains
32.1507 ounces, troy
35.274 ounces, avoirdupois
2.67023 pounds, troy
2.20462 pounds, avoirdupois
0.001102 ton
0.001 ton, metric
1,000 milligrams
100 centigrams
10 decigrams

1 kilogram per meter
0.056 pound per inch
0.67197 pound per foot
1.774 tons per mile
1 ton per kilometer, metric
10 grams per centimeter

1 kilogram per square centimeter
14.2234 pounds per square inch

2,048.17 pounds per square foot
1.02408 tons per square foot
0.96778 atmosphere, standard

1 kilogram per hectoliter
0.0003613 pound per cubic inch
0.62428 pound per cubic foot
0.08345 pound per liquid gallon
0.01 gram per cubic centimeter
10 kilograms per cubic meter

1 kilogram per cubic meter
0.00003613 pound per cubic inch
0.06243 pound per cubic foot
0.001 gram per cubic centimeter
0.1 kilogram per hectoliter

1 kilogram-meter
7.233 foot-pounds
0.009296 British thermal unit
0.000003653 horsepower-hour
0.000003704 horsepower-hour, metric
9.80597 joules
0.002342 calorie
0.000002724 kilowatt-hour

1 kilogram-meter per second
7.233 foot-pounds per second
0.01315 horsepower
0.01333 horsepower, metric
0.009296 British thermal unit per
  second
0.002342 calorie per second
9.80597 watts
0.009806 kilowatt

1 kilometer
39,370 inches
3,280.83 feet
1,000 meters

1 kilometer per hour
0.91134 foot per second
0.62137 mile per hour
0.27778 meter per second

1 kilometer per hour per second
0.91134 foot per second per second
0.623137 mile per hour per second
0.27778 meter per second per second

1 kilowatt
737.612 foot-pounds per second
1.34111 horsepower
1.35972 horsepower, metric

Table 1   Equivalents (*Continued*)

0.94796 British thermal unit per second
0.23888 calorie
1,000 watts
101.979 kilogram-meters per second

1 kilowatt-hour
2,655,403 foot-pounds
1.34111 horsepower-hours
1.35972 horsepower-hours, metric
3412.66 British thermal units
859.975 calories
3,600,000 joules
1,000 watt-hours
367,123 kilogram-meters

1 liter
61.023 cubic inches
1.0567 quarts
1,000 cubic centimeters

1 meter
39.37 inches
3.28083 feet
0.001 kilometer
1,000 millimeters
100 centimeters
10 decimeters

1 meter per second
3.28083 feet per second
2.23693 miles per hour
3.6 kilometers

1 meter per second per second
3.28083 feet per second per second
2.23693 miles per hour per second
3.6 kilometers per hour per second

1 meter, square
1,550 square inches
10.7639 square feet
10,000 square centimeters
100 square decimeters

1 meter, cubic
0.006102 cubic inch
0.000035 cubic foot
0.00156 liquid quart
0.000028 bushel
1,000,000 cubic centimeters
1,000 cubic decimeters

1 mile
63,360 inches
5,280 feet

1,760 yards
320 rods
80 chains
1,609.35 meters
1,60935 kilometers

1 mile, square
27,878,400 square feet
640 acres
2,589,999 square meters
2.59 square kilometers

1 mile per hour
1.46667 feet per second
0.44704 meter per second
1.60935 kilometers per hour

1 mile per hour per second
1.46667 feet per second
0.44704 meter per second per second
1.60935 kilometers per hour per second

1 ounce, troy
480 grains
1.09714 ounces, avoirdupois
0.08333 pound, troy
0.06857 pound, avoirdupois
0.0311 kilogram

1 ounce, avoirdupois
437.5 grains
0.91146 ounce, troy
0.07595 pound, troy
0.0625 pound, avoirdupois
0.02835 kilogram

1 pound, troy
12 ounces, troy
13.1657 ounces, avoirdupois
0.82286 pound, avoirdupois
0.0004114 ton
0.0003732 ton, metric
0.37324 kilogram

1 pound, avoirdupois
14.5833 ounces, troy
16 ounces, avoirdupois
1.21528 pounds, troy
0.0005 ton
0.0004536 ton, metric
0.45359 kilogram

1 pound per inch
12 pounds per foot
31.68 tons per mile

Table 1   Equivalents (*Continued*)

17.8579 tons, metric, per kilometer
178.579 grams per centimeter
17.8579 kilograms per meter

1 pound per square inch
144 pounds per square foot
0.072 ton per square foot
0.07031 kilogram per square centimeter
0.06804 atmosphere, standard

1 pound per cubic inch
1,728 pounds per cubic foot
27.6797 grams per cubic centimeter
2,767.97 kilograms per hectoliter
27,679.7 kilograms per cubic meter

1 pound per foot
0.08333 pound per inch
2.64 tons per mile
1.48816 tons, metric, per kilometer
14.8816 grams per centimeter
1.48816 kilograms per meter

1 pound per square foot
0.006944 pound per square inch
0.0004882 kilogram per square
   centimeter
0.0004725 atmosphere, standard

1 pound per cubic foot
0.0005787 pound per cubic inch
1.24446 pounds per bushel
0.01602 gram per cubic centimeter
1.60184 kilograms per hectoliter
16.0184 kilograms per cubic meter

1 pound per yard
0.02778 pound per inch
0.333 pound per foot
0.88 ton per mile
0.49605 ton, metric, per kilometer
4.96054 grams per centimeter
0.49605 kilogram per meter

1 pound per cubic yard
0.00002143 pound per cubic inch
0.03704 pound per cubic foot
0.04609 pound per bushel
0.0005933 gram per cubic centimeter
0.05933 kilogram per hectoliter
0.59327 kilogram per cubic meter

1 quart, liquid
57.75 cubic inches
0.03342 cubic foot
0.94636 cubic decimeter

1 quart, dry
67.2006 cubic inches
0.03889 cubic foot
0.25 dry gallon
0.03125 bushel
1.10123 cubic decimeters

1 radian per second
0.159155 revolution per second

1 rod
198 inches
16.5 feet
5.5 yards
0.25 chain
0.003125 mile
5.02921 meters
0.005029 kilometer

1 rod, square
39,204 square inches
272.25 square feet
0.00625 acre
0.000009766 square mile
25.293 square meters

1 ton
2,430.56 pounds, troy
2,000 pounds, avoirdupois
0.90719 ton, metric
907.185 kilograms

1 ton per mile
0.03157 pound per inch
0.37879 pound per foot
0.5637 ton, metric, per kilometer
5.63698 grams per centimeter
0.5637 kilogram per meter

1 ton, metric
2,679.23 pounds, troy
2,204.62 pounds, avoirdupois
1.10231 tons
1000 kilograms

1 ton, metric, per kilometer
0.056 pound per inch
0.67197 pound per foot
1.774 tons per mile
10 grams per centimeter
1 kilogram per meter

1 watt
0.73761 foot-pound per second
0.001341 horsepower
0.00136 horsepower, metric

Table 1 **Equivalents** (*Continued*)

0.000948 British thermal unit per second	0.9144 meter
0.0002389 calorie per second	1.0000029 yards, British
1 joule per second	
0.001 kilowatt	**1 yard, square**
0.10198 kilogram-meter per second	1,296 square inches
	9 square feet
	0.83613 square meter
	1 yard, cubic
1 yard	46,656 cubic inches
36 inches	27 cubic feet
3 feet	764.559 cubic decimeters

Table 2   Conversion Factors

Multiply	by*	to obtain
acres	0.404687	hectares
acres	4.04687×10^{-3}	square kilometers
acres	1076.39	square feet
board feet	144 sq. in. x 1 in.	cubic inches
board feet	0.0833	cubic feet
centimeters	3.28083×10^{-2}	feet
centimeters	0.3937	inches
cubic centimeters	3.53145×10^{-5}	cubic feet
cubic centimeters	6.102×10^{-2}	cubic inches
cubic feet	2.8317×10^{-4}	cubic centimeters
cubic feet	2.8317×10^{-2}	cubic meters
cubic feet	6.22905	gallons, British Imperial
cubic feet	28.3170	liters
cubic feet	2.38095×10^{-2}	tons, British Shipping
cubic feet	0.025	tons, U.S. Shipping
cubic inches	16.38716	cubic centimeters
cubic meters	35.3145	cubic feet
cubic meters	1.30794	cubic yards
cubic yards	0.764559	cubic meters
degrees, angular	0.0174533	radians
degrees, Fahrenheit (less 32° F.)	0.5556	degrees, centigrade
degrees, centigrade	1.8	degrees, Fahrenheit (less 32° F.)
foot-pounds	0.13826	kilogram-meters
feet	30.4801	centimeters
feet	0.304801	meters
feet	304.801	millimeters
feet	1.64468×10^{-4}	miles, nautical
gallons, British Imperial	0.160538	cubic feet
gallons, British Imperial	1.20091	gallons, U.S.
gallons, British Imperial	4.54596	liters
gallons, U.S.	0.832702	gallons, British Imperial
gallons, U.S.	0.13368	cubic feet
gallons, U.S.	231.0	cubic inches
gallons, U.S.	3.78543	liters
grams, metric	2.20462×10^{-3}	pounds, avoirdupois
hectares	2.47104	acres
hectares	1.076387×10^{5}	square feet
hectares	3.86101×10^{-3}	square miles
horsepower, metric	0.98632	horsepower, U.S.
horsepower, U.S.	1.01387	horsepower, metric
inches	2.54001	centimeters
inches	2.54001×10^{-2}	meters
inches	25.4001	millimeters
kilograms	2.20462	pounds
kilograms	9.84206×10^{-4}	long tons
kilograms	1.10231×10^{-3}	short tons
kilogram-meters	7.233	foot-pounds
kilograms per meter	0.671972	pounds per foot
kilograms per square centimeter	14.2234	pounds per square inch
kilograms per square meter	0.204817	pounds per square foot
kilograms per square meter	9.14362×10^{-5}	long tons per square foot

Table 2   Conversion Factors (*Continued*)

Multiply	by*	to obtain
kilograms per square millimeter . .	1422.34	pounds per square inch
kilograms per square millimeter . .	0.634973	long tons per square inch
kilograms per cubic meter	6.24283×10^{-2}	pounds per cubic foot
kilometers	0.62137	miles, statute
kilometers	0.53959	miles, nautical
liters	0.219976	gallons, British Imperial
liters	0.26417	gallons, U.S.
liters	3.53145×10^{-2}	cubic feet
meters	3.28083	feet
meters	39.37	inches
meters	1.09361	yards
miles, statute	1.60935	kilometers
miles, statute	0.8684	miles, nautical
miles, nautical	6080.204	feet
miles, nautical	1.85325	kilometers
miles, nautical	1.1516	miles, statute
millimeters	3.28083×10^{-3}	feet
millimeters	3.937×10^{-2}	inches
pounds, avoirdupois	453.592	grams, metric
pounds, avoirdupois	0.453592	kilograms
pounds, avoirdupois	4.464×10^{-4}	tons, long
pounds, avoirdupois	4.53592×10^{-4}	tons, metric
pounds per foot	1.48816	kilograms per meter
pounds per square foot	4.88241	kilograms per square meter
pounds per square inch	7.031×10^{-2}	kilograms per square centimeter
pounds per square inch	7.031×10^{-4}	kilograms per square millimeter
pounds per cubic foot	16.0184	kilograms per cubic meter
radians	57.29578	degrees, angular
square centimeters	0.1550	square inches
square feet	9.29034×10^{-4}	acres
square feet	9.29034×10^{-6}	hectares
square feet	0.0929034	square meters
square inches	6.45163	square centimeters
square inches	645.163	square millimeters
square kilometers	247.104	acres
square kilometers	0.3861	square miles
square meters	10.7639	square feet
square meters	1.19599	square yards
square miles	259.0	hectares
square miles	2.590	square kilometers
square millimeters	1.550×10^{-3}	square inches
square yards	0.83613	square meters
tons, long	1016.05	kilograms
tons, long	2240.0	pounds
tons, long	1.01605	tons, metric
tons, long	1.120	tons, short
tons, long, per square foot	1.09366×10^4	kilograms per square meter
tons, long, per square inch	1.57494	kilograms per square millimeter
tons, metric	2204.62	pounds
tons, metric	0.98421	tons, long
tons, metric	1.10231	tons, short

Table 2   Conversion Factors (*Continued*)

Multiply	by*	to obtain
tons, short	907.185	kilograms
tons, short	0.892857	tons, long
tons, short	0.907185	tons, metric
tons, British Shipping	42.00	cubic feet
tons, British Shipping	1.050	tons, U.S. Shipping
tons, U.S. Shipping	40.0	cubic feet
tons, U.S. Shipping	0.952381	tons, British Shipping
yards	0.914402 meters	meters

*The expressions $\times 10^{-2}$, $\times 10^{-3}$, $\times 10^{-4}$, $\times 10^{-5}$, and $\times 10^{-6}$, following certain multipliers, indicate that the decimal point in the product—of left-column value times multiplier—is to be moved respectively 2, 3, 4, 5, or 6 places to the left.

Table 3   Conversion of Millimeters to Inches

mm	in	mm	in	mm	in	mm	in
1 = 0.03937		26 = 1.02362		51 = 2.00787		76 = 2.99212	
2 = 0.07874		27 = 1.06299		52 = 2.04724		77 = 3.03149	
3 = 0.11811		28 = 1.10236		53 = 2.08661		78 = 3.07086	
4 = 0.15748		29 = 1.14173		54 = 2.12598		79 = 3.11023	
5 = 0.19685		30 = 1.18110		55 = 2.16535		80 = 3.14960	
6 = 0.23622		31 = 1.22047		56 = 2.20472		81 = 3.18897	
7 = 0.27559		32 = 1.25984		57 = 2.24409		82 = 3.22834	
8 = 0.31496		33 = 1.29921		58 = 2.28346		83 = 3.26771	
9 = 0.35433		34 = 1.33858		59 = 2.32283		84 = 3.30708	
10 = 0.39370		35 = 1.37795		60 = 2.36220		85 = 3.34645	
11 = 0.43307		36 = 1.41732		61 = 2.40157		86 = 3.38582	
12 = 0.47244		37 = 1.45669		62 = 2.44094		87 = 3.42519	
13 = 0.51181		38 = 1.49606		63 = 2.48031		88 = 3.46456	
14 = 0.55118		39 = 1.53543		64 = 2.51968		89 = 3.50393	
15 = 0.59055		40 = 1.57480		65 = 2.55905		90 = 3.54330	
16 = 0.62992		41 = 1.61417		66 = 2.59842		91 = 3.58267	
17 = 0.66929		42 = 1.65354		67 = 2.63779		92 = 3.62204	
18 = 0.70866		43 = 1.69291		68 = 2.67716		93 = 3.66141	
19 = 0.74803		44 = 1.73228		69 = 2.71653		94 = 3.70078	
20 = 0.78740		45 = 1.77165		70 = 2.75590		95 = 3.74015	
21 = 0.82677		46 = 1.81102		71 = 2.79527		96 = 3.77952	
22 = 0.86614		47 = 1.85039		72 = 2.83464		97 = 3.81889	
23 = 0.90551		48 = 1.88976		73 = 2.87401		98 = 3.85826	
24 = 0.94488		49 = 1.92913		74 = 2.91338		99 = 3.89763	
25 = 0.98425		50 = 1.96850		75 = 2.95275		100 = 3.93700	

Table 4   Fractions of an Inch and Decimal Equivalents

1/8	1/16	1/32	1/64	Decimal
			1/64	0.015625
		1/32	....	0.03125
			3/64	0.046875
	1/16	....	....	0.0625
			5/64	0.078125
		3/32	....	0.09375
			7/64	0.109375
1/8	....	....	....	0.125
			9/64	0.140625
		5/32	....	0.15625
			11/64	0.171875
	3/16	....	....	0.1875
			13/64	0.203125
		7/32	....	0.21875
			15/64	0.234375
1/4	....	....	....	0.250
			17/64	0.265625
		9/32	....	0.28125
			19/64	0.296875
	5/16	....	....	0.3125
			21/64	0.328125
		11/32	....	0.34375
			23/64	0.359375
3/8	....	....	....	0.375
			25/64	0.390625
		13/32	....	0.40625
	7/16	....	....	0.4375
		15/32	....	0.46875
1/2	....	....	....	0.500
		17/32	....	0.53125
	9/16	....	....	0.5625
		19/32	....	0.59375
5/8	....	....	....	0.625
		21/32	....	0.65625
	11/16	....	....	0.6875
		23/32	....	0.71875
3/4	....	....	....	0.750
		25/32	....	0.78125
	13/16	....	....	0.8125
		27/32	....	0.84375
7/8	....	....	...	0.875
		29/32	....	0.90625
	15/16	....	....	0.9375
		31/32	....	0.96875
1	....	....	....	1.000

Table 5 Converting Inches and Fractions of an Inch into Decimals of a Foot

		1 in	2 in	3 in	4 in	5 in	6 in	7 in	8 in	9 in	10 in	11 in	
		.0833	.1667	.2500	.3333	.4167	.5000	.5833	.6667	.7500	.8333	.9167	
1/16	.0052	.0885	.1719	.2552	.3385	.4219	.5052	.5885	.6719	.7552	.8385	.9219	1/16
1/8	.0104	.0938	.1771	.2604	.3438	.4271	.5104	.5938	.6771	.7604	.8438	.9271	1/8
3/16	.0156	.0990	.1823	.2656	.3490	.4323	.5156	.5990	.6823	.7656	.8490	.9323	3/16
1/4	.0208	.1042	.1875	.2708	.3542	.4375	.5208	.6042	.6875	.7708	.8542	.9375	1/4
5/16	.0260	.1094	.1927	.2760	.3594	.4427	.5260	.6094	.6927	.7760	.8594	.9427	5/16
3/8	.0313	.1146	.1979	.2813	.3646	.4479	.5313	.6146	.6979	.7813	.8646	.9479	3/8
7/16	.0365	.1198	.2031	.2865	.3698	.4531	.5365	.6198	.7031	.7865	.8698	.9531	7/16
1/2	.0417	.1250	.2083	.2917	.3750	.4583	.5417	.6250	.7083	.7917	.8750	.9583	1/2
9/16	.0469	.1302	.2135	.2969	.3802	.4635	.5469	.6302	.7135	.7969	.8802	.9635	9/16
5/8	.0521	.1354	.2188	.3021	.3854	.4688	.5521	.6354	.7188	.8021	.8854	.9688	5/8
11/16	.0573	.1406	.2240	.3073	.3906	.4740	.5573	.6406	.7240	.8073	.8906	.9740	11/16
3/4	.0625	.1458	.2294	.3125	.3958	.4792	.5625	.6458	.7292	.8125	.8958	.9792	3/4
13/16	.0677	.1510	.2344	.3177	.4010	.4844	.5677	.6510	.7344	.8177	.9010	.9844	13/16
7/8	.0729	.1563	.2396	.3229	.4063	.4896	.5729	.6563	.7396	.8229	.9063	.9896	7/8
15/16	.0781	.1615	.2448	.3281	.4115	.4948	.5781	.6615	.7448	.8281	.9115	.9948	15/16

Table 6 Draft and Power Requirements of Crop Machines

Machine	Normal Range
Tillage:	
Plow	5-12 lb/in^2 of furrow section
Lister	400-750 lb per row
One-way disk	150-350 lb per ft width
Single-disk harrow	40-130 lb per ft width
Tandem-disk harrow	80-160 lb per ft width
Tandem-disk harrow, 22-in diameter,	
9-in spacing	170-225 lb per ft width, or 90% of weight
Spike-tooth harrow	30-60 lb per ft width
Spring-tooth harrow	75-150 lb per ft width
Duck foot field cultivator	90-160 lb per ft width
Roller	30-60 lb per ft width
Subsoiler	80-160 lb per in of depth*
Planting:	
Grain drill	30-80 lb per ft width
Corn planter	80-120 lb per row
Cultivating:	
Rotary hoe	30-60 lb per ft width
Corn cultivator	22-95 lb per shovel
Spring-tooth weeder	25-35 lb per ft width
Rod weeder	80-110 lb per ft width
Harvesting:	
Mower	60-100 lb per ft width
Grain binder	65-150 lb per ft width
Thresher	0.8-1.2 hp per in cylinder width
Combine, 5 and 6 ft	2-4$\frac{1}{2}$ (pto) hp per ft of cutter bar
Combine, 8-12 ft	Engine with 2-3 net hp per ft of cutter bar
Corn picker, 2-row	2-5 (pto) hp per row
Stationary silage cutter	0.761-1.60 hp-h per ton
Husker-shredder	0.25-0.35 hp-h per bu
Pickup baler	Engine with 15-25 net hp
Forage harvester	1-3 (pto) hp-h per ton of grass silage at $\frac{1}{2}$-in cut

* In certain Southern and Far Western soils, draft figures ranging up to a maximum of approximately double the above have been recorded.

Source: Agr. Engin. Yearbook, 1962.

Table 7   Historical Dates in the Development of Farm Equipment
in the United States

Soil preparation			
1788	Thomas Jefferson applied mathematics to the plow moldboard	1895	Heavy-duty, deep-tillage chisel to penetrate subsoil of irrigated land
1797	Cast-iron moldboard plow patented	1898	High-lift, frame-type gang sulky plows
1814	Iron plow with replaceable parts patented	1900	Heavy-duty, deep-tillage implements in commercial production
1837	Steel plow industry began		
1847	Revolving plow moldboard patented (disk plow)	1910	Rod weeder for fallow land farming
1860s	Chisel cultivators came into use	1915	Unit frame tractor plows introduced
1864	Successful sulky (riding) plow	1923	Combination tool carrier for heavy-duty implements
1867	Walking gang plow supported on wheels	1924	Offset disk harrow
1868	Soft-center steel for plows	1927	One-way or wheatland disk plows first sold in large numbers
1869	Chilled plow process patented		
1869	Spring-tooth harrow patented	1935-1940	Lift-type mounted plow introduced
1877	Disk harrow with concave blades patented	1941	Hydraulic remote control of drawn implements
1880	Lister produced commercially	1949	Wheel disk harrow introduced
1884	First three-wheel riding plow	1953	Safety trip plow beams in use

Planting and cultivating			
1799	Seeding machine patented	1890s	Accumulative or hill-drop corn planter
1839	Two-row corn planter patented		
1840-1841	Grain drill feeding mechanism patents	1912	Rotary hoe produced commercially
1846	First wheel cultivator (one-horse)	1917	Two-row potato planters
1851	Force-feed grain drill	1921	Tool bar idea for mounting tillage tools on tractors
1857	Early-type broadcast seeder		
1860	Two-row corn planter with hand control drop for cross-checking	1924	Mounted-type tractor implements
		1929	Attachments for placing fertilizer in bands
1863	Commercially successful riding cultivator	1939	Single-seed beet planters made available commercially
1870s	Straddle-row, two-horse cultivators (one row)	1942	First commercial plantings of segmented beet seed
1875	Automatic check rowers for planting corn	1950	Sodland seeder introduced
		1957	Planter attachments to apply liquid fertilizers
1880	Potato planters in use		
1890	Transplanting machine for tobacco and vegetable plants	1973	Air-metering planter

Table 7   Historical Dates in the Development of Farm Equipment in the United States (*Continued*)

Harvesting

1831	Accepted birth date of the type of reaper that attained success	1892	Corn binder patented (self-binding)
1836	Combine built in Michigan	1904	First steel thresher
1837	Thresher patented which introduced principles of later machines	1909	Corn picker built commercially
1843	Vibrating principle in threshers patented	1926	Cotton strippers built commercially for the High Plains
1846-1847	Reaper put into quantity production	1928	Two-row PTO-operated corn picker and one-row mounted picker
1848-1849	Push-type headers invented	1929	Two-row mounted corn picker
1854	Self-rake reaper produced in quantity	1935	One-man power-takeoff combine
1858	First successful harvester on which men could ride to bind	1938	Self-propelled combine in America
1863	Wire binder for harvester produced in large numbers	1940-1945	Potato harvester
1871	Horse-drawn potato digger	1941	Successful cotton picker built
1872	Wire binder adapted to Marsh-type harvester	1943	First appreciable use of commercial sugar beet harvesters
1878	Successful twine binder for Marsh-type harvester	1943	Tractor-mounted cotton strippers
1880	Factory production of combines started on Pacific Coast	1946	Self-propelled corn picker
1880	First important corn picker patent	1949	Portable batch dryer available
1885	Corn husker-shredder appeared	1950	Self-propelled windrower
1886	Potato diggers built commercially	1954	Corn head attachment for combine available commercially
		1958	Shelling attachment for corn pickers

Hay and forage

1820	Horse-drawn revolving hay rake	1867	Hay carrier patented
1822	Mower or grass-cutting machine patented	1872	Continuous hay press invented
1850	Hand dump rake with iron or steel teeth	1874	Successful mechanical hay loader
1853	Early-type hay press	1876	Ensilage cutters appeared
1853-1860	Two-wheeled mowers with flexible or hinged bars	1882	Portable hay stacker
1855-1870	Seven patents on iron or steel tedders with seat for driver	1885	Sweep rakes in use
		1890	Hay loader reaches commercial importance
1864	Harpoon-type hay fork patented	1893	Early-type side-delivery rake
		1914	Reel-type side-delivery rake
		1932	Field pickup baler introduced

Table 7   Historical Dates in the Development of Farm Equipment in the United States (*Continued*)

Hay and forage			
1936	Forage harvester built commercially	1945	Hay crusher built commercially
1940	Automatic, self-tying pickup baler	1946	Silo unloader introduced
		1952	Flail-type forage harvester
1940	Power-takeoff-driven side delivery rake	1953	Dynamically balanced mower introduced
1945	Mow hay finishers built commercially	1958	Self-propelled baler

Specialized equipment			
1840	Wooden hand pump—first use of suction to lift water	1900	Riding, single-row stalk cutter
1842	First grain elevator constructed	1902	First commercially successful cylinder corn shellers
1844	First American incubator invented	1902	Steel stanchions for dairy cows
1850	Feed cutters—self-feeding, knife cutting against steel	1902	Portable grain elevators
1854	Self-regulating windmill	1904-1905	Steel stalls and alignment device to make stanchions adjustable
1855	Portable grinding mills with metallic plates	1910	Electric water systems produced in quantity
1850s	Picker-wheel-type sheller, hand and power	1910	Colony-type heated brooder for baby chicks
1850s	Specialized garden implements	1911	Pressure regulator and air chamber for power sprayers
1860	Self-feeding device for corn shellers	1912	Automatic, individually operated water bowls for cows
1865	First manure spreader, wagon type	1914	Land leveler for irrigation farming
1877	First commercially known spreader—endless apron	1914-1917	Power-operated milking machines introduced
1878	First milking machine using vacuum principle	1916	Garden tractors introduced
1879	Centrifugal cream separator introduced	1920	Hammer mill introduced
1883	Steel windmill and tower	1924	Electric ventilating system for dairy barns
1885	First American factory-made incubator	1925	Two-row, tractor-drawn rolling stalk cutter
1887	First traction sprayer	1925	Air-blast-type sprayer introduced
1890	Gasoline engines for stationary power	1926	Jet-type pumps introduced
1895	First milking machine with intermittent suction	1930	Portable sprinkler irrigation
1897	Litter carrier and feed carrier introduced	1937	Automatic barn cleaner built commercially
1900	First power sprayer operated by gasoline engine	1938	Tractor loader built commercially

Table 7   Historical Dates in the Development of Farm Equipment in the United States (*Continued*)

Specialized equipment

1941	Precision planting of vegetable seed	1948	Submersible pump (water well)
1944	Dusters with attachments to inject liquid	1948	Self-propelled sprayer
1944	Low-pressure, low-volume sprayers	1948-1950	Pipeline milking important commercially
1947	Equipment to apply fertilizer in vapor form (anhydrous ammonia)	1950s	Automatic feeding of beef and poultry

Tractor power

1850	Portable steam engines for farm use	1931	First diesel-powered track-type tractor
1876	Steam traction engine	1932	Low-pressure rubber tires introduced for farm tractors
1892-1901	Experimental models, gas traction engines	1935	Factory-built, high-compression tractors to burn leaded gasoline
1903	Gas traction engines produced commercially	1935	Hydraulic lift equipment on the tractor
1904	Track-type traction engine	1939	Weight transfer for increased traction
1906-1907	The word "tractor" came into use	1941	First factory-built LP gas tractors
1908	First gasoline track-type tractor	1944	Power-takeoff and drawbar dimensions standardized
1913	Frameless or unit design in farm tractors	1946	Transmission clutch for live power takeoff
1915	"Motor cultivator" type tractor	1949	Hydraulic remote-control cylinder dimensions standardized
1919	Farm tractor equipped with power takeoff	1954-1958	"On-the-go" and automatic shifting of tractor transmissions
1920	Starter and lights for tractors		
1924	Cultivating or tricycle-type tractor		
1924	Mounted-type tractor implements introduced		
1930	Farm tractor equipped with power lift		

Source: Land of Plenty, Farm Equipment Institute, 1959.

Table 8   Miles per Hour for Given Tire Size

r/min	12 x 38	11 x 38	10 x 38	9 x 38	9 x 36	9 x 32	11 x 28	10 x 28	10 x 24	9 x 24
12	2.1	2.0	1.9	1.9	1.8	1.6	1.6	1.6	1.4	1.4
13	2.2	2.1	2.1	2.0	2.0	1.8	1.8	1.7	1.5	1.5
14	2.4	2.3	2.2	2.2	2.1	1.9	1.9	1.8	1.7	1.6
15	2.6	2.5	2.4	2.3	2.2	2.1	2.0	2.0	1.8	1.7
16	2.7	2.6	2.6	2.5	2.4	2.2	2.2	2.1	1.9	1.8
17	2.9	2.8	2.7	2.6	2.5	2.3	2.3	2.2	2.0	2.0
18	3.1	3.0	2.9	2.8	2.7	2.5	2.4	2.4	2.1	2.1
19	3.2	3.1	3.0	2.9	2.8	2.6	2.6	2.5	2.2	2.2
20	3.4	3.3	3.2	3.1	3.0	2.7	2.7	2.6	2.4	2.3
21	3.6	3.5	3.4	3.3	3.1	2.9	2.8	2.8	2.5	2.4
22	3.8	3.6	3.5	3.4	3.3	3.0	3.0	2.9	2.6	2.5
23	3.9	3.8	3.7	3.5	3.4	3.1	3.1	3.0	2.7	2.6
24	4.1	4.0	3.8	3.6	3.6	3.3	3.3	3.1	2.8	2.8
25	4.3	4.1	4.0	3.8	3.7	3.4	3.4	3.3	3.0	2.9
26	4.4	4.3	4.2	3.9	3.9	3.6	3.5	3.4	3.1	3.0
27	4.6	4.5	4.3	4.1	4.0	3.7	3.7	3.5	3.2	3.1
28	4.8	4.6	4.5	4.2	4.2	3.8	3.8	3.7	3.3	3.2
29	4.9	4.8	4.6	4.5	4.3	4.0	3.9	3.8	3.4	3.3
30	5.1	5.0	4.8	4.6	4.5	4.1	4.1	3.9	3.5	3.4
31	5.3	5.1	5.0	4.8	4.6	4.2	4.2	4.1	3.7	3.6
32	5.5	5.3	5.1	5.0	4.8	4.4	4.3	4.2	3.8	3.7
33	5.6	5.5	5.3	5.1	4.9	4.5	4.5	4.3	3.9	3.8
34	5.8	5.6	5.4	5.3	5.1	4.6	4.6	4.4	4.0	3.9
35	6.0	5.8	5.6	5.4	5.2	4.8	4.7	4.6	4.1	4.0
36	6.1	5.9	5.8	5.6	5.4	4.9	4.9	4.7	4.3	4.1
37	6.3	6.1	5.9	5.7	5.5	5.1	5.0	4.8	4.4	4.2
38	6.5	6.3	6.1	5.9	5.7	5.2	5.1	5.0	4.5	4.4
39	6.7	6.4	6.2	6.0	5.8	5.3	5.3	5.1	4.6	4.5
40	6.8	6.6	6.4	6.2	6.0	5.5	5.4	5.2	4.7	4.6
41	7.0	6.8	6.6	6.3	6.1	5.6	5.5	5.4	4.9	4.7
42	7.2	7.0	6.7	6.5	6.3	5.8	5.7	5.5	5.0	4.8
43	7.3	7.1	6.9	6.7	6.4	5.9	5.8	5.6	5.1	4.9
44	7.5	7.3	7.1	6.8	6.6	6.0	6.0	5.8	5.2	5.1
45	7.7	7.4	7.2	7.0	6.7	6.2	6.1	5.9	5.3	5.2
46	7.8	7.6	7.4	7.1	6.9	6.3	6.2	6.0	5.4	5.3
47	8.0	7.8	7.5	7.3	7.0	6.4	6.4	6.1	5.6	5.4
48	8.2	7.9	7.7	7.4	7.2	6.6	6.5	6.3	5.7	5.5
49	8.4	8.1	7.8	7.6	7.3	6.7	6.6	6.4	5.8	5.6
50	8.5	8.3	8.0	7.7	7.5	6.8	6.8	6.5	5.9	5.7
51	8.7	8.4	8.2	7.9	7.6	7.0	6.9	6.7	6.0	5.9
52	8.9	8.6	8.3	8.1	7.8	7.1	7.0	6.8	6.2	6.0
53	9.0	8.8	8.5	8.2	7.9	7.3	7.2	6.9	6.3	6.1
54	9.2	8.9	8.6	8.4	8.1	7.4	7.3	7.1	6.4	6.2
55	9.4	9.1	8.8	8.5	8.2	7.5	7.4	7.2	6.5	6.3
56	9.5	9.2	9.0	8.7	8.3	7.7	7.6	7.3	6.6	6.4

Table 9 **Tractor Speed in Miles and Kilometers per Hour**

mi/h	km/h
2.0	3.2
2.5	4.0
3.0	4.8
3.5	5.6
4.0	6.4
4.5	7.2
5.0	8.0
5.5	8.6
6.0	9.7
6.5	10.5
7.0	11.2
7.5	12.1
8.0	12.9
8.5	13.7
9.0	14.5
9.5	15.3
10.0	16.1

Table 10   Distance Traveled in Tilling an Acre of Land with Various Widths of Cut

Width of cut		Distance traveled	
in	cm	mi	km
10	25.4	9.9	15.9
11	27.9	9.0	14.5
12	30.5	8.2	13.2
13	33.0	7.6	12.2
14	35.6	7.0	11.3
15	38.1	6.6	10.6
16	40.6	6.2	10.0
24	61.0	4.1	6.6
28	71.1	3.5	5.6
32	81.3	3.1	5.0
36	91.4	2.8	4.5
40	101.6	2.5	4.0
42	106.7	2.4	3.9
48	121.9	2.1	3.4
56	142.2	1.8	2.9
64	162.6	1.6	2.6
72	182.9	1.4	2.3
80	203.2	1.24	2.0
84	213.4	1.18	1.9
96	243.8	1.03	1.7
108	274.3	0.92	1.5
120	304.8	0.82	1.31
132	335.3	0.75	1.21
144	365.8	0.69	1.11

Table 11   Miles Traveled in Planting or Cultivating an Acre for 40-in (3.33-ft) Row Spacing

Number of rows	Miles traveled
1	2.48
2	1.24
4	0.62
6	0.41
8	0.31
10	0.25
12	0.12

Table 12   Plants per Acre for 40-in (1.02-m) Row Spacing

Distance apart in row, in	Plants per ft	Plants per acre
12	1	13,081
6	2	26,162
4	3	39,243
3	4	52,324
2.4	5	65,405
2.0	6	78,486
1.7	7	91,567
1.5	8	104,480
1.3	9	117,772
1.2	10	130,081
1.09	11	143,891
1.00	12	156,972

Table 13   Multiplying Factor for Converting Pounds per Row to Pounds per Acre

Row spacing, in	Row length, ft																	Linear ft per acre
	10	20	30	40	50	60	70	80	90	100	120	140	160	180	200	300	400	
10	5,227.2	2,613.6	1,742.4	1,306.8	1,045.4	871.2	746.7	653.4	580.8	522.7	435.6	373.4	326.7	290.4	261.4	174.2	130.7	52,272
11	4,752.0	2,376.0	1,584.0	1,188.0	950.4	792.0	678.9	594.0	528.0	475.2	396.0	339.4	297.0	264.0	237.6	158.4	118.8	47,520
12	4,356.0	2,178.0	1,452.0	1,089.0	871.2	726.0	622.3	544.5	484.0	435.6	363.0	311.1	272.3	242.0	217.8	145.2	108.9	43,560
13	4,020.9	2,010.5	1,340.3	1,005.2	804.2	670.2	574.4	502.6	446.8	402.1	335.1	287.2	251.3	223.4	201.1	134.0	100.5	40,209
14	3,733.7	1,866.9	1,244.6	933.4	746.7	622.3	533.4	466.7	414.9	373.4	311.1	266.7	233.4	207.4	186.7	124.5	93.3	37,337
15	3,484.8	1,742.4	1,161.6	871.2	697.0	580.8	497.8	435.6	387.2	348.5	290.4	248.9	217.8	193.6	174.2	116.2	87.1	34,848
16	3,267.0	1,633.5	1,089.0	816.8	653.4	544.5	466.7	408.4	363.0	326.7	272.3	233.4	204.2	181.5	163.4	108.9	81.7	32,670
17	3,074.8	1,537.4	1,024.9	768.7	615.0	512.5	439.3	384.4	341.6	307.5	256.2	219.6	192.2	170.8	153.7	102.5	76.9	30,748
18	2,904.0	1,452.0	968.0	726.0	580.8	484.0	414.9	363.0	322.7	290.4	242.0	207.4	181.5	161.3	145.2	96.8	72.6	29,040
19	2,751.2	1,375.6	917.1	687.8	550.2	458.5	393.0	343.9	305.7	275.1	229.3	196.5	172.0	152.8	137.6	91.7	68.8	27,512
20	2,613.6	1,306.8	871.2	653.4	522.7	435.6	373.4	326.7	290.4	261.4	217.8	186.7	163.4	145.2	130.7	87.1	65.3	26,136
21	2,489.1	1,244.6	829.7	622.3	497.8	414.9	355.6	311.1	276.6	248.9	207.4	177.8	155.6	138.3	124.5	83.0	62.2	24,891
22	2,376.0	1,188.0	792.0	594.0	475.2	396.0	339.4	297.0	264.0	237.6	198.0	169.7	148.5	132.0	118.8	79.2	59.4	23,760
23	2,272.7	1,136.4	757.6	568.2	454.4	378.8	324.7	284.1	252.5	227.3	189.4	162.3	142.0	126.3	113.6	75.8	56.8	22,727
24	2,178.0	1,089.0	726.0	544.5	435.6	363.0	311.1	272.3	242.0	217.8	181.5	155.6	136.1	121.0	108.9	72.6	54.5	21,780
25	2,090.9	1,045.5	697.0	522.7	418.2	348.5	298.7	261.4	232.3	209.1	174.2	149.4	130.7	116.2	104.6	69.7	52.3	20,909
26	2,010.5	1,005.3	670.2	502.6	402.1	335.1	287.2	251.3	223.4	201.1	167.5	143.6	125.7	111.7	100.5	67.0	50.3	20,105
27	1,936.0	968.0	645.3	484.0	387.2	322.7	276.6	242.0	215.1	193.6	161.3	138.3	121.0	107.6	96.8	64.5	48.4	19,360
28	1,866.9	933.5	622.3	466.7	373.4	311.2	266.7	233.4	207.4	186.7	155.6	133.4	116.7	103.7	93.4	62.2	46.7	18,669
29	1,802.5	901.3	600.8	450.6	360.5	300.4	257.5	225.3	200.3	180.3	150.2	128.8	112.7	100.1	90.1	60.1	45.1	18,025
30	1,742.4	871.2	580.8	435.6	348.5	290.4	248.9	217.8	193.6	174.2	145.2	124.5	108.9	96.8	87.1	58.1	43.6	17,424
31	1,686.3	843.1	562.1	421.6	337.2	281.0	240.9	210.8	187.4	168.6	140.5	120.4	105.4	93.7	84.3	56.2	42.2	16,862
32	1,633.5	816.8	544.5	408.4	326.7	272.3	233.4	204.2	181.5	163.4	136.1	116.7	102.1	90.8	81.7	54.5	40.8	16,335
33	1,584.0	792.0	528.0	396.0	316.8	264.0	226.3	198.0	176.0	158.4	132.0	113.1	99.0	88.0	79.2	52.8	39.6	15,840

Table 13   Multiplying Factor for Converting Pounds per Row to Pounds per Acre (*Continued*)

Row spacing, in	Row length, ft																	Linear ft per acre
	10	20	30	40	50	60	70	80	90	100	120	140	160	180	200	300	400	
34	1,537.4	768.7	512.5	384.4	307.5	256.2	219.6	192.2	170.8	153.7	128.1	109.8	96.1	85.4	76.9	51.3	38.4	15,374
35	1,493.5	746.8	497.8	374.4	298.7	248.9	213.4	186.7	165.9	149.4	124.5	106.7	93.3	83.0	74.7	49.8	37.4	14,935
36	1,452.0	726.0	484.0	363.0	290.4	242.0	207.4	181.5	161.3	145.2	121.0	103.7	90.8	80.7	72.6	48.4	36.3	14,520
37	1,412.8	706.4	470.9	353.2	282.6	235.5	201.8	176.6	157.0	141.3	117.7	100.9	88.3	78.5	70.6	47.1	35.3	14,128
38	1,375.6	687.8	458.5	343.9	275.1	229.3	196.5	172.0	152.8	137.6	114.6	98.3	86.0	76.4	68.8	45.9	34.4	13,756
39	1,340.3	670.2	446.8	335.1	268.1	223.4	191.5	167.5	148.9	134.0	111.7	95.7	83.8	74.5	67.0	44.7	33.5	13,403
40	1,306.8	653.4	435.6	326.7	261.4	217.8	186.7	163.4	145.2	130.7	108.9	93.3	81.7	72.6	65.3	43.6	32.7	13,068
41	1,274.9	637.5	425.0	318.7	255.0	212.5	182.1	159.4	141.7	127.5	106.2	91.1	79.7	70.8	63.8	42.5	31.9	12,749
42	1,244.6	622.3	414.9	311.2	248.9	207.4	177.8	155.6	138.3	124.5	103.7	88.9	77.8	69.1	62.2	41.5	31.1	12,446
45	1,161.6	580.8	387.2	290.4	232.3	193.6	165.9	145.2	129.1	116.2	96.8	83.0	72.6	64.5	58.1	38.7	29.0	11,616
48	1,089.0	544.5	363.0	272.3	217.8	181.5	155.6	136.1	121.0	108.9	90.8	77.8	68.1	60.5	54.5	36.3	27.2	10,890
54	968.0	484.0	322.7	242.0	193.6	161.3	138.3	121.0	107.6	96.8	80.7	69.1	60.5	53.8	48.4	32.3	24.2	9,680
60	871.2	435.6	290.4	217.8	174.2	145.2	124.5	108.9	96.8	87.1	72.6	62.2	54.5	48.4	43.6	29.0	21.8	8,712
66	792.0	396.0	264.0	198.0	158.4	132.0	113.1	99.0	88.0	79.2	66.0	56.6	49.5	44.0	39.6	26.4	19.8	7,920
72	726.0	363.0	242.0	181.5	145.2	121.0	103.7	90.8	80.7	72.6	60.5	51.9	45.4	40.3	36.3	24.2	18.2	7,260

Source: U.S. Dept. Agr., Agricultural Engineering Branch.

Table 14   Temperature Conversion Table*

-100 to 23			24 to 57			58 to 91			92 to 330			340 to 670			680 to 1000		
C	C or F	F	C	C or F	F	C	C or F	F	C	C or F	F	C	C or F	F	C	C or F	F
-73	-100	-148	-4.4	24	75.2	14.4	58	136.4	33.3	92	197.6	171	340	644	360	680	1256
-68	-90	-130	-3.9	25	77.0	15.0	59	138.2	33.9	93	199.4	177	350	662	366	690	1274
-62	-80	-112	-3.3	26	78.8	15.6	60	140.0	34.4	94	201.2	182	360	680	371	700	1292
-57	-70	-94	-2.8	27	80.6	16.1	61	141.8	35.0	95	203.0	188	370	698	377	710	1310
-51	-60	-76	-2.2	28	82.4	16.7	62	143.6	35.6	96	204.8	193	380	716	382	720	1328
-46	-50	-58	-1.7	29	84.2	17.2	63	145.4	36.1	97	206.6	199	390	734	388	730	1346
-40	-40	-40	-1.1	30	86.0	17.8	64	147.2	36.7	98	208.4	204	400	752	393	740	1364
-34.4	-30	-22	-.6	31	87.8	18.3	65	149.0	37.2	99	210.2	210	410	770	399	750	1382
-28.9	-20	-4	0	32	89.6	18.9	66	150.8	37.8	100	212.0	216	420	788	404	760	1400
-23.3	-10	14	0.6	33	91.4	19.4	67	152.6	38	100	212	221	430	806	410	770	1418
-17.8	0	32	1.1	34	93.2	20.0	68	154.4	43	110	230	227	440	824	416	780	1436
-17.2	1	33.8	1.7	35	95.0	20.6	69	156.2	49	120	248	232	450	842	421	790	1454
-16.7	2	35.6	2.2	36	96.8	21.1	70	158.0	54	130	266	238	460	860	427	800	1472
-16.1	3	37.4	2.8	37	98.6	21.7	71	159.8	60	140	284	243	470	878	432	810	1490
-15.6	4	39.2	3.3	38	100.4	22.2	72	161.6	66	150	302	249	480	896	438	820	1508
-15.0	5	41.0	3.9	39	102.2	22.8	73	163.4	71	160	320	254	490	914	443	830	1526
-14.4	6	42.8	4.4	40	104.0	23.3	74	165.2	77	170	338	260	500	932	449	840	1544
-13.9	7	44.6	5.0	41	105.8	23.9	75	167.0	82	180	356	266	510	950	454	850	1562
-13.3	8	46.4	5.6	42	107.6	24.4	76	168.8	88	190	374	271	520	968	460	860	1580
-12.8	9	48.2	6.1	43	109.4	25.0	77	170.6	93	200	392	277	530	986	466	870	1598
-12.2	10	50.0	6.7	44	111.2	25.6	78	172.4	99	210	410	282	540	1004	471	880	1616
-11.7	11	51.8	7.2	45	113.0	26.1	79	174.2	100	212	414	288	550	1022	477	890	1634
-11.1	12	53.6	7.8	46	114.8	26.7	80	176.0	104	220	428	293	560	1040	482	900	1652
-10.6	13	55.4	8.3	47	116.6	27.2	81	177.8	110	230	446	299	570	1058	488	910	1670

Table 14   Temperature Conversion Table*

−100 to 23			24 to 57			58 to 91			92 to 330			340 to 670			680 to 1000		
C	C or F	F	C	C or F	F	C	C or F	F	C	C or F	F	C	C or F	F	C	C or F	F
−10.0	14	57.2	8.9	48	118.4	27.8	82	179.6	116	240	464	304	580	1076	493	920	1688
− 9.4	15	59.0	9.4	49	120.2	28.3	83	181.4	121	250	482	310	590	1094	499	930	1706
− 8.9	16	60.8	10.0	50	122.0	28.9	84	183.2	127	260	500	316	600	1112	504	940	1724
− 8.3	17	62.6	10.6	51	123.8	29.4	85	185.0	132	270	518	321	610	1130	510	950	1742
− 7.8	18	64.4	11.1	52	125.6	30.0	86	186.8	138	280	536	327	620	1148	516	960	1760
− 7.2	19	66.2	11.7	53	127.4	30.6	87	188.6	143	290	554	332	630	1166	521	970	1778
− 6.7	20	68.0	12.2	54	129.2	31.1	88	190.4	149	300	572	338	640	1184	527	980	1796
− 6.1	21	69.8	12.8	55	131.0	31.7	89	192.2	154	310	590	343	650	1202	532	990	1814
− 5.6	22	71.6	13.3	56	132.8	32.2	90	194.0	160	320	608	349	660	1220	538	1000	1832
− 5.0	23	73.4	13.9	57	134.6	32.8	91	195.8	166	330	626	354	670	1238			

*The numbers in italics in the center column refer to the temperature C in either Celsius or Fahrenheit which one desires to convert to the other scale. If converting Fahrenheit to Celsius, the equivalent temperature will be found in the left column. If converting Celsius to Fahrenheit, the equivalent temperature will be found in the column on the right.

Source: Reprinted by permission of Fisher Scientific Company.

Table 15   Relative Humidity Table

Difference between wet- and dry-bulb thermometers, °F

Temperature of dry-bulb	1	2	3	4	5	6	7	8	9	10	11	12	13	14	15	16	17	18	19	20	21	22	23	24	25	26	27	28	29	30	31	32	33	34	35	36	37	38	39	40
60	94	89	83	78	73	68	63	58	53	48	44	39	34	30																										
70	95	90	85	81	76	72	68	64	59	56	51	47	43	40	37	33	29																							
80	95	91	87	83	79	75	71	68	64	61	57	54	50	47	44	41	38	35	32	29																				
90	96	92	88	85	81	77	74	71	67	65	61	58	55	52	49	47	44	41	39	36	34	31	29	26	25															
100	96	93	89	86	83	80	76	73	70	68	64	62	59	57	54	51	48	46	44	42	39	37	35	33	31	29	27													
102	96	93	89	86	83	80	77	74	71	69	65	62	59	57	54	52	49	47	45	43	40	38	36	34	32	30	28													
104	96	93	90	86	83	80	77	74	71	69	65	63	60	58	55	52	50	47	46	43	41	39	37	35	33	31	29	27												
106	96	93	90	87	83	80	77	74	72	69	66	63	60	58	56	53	51	48	46	44	42	40	38	36	34	32	30	28												
108	96	93	90	87	84	81	78	75	72	70	66	64	61	59	56	54	51	49	47	45	43	41	39	37	35	33	31	29	28											
110	96	93	90	87	84	81	78	75	72	70	67	64	62	60	57	55	52	50	48	46	44	41	39	37	36	34	32	30	28	27										
112	96	93	90	87	84	81	78	75	73	70	67	65	62	60	57	55	53	51	49	47	44	42	40	38	37	35	33	31	29	28										
114	97	93	90	87	84	81	78	75	73	71	68	65	63	61	58	56	53	51	49	47	45	43	41	39	38	36	34	32	30	28	27									
116	97	93	90	88	84	82	79	76	74	71	68	66	63	61	59	56	54	52	50	48	46	44	42	40	38	36	34	33	31	29	28	27								
118	97	93	91	88	85	82	79	76	74	71	68	66	64	62	59	57	54	53	51	49	46	44	42	41	39	37	35	34	32	30	29	28	27							
120	97	94	91	88	85	82	79	77	74	72	69	66	64	62	60	57	55	53	51	49	47	45	43	41	40	38	36	34	33	31	30	28	27							
122	97	94	91	88	85	82	79	77	75	72	69	67	65	62	60	58	56	54	52	50	48	46	44	42	40	38	37	35	33	32	30	29	27							
124	97	94	91	88	85	83	80	77	75	72	70	67	65	63	60	58	56	54	52	51	48	46	44	43	41	39	38	36	34	33	31	29	28	27						
126	97	94	91	88	86	83	80	78	75	73	70	67	65	63	61	59	57	55	53	51	49	47	45	43	42	40	38	37	35	34	32	31	29	28	27					
128	97	94	91	89	86	83	80	78	76	73	71	68	66	64	61	59	57	55	53	51	49	47	46	44	42	40	39	37	36	34	33	31	30	28	27					
130	97	94	91	89	86	83	80	78	76	73	71	68	66	64	62	60	58	56	54	52	50	48	46	44	43	41	39	38	36	35	33	32	30	29	28	27				
132	97	94	92	89	86	83	81	78	76	74	71	69	67	65	62	60	58	56	54	53	51	49	47	45	43	42	40	39	37	36	34	33	31	30	29	27				
134	97	94	92	89	86	84	81	79	76	74	72	69	67	65	63	61	59	57	55	53	51	49	48	46	44	43	41	40	38	37	35	34	32	31	29	28	27			
136	97	94	92	89	86	84	81	79	77	74	72	69	67	65	63	61	59	57	55	53	51	50	48	46	45	43	42	40	39	37	36	35	33	32	30	29	28	27		
138	97	94	92	89	86	84	81	79	77	74	72	70	68	66	63	61	59	58	56	54	52	50	48	47	45	43	42	41	39	38	36	35	34	33	31	30	28	27		
140	97	94	92	89	87	84	81	79	77	75	72	70	68	66	64	62	60	58	56	54	52	51	49	47	46	44	42	41	39	38	36	35	34	33	31	30	28	27		
142	97	94	92	89	87	84	82	80	77	75	73	70	68	66	64	62	60	58	57	55	53	51	49	48	46	44	43	42	40	39	37	36	34	33	32	30	29	28	27	
144	97	95	92	89	87	84	82	80	78	75	73	71	69	67	65	63	61	59	57	55	53	52	50	48	47	45	44	42	41	39	38	36	35	34	32	31	30	29	28	
146	97	95	92	90	87	85	82	80	78	75	73	71	69	67	65	63	61	59	57	56	54	52	50	49	47	45	44	43	41	40	38	37	36	35	33	32	30	29	28	27

Table 15 Relative Humidity Table (*Continued*)

Difference between wet- and dry-bulb thermometers, °F

Temperature of dry-bulb	1	2	3	4	5	6	7	8	9	10	11	12	13	14	15	16	17	18	19	20	21	22	23	24	25	26	27	28	29	30	31	32	33	34	35	36	37	38	39	40
148	97	95	92	90	87	85	82	80	78	76	73	71	69	67	65	63	61	60	58	56	54	53	51	49	48	46	45	43	42	40	39	38	36	35	34	32	31	30	28	28
150	98	95	92	90	87	85	82	80	78	76	74	72	70	68	66	64	62	60	58	57	55	53	51	49	48	46	45	43	42	41	39	38	37	36	34	33	31	30	29	28
152	98	95	93	90	88	85	83	81	79	76	74	72	70	68	66	64	62	60	59	57	55	53	52	50	48	47	46	44	42	41	40	39	37	36	35	33	32	31	30	29
154	98	95	93	90	88	85	83	81	79	77	74	72	70	68	66	65	63	61	59	57	56	54	52	50	49	47	46	44	43	42	40	39	38	37	35	34	33	32	30	29
156	98	95	93	90	88	85	83	81	79	77	74	72	71	69	67	65	63	61	59	58	56	54	53	51	49	48	46	45	43	42	41	40	38	37	36	34	33	32	31	30
158	98	95	93	90	88	86	83	81	79	77	75	73	71	69	67	65	63	61	60	58	56	55	53	51	50	48	47	45	44	43	41	40	39	38	36	35	34	33	31	30
160	98	95	93	90	88	86	83	81	79	77	75	73	71	69	67	65	64	62	60	58	57	55	53	52	50	49	47	46	44	43	42	41	39	38	37	35	34	33	32	31
162	98	95	93	90	88	86	84	82	80	77	75	73	71	69	68	66	64	62	60	59	57	55	54	52	51	49	48	46	45	44	42	41	40	39	37	36	35	34	32	31
164	98	95	93	91	88	86	84	82	80	78	75	73	72	70	68	66	64	62	61	59	58	56	54	52	51	49	48	47	45	44	43	41	40	39	38	36	35	34	33	32
166	98	95	93	91	88	86	84	82	80	78	76	74	72	70	68	66	64	63	61	59	58	56	54	53	51	50	48	47	46	44	43	42	41	39	38	37	36	35	33	32
168	98	95	93	91	88	86	84	82	80	78	76	74	72	70	68	67	65	63	61	60	58	56	55	53	52	50	49	47	46	45	43	42	41	40	39	37	36	35	34	33
170	98	95	93	91	89	86	84	82	80	78	76	74	72	70	69	67	65	63	62	60	59	57	55	53	52	50	49	48	47	45	44	43	41	40	39	38	37	36	35	33
172	98	95	93	91	89	86	84	82	80	78	76	74	73	71	69	67	66	64	62	60	59	57	55	54	52	51	49	48	47	45	44	43	41	40	39	38	37	36	35	34
174	98	95	93	91	89	87	84	83	81	78	76	74	73	71	69	67	66	64	62	61	59	57	55	54	53	51	50	48	47	46	45	43	42	41	39	38	37	36	35	34
176	98	96	94	91	89	87	85	83	81	79	77	75	73	71	70	68	66	64	63	61	60	58	56	55	53	51	50	49	47	46	45	44	42	41	40	39	38	37	36	35
178	98	96	94	91	89	87	85	83	81	79	77	75	73	72	70	68	66	64	63	61	60	58	56	55	54	52	51	49	48	47	45	44	43	42	41	39	38	37	36	35
180	98	96	94	91	89	87	85	83	81	79	77	75	73	72	70	68	67	65	63	62	60	58	57	55	54	52	51	50	48	47	46	45	43	42	41	40	39	38	36	35
182	98	96	94	91	89	87	85	83	81	79	77	75	74	72	70	68	67	65	63	62	60	59	57	56	54	53	51	50	49	48	46	45	44	43	42	40	39	38	37	36
184	98	96	94	92	89	87	85	83	81	79	77	76	74	72	70	69	67	65	64	62	60	59	57	56	54	53	52	50	49	48	46	45	44	43	42	40	39	38	37	36
186	98	96	94	92	90	87	85	83	82	80	78	76	74	72	71	69	67	66	64	62	61	59	58	56	55	53	52	51	49	48	47	46	44	43	42	41	39	38	37	36
188	98	96	94	92	90	87	85	83	82	80	78	76	74	73	71	69	68	66	64	63	61	59	58	57	55	54	52	51	49	48	47	46	45	44	42	41	40	39	38	37
190	98	96	94	92	90	88	86	84	82	80	78	76	75	73	71	69	68	66	65	63	62	60	59	57	56	54	53	51	50	49	48	46	45	44	43	41	40	39	38	37
200	98	96	94	92	90	88	86	84	82	80	79	77	75	74	72	70	69	67	66	64	63	61	60	58	57	55	54	53	51	50	49	48	47	46	45	43	42	41	40	39

*To find the relative humidity, assume the reading of the dry-bulb is 90° and the wet-bulb is 85°; subtract the wet-bulb reading (85°) from the dry-bulb reading (90°), which gives a difference of 5°. Follow down the column headed (5°) and read across the table from dry-bulb reading (90°). We find the relative humidity is 81 percent where the two lines intersect.
Temperature readings in degrees Fahrenheit. Relative humidity readings in percent. Barometric pressure 29.92 in.

Index